NUMBER SYSTEMS
AND THE FOUNDATIONS
OF ANALYSIS

NUMBER SYSTEMS
AND THE FOUNDATIONS
OF ANALYSIS

ELLIOTT MENDELSON

Queens College
The City University of New York

ACADEMIC PRESS New York and London
A Subsidiary of Harcourt Brace Jovanovich, Publishers

ACADEMIC PRESS, INC.
111 Fifth Avenue, New York, New York 10003

United Kingdom Edition published by
ACADEMIC PRESS, INC. (LONDON) LTD.
24/28 Oval Road, London NW1

LIBRARY OF CONGRESS CATALOG CARD NUMBER: 72-82629

AMS (MOS) 1970 Subject Classifications: 00-01, 06A70, 10M15,
12D99, 12J15, 26A03

February 1973, First Printing
August 1973, Second Printing

PRINTED IN THE UNITED STATES OF AMERICA

To my mother, Helen,
and to the memory of my father, Joseph Mendelson

CONTENTS

Chapter 2 **The Natural Numbers**

Chapter 3 **The Integers**

Chapter 4 **Rational Numbers and Ordered Fields**

Chapter 5 **The Real Number System**

PREFACE

The aim of this book is to study the basic number systems of mathematics: natural numbers, integers, rational numbers, real numbers, and complex numbers. Although the principal users of this text probably will be undergraduate students of mathematics and related subjects, as well as beginning graduate students, we always have kept a wider audience in mind.

No presupposition has been made as to previous mathematical training. In fact, we have assumed that the reader has no experience in abstract mathematical thinking. For this reason, many of the proofs have been presented in great, sometimes excruciating, detail. Not only does this give the beginner a better chance to follow kinds of abstract reasoning which may be new to him, but it also will allow an instructor to devote more time in class to explanation and discussion. Only in the last three sections of the last chapter do the proofs begin to resemble the standard abbreviated proofs found in most textbooks. By that time, the reader should be able to stand on his own feet.

A course in number systems usually runs for only one semester. Experience has taught me that, in order to reach the most important part of the subject, namely the real number system, certain economies have to be practiced in the earlier chapters. The following suggestions may help the inexperienced instructor or the independent reader.

a. The basic ideas and notation of Chapter 1 should be summarized by the instructor in at most two lectures. Any material not thoroughly covered can be reviewed later, when needed. The student should be asked to read the chapter at a slower pace and to do many of the exercises.

b. There is very little fat that can be eliminated from Chapter 2. Most of Chapters 3–4 is essential, but there are various specific results, such as the characterizations of the integers (for example, Theorem 3.9.5), which can be harmlessly omitted.

c. Sections 5.1–5.6 form the core of Chapter 5, while Section 5.7 is recommended, but not essential. Sections 5.7–5.9 illustrate how the further development of analysis follows from the basic properties of a complete ordered field. The exposition in Sections 5.7–5.9 is very compact and is in no way intended to be a thorough introduction to analysis. Nevertheless, it may be enlightening to readers who have previously studied calculus. However, the author would suggest that it is unreasonable to expect to get beyond Section 5.6 in a one-semester course. The material up through Section 5.6 will provide sufficient work for a thorough course on number systems.

Our approach to the number systems is to start with an axiomatic presentation of the simplest structure, the system of natural numbers, and then, by set-theoretic methods, to construct in succession the integers, rational numbers, and the real numbers. Although this procedure consists of adding the negative integers, then the proper fractions, and finally the "irrational" numbers, the actual methods of construction are not straightforward. This is the case, paradoxically, for reasons of simplicity. The natural, psychologically plausible methods turn out to be much more complicated, in the long run, than the rather sophisticated procedures that we shall employ. In any case, the crucial point is that there is no recourse to spatial or temporal intuition. The whole fantastic hierarchy of number systems is built up by purely set-theoretic means from a few simple assumptions about natural numbers.

Notation.   (i)   References to the Bibliography are given by author and year of publication.

(ii)   Theorems are numbered consecutively within each section of a chapter. References to a theorem in the same chapter use the numbers of the section and the theorem. Thus, "Theorem 5.3" refers to the third theorem of Section 5. However, if the reference is to a theorem in another chapter, then the number of the chapter will also be mentioned. Thus, "Theorem 2.3.7" means the seventh theorem of Section 3 of Chapter 2.

(iii)   Difficult exercises are indicated by a capital D as a superscript after the number of the exercise. Exercises that are crucial to later work are indicated by a capital C as a superscript after the number of the exercise. Crucial exercises should be considered an essential part of the text.

CHAPTER 1

BASIC FACTS AND NOTIONS OF LOGIC AND SET THEORY

This chapter is a very brief introduction to the basic concepts and terminology of logic and set theory. We shall emphasize the meaning of the concepts and ease of translation between ordinary English and the mathematical formalism.

1.1 LOGICAL CONNECTIVES

Logic is concerned with ways in which new sentences can be constructed from given sentences and with the deducibility relation between sentences. The principal task of logic is to determine which sentences *follow from* other given sentences. In this introductory chapter, we shall deal only with the basic logical notions and their meaning.

The simplest way of constructing new sentences is by means of **connectives**. One of the most easily understood connectives is *and*. Given sentences A and B, we may form a new sentence A *and* B, which we shall write $A \wedge B$. The relation between the new sentence $A \wedge B$ and the component sentences A, B can be made clear by means of a so-called truth table:

A	B	$A \wedge B$
T	T	T
F	T	F
T	F	F
F	F	F

In this table, as in all tables to follow, T stands for "true" and F for "false." Each row of the table corresponds to a possible assignment of truth values to the component sentences A and B. The entry under $A \wedge B$ gives the corresponding truth value of $A \wedge B$. Thus, $A \wedge B$ is true only in the first row, where both A and B are true. The sentence $A \wedge B$ is called the **conjunction** of A and B.

An even simpler connective is **negation**. If A is a sentence, then *not*-A is also a sentence, which we shall denote $\neg A$. The corresponding truth table is

A	$\neg A$
T	F
F	T

$\neg A$ is T when A is F, and F when A is T. $\neg A$ is called the **denial** or **negation** of A.

A somewhat less clear connective is *or*. If A and B are sentences, we may form a new sentence A *or* B, which we shall abbreviate $A \vee B$. This is understood in the **inclusive sense**, namely, $A \vee B$ is true when A is true or B is true or both. This is the sense of the expression *and/or* used in legal documents. The corresponding truth table is

A	B	$A \vee B$
T	T	T
F	T	T
T	F	T
F	F	F

$A \vee B$ is called the **disjunction** of A and B.

EXAMPLES

1. If A stands for "John is tired" and B stands for "John looks tired," let us translate the following formulas into ordinary English.

a. $\neg A$.   John is not tired.

b. $\neg B$.   John does not look tired.

c. $A \wedge B$.   John is tired and looks tired.

d. $A \wedge (\neg B)$.   John is tired but he does not look tired.

e. $(\neg A) \wedge B$.   John is not tired but he looks tired.

f. $A \vee B$.   John is tired or he looks tired.

Notice that "but" has the same meaning as "and," with an element of surprise.

2.   Using letters to stand for basic sentences, let us translate the following sentences into formulas.

a.   The street is wet but it didn't rain.

$A \wedge (\neg B)$, where A stands for "The street is wet" and B stands for "It rained."

b.   John is lucky in love and lucky in cards.

$A \wedge B$, where A stands for "John is lucky in love" and B stands for "John is lucky in cards."

c. x is either rational or irrational.

$A \vee \neg A$, where A stands for "x is rational."

d.   Neither 1 nor 2 is transcendental.

$(\neg A) \wedge (\neg B)$, where A stands for "1 is transcendental" and B stands for "2 is transcendental."

3.   Let us show that the following pairs of formulas are equivalent, in the sense that the formulas always take the same truth values.

a. $A \vee A$ and A.

When A is T, so is $A \vee A$, and vice versa.

b. $A \wedge A$ and A.

A is T if and only if $A \wedge A$ is T.

c. $A \vee (B \wedge A)$ and A.

When A is T, so is $A \vee (B \wedge A)$. If A is F, then both A and $(B \wedge A)$ are F, and, therefore, $A \vee (B \wedge A)$ is F. An alternative demonstration can be given by means of the following table:

A	B	$B \wedge A$	$A \vee (B \wedge A)$
T	T	T	T
F	T	F	F
T	F	F	T
F	F	F	F

It is only necessary to observe that the columns under A and $A \vee (B \wedge A)$ are identical.

d. $A \wedge (B \vee A)$ and A.

A	B	$B \vee A$	$A \wedge (B \vee A)$
T	T	T	T
F	T	T	F
T	F	T	T
F	F	F	F

EXERCISES

1. Write the following English sentences in terms of our new notation. Use letters A, B, C, ... to stand for the basic sentences.

   a. I will go to the movies or to the beach.
   b. Peter is handsome but Paul is not.
   c. Either x has a square root or x is not positive.
   d. It is not the case that x is both positive and negative.
   e. Neither John nor James is here.

2. Render the following into idiomatic English.

   a. $x > 2 \vee x < 2$.
   b. $\neg(y$ is rational $\vee$ y is transcendental).
   c. $e > 2 \wedge e < 3$.
   d. $e > 2 \wedge e \not> 3$.

3. Show that the following pairs of formulas always take the same truth values, and, therefore, are, in a certain sense, equivalent:

   a. $\neg(A \vee B)$ and $(\neg A) \wedge (\neg B)$.
   b. $\neg(A \wedge B)$ and $(\neg A) \vee (\neg B)$.
   c. $\neg\neg B$ and B.
   d. $A \wedge (B \vee C)$ and $(A \wedge B) \vee (A \wedge C)$.
   e. $A \vee (B \wedge C)$ and $(A \vee B) \wedge (A \vee C)$.

1.2 CONDITIONALS

A very important connective, but one which causes a great deal of confusion, is "If ____, then ...." If A and B are sentences, we may form a new sentence "If A then B," which we shall denote $A \Rightarrow B$. The sentence $A \Rightarrow B$ is called a **conditional**, with antecedent A and **consequent** B. The truth table which we shall adopt for this connective is

A	B	$A \Rightarrow B$
T	T	T
F	T	T
T	F	F
F	F	T

There is one line of this table which is clear, the third. For, if A is true and B is false, then "If A then B" is obviously false. However, there is no obvious reason for the entries in the other three rows of the table. For example, the sentences

i. If $2 + 2 = 4$, then there are 31 days in October
ii. If $2 + 2 \neq 4$, then there are 31 days in October
iii. If $2 + 2 \neq 4$, then there are 30 days in October

cannot be assigned a truth value on the basis of ordinary usage. In fact, it is doubtful that these sentences would be considered meaningful. Thus, the truth table we have established for $\Rightarrow$ is an *extension* of ordinary usage, but in no way does it *contradict* ordinary usage. It also should be emphasized that the meaning we have assigned to $A \Rightarrow B$ is precisely the meaning that it has in contemporary logic and mathematics: $A \Rightarrow B$ is false when and only when A is true and B is false.

Two peculiar properties of $\Rightarrow$ sometimes cause trouble:

I. If A is false, then $A \Rightarrow B$ is true (no matter what B is).
II. If B is true, then $A \Rightarrow B$ is true (no matter what A is).

In this connection, it is customary in mathematics to say that $A \Rightarrow B$ is **trivially true** in Cases I or II.

Aside from these peculiar properties of $\Rightarrow$, we must emphasize the two most important properties of $\Rightarrow$:

a. When A and $A \Rightarrow B$ are both true, then B must be true.
b. In order to prove $A \Rightarrow B$, it suffices to assume A true and then show that B must also be true.

EXAMPLES

1. The sentence

$$1 + 1 \neq 2 \Rightarrow \text{Fermat's Last Theorem is correct}$$

is trivially true, since $1 + 1 \neq 2$ is false. We do not know the truth value of "Fermat's Last Theorem is correct."

2.   The sentence

$$1 + 1 \neq 2 \Rightarrow 7 < 4$$

is also trivially true, since $1 + 1 \neq 2$ is false.

3.   The sentence

$$\text{Fermat's Last Theorem is correct} \Rightarrow 2 + 2 = 4$$

is trivially true, since $2 + 2 = 4$ is true.

4.   The sentence

$$1 + 1 = 2 \Rightarrow 6 = 7$$

is false, since $1 + 1 = 2$ is true while $6 = 7$ is false.

5.   ($1 + 1 = 2 \Rightarrow$ London is the capital of England) is true, since both antecedent and consequent are true. Notice that we are not asserting any meaningful connection between "$1 + 1 = 2$" and "London is the capital of England." We are only interested in the truth values of the sentences.

6.   The sentence "The truth of A is a **necessary condition** for the truth of B" means that, if B is true, then A also must be true, and, thus, should be translated as $B \Rightarrow A$. For instance,

> a necessary condition for x to be greater than 10 is that x is greater than 5

should be written as $x > 10 \Rightarrow x > 5$.

7.   The sentence "The truth of A is a **sufficient condition** for the truth of B" means that, if A is true, B also must be true, and, thus, should be translated as $A \Rightarrow B$. For instance,

> a sufficient condition for x to be odd is that x is the square of an odd integer

should be written as $A \Rightarrow B$, where A stands for "x is the square of an odd integer" and B stands for "x is odd."

EXERCISES

1. Translate into our symbolism the following English sentences.

    a. If John is not sick and it is not raining, we shall go on a picnic.
    b. Sam goes to work only if the sun is not shining.
    c. A sufficient condition for f to be continuous is that f is differentiable.
    d. A necessary condition for s to converge is that s is bounded.

2. Show that following sentences always have the same truth values and are, therefore, equivalent.

    a. $A \Rightarrow B$.
    b. $(\neg A) \vee B$.
    c. $\neg(A \wedge (\neg B))$.

3. Verify the following justification for our truth table for $A \Rightarrow B$. It is ob· ›us that $(C \wedge D) \Rightarrow C$ must always be true. Hence: (i) When C and D are both true, we have $(T \Rightarrow T) = T$. This gives the first row of the table for $\Rightarrow$. (ii) When C is false and D is true, we have $(F \Rightarrow T) = T$, which is the second line of the table. (iii) When D is false, we have $(F \Rightarrow F) = T$, which is the fourth line of the table.

4. Show that the following pairs of sentences are equivalent, that is, always take the same truth values:

    a. $\neg(A \Rightarrow B)$ and $A \wedge (\neg B)$.
    b. $A \Rightarrow (B \Rightarrow C)$ and $(A \wedge B) \Rightarrow C$.
    c. $A \Rightarrow B$ and $(\neg B) \Rightarrow (\neg A)$ (Law of Contrapositives).

5. a. Show that, if $\neg A \Rightarrow (B \wedge (\neg B))$ is true, then A is true. (This justifies proofs by *reductio ad absurdum*: To prove A, show that $\neg A$ leads to a contradiction.)
    b. Prove by reduction ad absurdum that $A \Rightarrow A$ is always true.
    Hint: Use Exercise 4a.

1.3 BICONDITIONALS

The sentence

$$A \text{ if and only if } B$$

is true precisely when A and B have the same truth value. We shall use $A \Leftrightarrow B$ to stand for "A if and only if B." The truth table reads

A	B	$A \Leftrightarrow B$
T	T	T
F	T	F
T	F	F
F	F	T

Another way of describing $A \Leftrightarrow B$ is to say that, whenever A is true, B is also true, and vice versa.

EXERCISES

1. Translate into our symbolism:

    a. A necessary and sufficient condition for s to be convergent is that it is Cauchy convergent.

    b. Horatio is happy if and only if the sun is shining and business is good.

    c. f is differentiable when and only when it is continuous.

2. Show that $A \Leftrightarrow B$ always has the same truth values as $(A \Rightarrow B) \wedge (B \Rightarrow A)$. (Hence, to prove $A \Leftrightarrow B$, is suffices to prove $A \Rightarrow B$ and $B \Rightarrow A$.)

3. Assume that $A \vee B$ is true. What, if anything, can be said of the truth values of the following?

    a. A b. $A \Rightarrow B$ c. $\neg A \Rightarrow B$ d. $\neg[(\neg A) \wedge (\neg B)]$

    e. $(A \wedge C) \vee (B \wedge (\neg C))$.

4. Assume that $A \Rightarrow B$ is true. What, if anything, can be said of the truth values of the following?

    a. B b. $A \vee B$ c. $(\neg A) \vee B$ d. $B \Rightarrow A$ e. $(\neg B) \Rightarrow (\neg A)$.

5. Assume that A is true, B is false, and C is true. What are the truth values of:

    a. $A \Rightarrow B$

    b. $B \Rightarrow C$

    c. $(A \vee \neg C) \vee B$

    d. $(A \Leftrightarrow C) \Leftrightarrow \neg B$

    e. $A \Rightarrow (B \Rightarrow A)$?

6. For each of the following pairs of sentences C and D, show that C is equivalent to D, in the sense that they always take the same truth values.

	C	D
a.	$A \wedge (B \vee \neg A)$	$A \wedge B$
b.	$A \vee (B \wedge \neg A)$	$A \vee B$
c.	$A \vee (B \wedge A)$	A
d.	$A \wedge (B \vee A)$	A
e.	$\neg(A \vee B)$	$(\neg A) \wedge (\neg B)$
f.	$\neg(A \wedge B)$	$(\neg A) \vee (\neg B)$.

1.4   QUANTIFIERS

The sentence

(1)   All men are mortal

can be reformulated as

(2)   For all x, if x is a man, then x is mortal.

If we let $M(x)$ stand for *x is a man* and $D(x)$ for *x is mortal*, then (2) can be rewritten as:

(3)   For all x $(M(x) \Rightarrow D(x))$.

Finally, we agree to abbreviate "For all x" by $(\forall x)$, obtaining

(4) $(\forall x)(M(x) \Rightarrow D(x))$.

The expression $(\forall x)$ is called a **universal quantifier.** It usually replaces the English expressions *for all x, for every x, for each x, for any x.*

Any English sentence of the form *All A's are B's* can be written in the form $(\forall x)(A(x) \Rightarrow B(x))$.

EXAMPLES

1.   "Any student who has patience can learn mathematics" can be formulated as $(\forall x)([S(x) \land P(x)] \Rightarrow M(x))$. Here, $S(x)$, $P(x)$, $M(x)$ stand for *x is a student, x has patience, x can learn mathematics*, respectively.

2.   The sentence

No even integer greater than 2 is a prime

can be rewritten as

$(\forall x)([E(x) \land x > 2] \Rightarrow \neg P(x))$.

Here, $E(x)$ and $P(x)$ stand for *x is an even integer* and *x is a prime*, respectively.

3.   The sentence

Every moment of life is dear

can be reformulated as

$(\forall x)(M(x) \Rightarrow D(x))$.

Here, $M(x)$ and $D(x)$ stand for *x is a moment of life* and *x is dear*, respectively.

There is another kind of quantifier occurring often in mathematics. Consider the sentence

Some continuous functions are not differentiable.

This can be written as

For some f, f is a continuous function and f is not differentiable.

This can be formulated as

For some f, $C(f) \land \neg D(f)$,

where $C(f)$ stands for *f is a continuous function*, and $D(f)$ stands for *f is a differentiable function*.

Finally, we introduce $(\exists f)$ as an abbreviation for *for some f*, obtaining

$$(\exists f)(C(f) \land \neg D(f)).$$

$(\exists f)$ is called an **existential quantifier**. It can be used to translate the English expressions *for some f, there exists an f such that, for at least one f*. A sentence of the form *Some A's are B's* can be rewritten as

$$(\exists x)(A(x) \land B(x)).$$

EXAMPLES

4. The sentence

Some swans are not white

can be rendered as

$$(\exists x)(S(x) \land \neg W(x)),$$

where $S(x)$ and $W(x)$ stand for *x is a swan* and *x is white*, respectively.

5. The sentence

There is no largest integer

can be written as

$$\neg(\exists x)(I(x) \land (\forall y)(I(y) \Rightarrow y \leq x)).$$

Here, $I(x)$ stands for *x is an integer*. We can read this expression in a long-winded way as

It is not the case that there exists an x such that x is an integer and all integers are less than or equal to x.

An equivalent sentence is

$$(\forall x)(I(x) \Rightarrow (\exists y)(I(y) \wedge x < y)),$$

that is, for every integer x, there is a larger integer y.

6. The sentence

Everyone loves somebody

can be written as

$$(\forall x)(P(x) \Rightarrow (\exists y)(P(y) \wedge L(x, y))).$$

Here, $P(x)$ and $L(x, y)$ stand for *x is a person* and *x loves y*, respectively.

7. Notice that $\neg(\exists x)A(x)$ has the same meaning as $(\forall x)\neg A(x)$. Consider the sentence

No integer is bigger than itself.

This can be rendered as $\neg(\exists x)(I(x) \wedge x < x)$, which is equivalent to $(\forall x)\neg(I(x) \wedge x < x)$. Since $\neg(A \wedge B)$ is equivalent to $A \Rightarrow \neg B$, we obtain

$$(\forall x)(I(x) \Rightarrow x \not< x).$$

Similarly, $\neg(\forall x)A(x)$ has the same meaning as $(\exists x)\neg A(x)$.

8. Consider the sentence

Not all primes are odd.

This can be written as $\neg(\forall x)(P(x) \Rightarrow O(x))$, which is equivalent to $(\exists x)\neg(P(x) \Rightarrow O(x))$. But, $\neg(A \Rightarrow B)$ is equivalent to $A \wedge \neg B$. Hence, we obtain $(\exists x)(P(x) \wedge \neg O(x))$, that is, some prime is not odd.

9. The sentence

No one loves a loser

can be translated as

$$(\forall x)(L(x) \Rightarrow (\forall y)(P(y) \Rightarrow \neg L(y, x))),$$

where $L(x)$, $P(y)$, $L(y, x)$ stand for *x is a loser*, *y is a person*, and *y loves x*, respectively.

10.   The sentence

> If anyone can solve this problem, John can

can be formulated as $[(\exists x)S(x)] \Rightarrow S(j)$, where $S(x)$ stands for *x can solve this problem*, and j denotes John. An equivalent formulation would be $(\forall x)(S(x) \Rightarrow S(j))$.

11.   The sentence

> A sequence converges only if it is bounded

can be rewritten as $(\forall x)([S(x) \wedge C(x)] \Rightarrow B(x))$, where $S(x)$, $C(x)$, $B(x)$ stand for *x is a sequence*, *x converges*, and *x is bounded*, respectively.

12.   The sentence

> The sequence s converges to b

can be written as

$$(\forall \varepsilon)(\varepsilon > 0 \Rightarrow (\exists n_0)(I(n_0) \wedge (\forall n)(I(n) \wedge n \geq n_0 \Rightarrow |s_n - b| < \varepsilon)).$$

Here, $I(x)$ stands for *x is a positive integer* and the other symbols have their usual meaning. This can be roughly paraphrased in English as: No matter what positive ε we take, if we go far enough along (past the n_0th term) in the sequence, all the terms from that term on will be within ε of b. The negation of this formula is equivalent to

$$(\exists \varepsilon)(\varepsilon > 0 \wedge \neg (\exists n_0)(I(n_0) \wedge (\forall n)(I(n) \wedge n \geq n_0 \Rightarrow |s_n - b| < \varepsilon))).$$

(Remember that $\neg (A \Rightarrow B)$ is equivalent to $A \wedge \neg B$.) By reformulating the inner negated formula, and remembering that $\neg (A \wedge B)$ is equivalent to $A \Rightarrow \neg B$, we obtain

$$(\exists \varepsilon)(\varepsilon > 0 \wedge (\forall n_0)(I(n_0) \Rightarrow \neg (\forall n)(I(n) \wedge n \geq n_0 \Rightarrow |s_n - b| < \varepsilon))).$$

Again transforming $\neg (\forall n)$ into $(\exists n)\neg$, we have

$$(\exists \varepsilon)(\varepsilon > 0 \wedge (\forall n_0)I(n_0) \Rightarrow (\exists n)(I(n) \wedge n \geq n_0 \wedge |s_n - b| \not< \varepsilon))$$

A rough English version would read: There is some positive ε such that no matter how far we go along in the sequence there will still be a later term which is at a distance at least ε away from b.

Sometimes it is convenient to apply a special uniqueness symbol: $(\exists!x)A(x)$ shall mean that *there is one and only one x such that $A(x)$.* Equivalent English sentences are: *there is a unique x such that $A(x)$; there is exactly one x such that $A(x)$; there is precisely one x such that $A(x)$.* We can define $(\exists!x)(A(x)$ as follows

$$(\exists x)[A(x) \wedge (\forall z)(A(z) \Rightarrow z = x)].$$

EXAMPLES

13. Consider the sentence

For any integer x, there is a unique integer y such that $x + y = 0$.

This can be formulated as

$$(\forall x)(I(x) \Rightarrow (\exists!y)(I(y) \wedge x + y = 0)).$$

14. The sentence

Every person has one and only one father

can be written as $(\forall x)(P(x) \Rightarrow (\exists!y)(F(y, x)))$, where $P(x)$ and $F(y, x)$ stand for *x is a person* and *y is the father of x*, respectively.

15. Notice that $(\exists!y)A(y)$ can be written in various equivalent forms:
  i. $[(\exists y)A(y)] \wedge [(\forall u)(\forall v)(A(u) \wedge A(v) \Rightarrow u = v)].$
  ii. $(\exists y)(\forall z)(A(z) \Leftrightarrow z = y).$

If we read the sentence of Example 14 in versions (i) and (ii), we have:
  i.   For every person x, there is a father of x, and any two fathers of x are identical.
  ii. For every person x, there is a person y such that any person z is a father of x if and only if z is identical with y.

16. The sentence

Every cat has more than one life

can be written as

$$(\forall x)[C(x) \Rightarrow (\exists y)(\exists z)(y \neq z \wedge L(y, x) \wedge L(z, x))].$$

Here, $C(x)$ and $L(u, v)$ mean *x is a cat* and *u is a life of v*, respectively.

17. The sentence

Monotheism is false

can be expressed as $\neg(\exists!x)G(x)$, where $G(x)$ means *x is a god*.

18. Euclid's Parallel Postulate may be written

$$(\forall x)(\forall y)(L(x) \wedge P(y) \wedge \neg I(x, y)$$
$$\Rightarrow (\exists!z)(L(z) \wedge I(z, y) \wedge \neg(\exists w)(P(w) \wedge I(z, w) \wedge I(x, w)))).$$

Here, $L(x)$, $P(x)$, $I(x, y)$ stand for *x is a line*, *x is a point*, and *y lies on x*, respectively.

EXERCISES

1. Translate the following English sentences into our symbolism.
    a. Not all American girls are beautiful.
    b. All American girls are not beautiful.
    c. There is a smallest positive integer.
    d. There is a rational number between any two real numbers.
    e. Not all drunkards are truculent.
    f. No one loves everybody.
    g. For every positive real number x, there is a unique real number y such that $2^y = x$.
    h. At least somebody cares about me.
    i. If a sequence has a limit, it has only one.
    j. Two nonparallel lines have exactly one point in common.
    k. All people are honest or no one is honest.
    l. Some people are honest and some people aren't honest.

2. Give idiomatic English versions of the following formulas.
    a. $(\forall x)(P(x) \Rightarrow (\exists y)(C(y) \wedge (\exists z)(T(z) \wedge S(x, y, z))))$,
        where $P(x)$, $C(x)$, $T(x)$, $S(x, y, z)$ mean *x is a student*, *x is a course*, *x is a bad teacher*, and *x studies y with z*, respectively.
    b. $(\forall x)(C(x) \Rightarrow (M(x) \Rightarrow N(x)))$,
        where $C(x)$, $M(x)$, $N(x)$ mean *x is a good chess player*, *x is a male*, and *x is neurotic*, respectively.
    c. $(\exists y)(P(y) \wedge (\exists x)(P(x) \wedge L(y, x)))$,
        where $P(y)$ and $L(x, y)$ mean *y is a person* and *x loves y*, respectively.
    d. $\neg(\exists x)[(\forall y)((D(y) \vee C(y)) \Rightarrow H(x, y)) \wedge B(x)]$,
        where $D(y)$, $C(y)$, $H(x, y)$, $B(x)$ mean *y is a dog*, *y is a child*, *x hates y*, and *x is all bad*, respectively.
    e. $(\exists!x)(A(x) \wedge \neg 0(x) \wedge S(x))$,
        where $A(x)$, $0(X)$, $S(x)$ mean *x is a real number*, *x = 0*, and *x = x²*, respectively.

f. $(\forall z)[([(\forall u)(E(u, z) \Rightarrow I(u))] \wedge (\exists u)E(u, z))$
$\Rightarrow (\exists!u)(E(u, z) \wedge (\forall v)(E(v, z) \Rightarrow L(u, v)))],$
where $E(u, z)$, $I(u)$, $L(u, v)$ mean $u \in z,$[†] u is a positive integer, and $u \le v$, respectively.

g. $(\forall x)(P(x) \wedge O(x) \Rightarrow \neg T(x)),$
where $P(x)$, $O(x)$, $T(x)$ mean x is a person, x is over 30, and x can be trusted, respectively.

h. $(\forall x)(\forall y)(P(x) \wedge P(y) \wedge \neg E(x, y) \Rightarrow (\exists!z)(L(z) \wedge I(z, x) \wedge I(z, y))),$
where $P(x)$, $E(x, y)$, $L(z)$, $I(z, x)$ mean x is a point, $x = y$, x is a line, and x lies on z, respectively.

3. The definition of f is continuous at x_0 is

$$(\forall \varepsilon)(\varepsilon > 0 \Rightarrow (\exists \delta)(\delta > 0 \wedge (\forall x)(|x - x_0| < \delta \Rightarrow |f(x) - f(x_0)| < \varepsilon))).$$

Find a simple formula for the denial: f is not continuous at x_0. Negation signs should apply only to the basic sentences, not to compound sentences.

1.5 SETS

By a **set** we mean any collection of objects.

Examples of sets: (i) the set of all positive real numbers; (ii) the set of all living American males; (iii) the set of all Kings of France.

The objects of which a set is composed are called the **elements** or **members** of the set. The elements of a set are said to **belong** to the set.

Of course, we have not given a precise definition of **set**, since the word **collection** has not been defined. Terms sometimes used as synonyms for **set** are **class**, **totality**, and **family**. It seems to be impossible to define these notions in terms of simpler ideas. Here, we shall assume that the reader grasps the intuitive meaning of the notion of *set*. (There are difficulties surrounding the usual naïve notion of *set*. In fact, contradictions can be derived from some of the classically accepted assumptions about sets. For this reason, an axiomatic treatment of set theory seems to be safest. Such an axiomatic approach is presented in Appendix E.)

Consider any property $P(x)$. We shall use

$$\{x : P(x)\}$$

to denote the set of all objects x such that $P(x)$ is true. In the case where a set consists of a finite number of elements $a_1, a_2, \ldots, a_n$, we shall also denote this set by $\{a_1, a_2, \ldots, a_n\}$. In particular, $\{b\}$ is a set whose only element is b, and $\{b, c\}$ is a set whose only elements are b and c.

[†] $u \in z$ means that z is a collection of objects and u is a member of this collection.

EXAMPLES

1. $\{b, c\} = \{x : x = b \lor x = c\}$.

2. $\{b, b\} = \{b\}$.

3. $\{x : x \text{ is an integer} \land x^2 = 1\} = \{-1, 1\}$.

4. $\{x : x \text{ is a real number} \land x^2 + 2x - 2 = 0\} = \{\sqrt{3} - 1, -\sqrt{3} - 1\}$.

5. $\{x : x \text{ has been a President of the United States}\}$
    $= \{\text{Washington, Adams, } \dots, \text{ Nixon}\}$.

6. $\{b, c\} = \{c, b\}$.

7. $\{b, c, d\} = \{c, b, d\}$.

1.6 MEMBERSHIP. EQUALITY AND INCLUSION OF SETS

To denote the fact that an object x is a member of a set A we use the customary notation

$$x \in A.$$

The negation of $x \in A$ will be written: $x \notin A$. Thus, $x \notin A$ stands for $\neg(x \in A)$.

EXAMPLES

1. $2 \in \{1, 2\}$.

2. Hume $\in \{x : x \text{ is a philosopher}\}$.

3. $3 \notin \{x : x^2 - x - 2 = 0\}$.

The basic property of **equality** of sets should be made explicit. Two sets are equal if and only if they contain the same members. Thus,

$$A = B \Leftrightarrow (\forall x)(x \in A \Leftrightarrow x \in B).$$

This means that equality of sets does not depend upon how they are defined but only on whether or not they happen to have the same members.

Another important notion is that of **inclusion** of sets.

Let A and B be sets. We say that A is **included** in B if and only if every member of A is a member of B. We shall use $A \subseteq B$ to denote the fact that

A is included in B. Thus,

$$A \subseteq B \Leftrightarrow (\forall x)(x \in A \Rightarrow x \in B).$$

Another way of asserting that A is included in B is to say that A is a **subset** of B.

Notice that any set A is a subset of itself, since every member of A is trivially again a member of A. Any subset of A which is different from A itself is called a **proper subset** of A. To denote that a set C is a proper subset of A, we use the notation $C \subsetneqq A$. Thus,

$$C \subsetneqq A \Leftrightarrow (C \subseteq A \wedge C \neq A).$$

It is clear that two sets are equal if and only if every element of one is an element of the other and vice versa. Thus,

$$A = B \Leftrightarrow (A \subseteq B \wedge B \subseteq A).$$

EXAMPLES

4. $\{1\} \subsetneqq \{1, 2\}$.
5. $\{x : x \text{ lives in New York}\} \subsetneqq \{x : x \text{ lives in the United States}\}$.
6. $\{x : x \text{ is an integer}\} \subsetneqq \{x : x \text{ is a rational number}\}$.

EXERCISES

Prove the following assertions.

1. $C \subsetneqq A \Leftrightarrow (C \subseteq A \wedge (\exists y)(y \in A \wedge y \notin C))$.
2. $A \neq B \Leftrightarrow (\exists y)([y \in B \wedge y \notin A] \vee [y \in A \wedge y \notin B])$.
3. $\{b, c\} = \{d\} \Leftrightarrow (b = c \wedge c = d)$.
4. $\{a, b\} = \{c, d\} \Leftrightarrow [(a = c \wedge b = d) \vee (a = d \wedge b = c)]$.

1.7 THE EMPTY SET

If we consider the set $\{x : x \neq x\}$, it is obvious that this set has no members.[†] It is called the **empty set** or **null set** and is denoted $\emptyset$.[‡] Thus,

$$(\forall x)(x \notin \emptyset).$$

---

† We take it for granted that $x \neq x$ is always false. An explanation of our use of equality is given in Appendix A.

‡ This is a letter of the Danish alphabet.

Observe that there is only one empty set. For, if two sets A and B both contain no members, then they have precisely the same members (namely, none at all), and, therefore, $A = B$.

Notice also that the empty set is a subset of every set:

$$(\forall B)(\varnothing \subseteq B).$$

For, the sentence $x \in \varnothing \Rightarrow x \in B$ is trivially true. (Remember that a sentence $P \Rightarrow Q$ is considered true whenever P is false.) Thus, $(\forall x)(x \in \varnothing \Rightarrow x \in B)$ is true, which means that $\varnothing \subseteq B$.

EXAMPLES

1. $\{x : x = 1 \wedge x \neq 1\} = \varnothing$.

2. $\{x : x \text{ is a married bachelor}\} = \varnothing$.

3. $\{x : x \text{ is an integer} \wedge 0 < x \wedge x < 1\} = \varnothing$.

EXERCISES

1. Prove: $\varnothing \in \{\varnothing\}$.

2. Prove: $\varnothing \notin \{1, 2\}$. (Thus, although $\varnothing$ must be a subset of every set, it is not necessarily a member of any given set.)

3. Verify the observation above that

$$[(\forall x)(x \notin B) \wedge (\forall x)(x \notin A)] \Rightarrow A = B.$$

1.8   UNION AND INTERSECTION

Consider any two sets A and B. By $A \cap B$ we mean the set of all objects which belong to both A and B. Thus,

$$A \cap B = \{x : x \in A \wedge x \in B\}.$$

$A \cap B$ is called the **intersection** of A and B.

EXAMPLES

1. $\{1, 2, 4\} \cap \{2, 3, 4\} = \{2, 4\}$.

2. $\{x : x \text{ is an adult male}\} \cap \{x : x \text{ is unmarried}\}$
    $= \{x : x \text{ is a bachelor}\}$.

3. $\{x : x \text{ is an integer} \wedge x \geq 0\} \cap \{x : x \text{ is an integer} \wedge x \leq 0\} = \{0\}$.

4. $\{1, 3, 5\} \cap \{2, 4\} = \varnothing$.

Observe the following obvious facts about intersections.

i. $A \cap A = A$.

ii. $A \cap B = B \cap A$.

iii. $A \cap \varnothing = \varnothing$.

Terminology When $A \cap B = \varnothing$, we say that A and B are **disjoint**. Thus, A and B are disjoint if and only if A and B have no elements in common.

Let A and B, be sets. By $A \cup B$ we mean the set of all objects which belong to A or B or both. Thus,

$$A \cup B = \{x : x \in A \vee x \in B\}.$$

$A \cup B$ is called the **union** of A and B.

EXAMPLES

5. $\{1, 2, 3\} \cup \{2, 4, 5\} = \{1, 2, 3, 4, 5\}$.

6. $\{x : x \text{ is an integer} \wedge x > 2\} \cup \{x : x \text{ is an integer} \wedge x < 5\}$
    $= \{x : x \text{ is an integer}\}$.

7. $\{1, 2\} \cup \varnothing = \{1, 2\}$.

We leave the following simple facts to be checked by the reader.

iv. $A \cup A = A$.

v. $A \cup B = B \cup A$.

vi. $A \cup \varnothing = A$.

Facts about the operations $\cap$ and $\cup$ usually can be proved by using corresponding logical laws.

EXAMPLE

8. Let us prove a **distributive law** for $\cap$ with respect to $\cup$:

$$A \cap (B \cup C) = (A \cap B) \cup (A \cap C).$$

Proof

$$x \in A \cap (B \cup C) \Leftrightarrow [x \in A \land x \in (B \cup C)]$$
$$\Leftrightarrow x \in A \land [x \in B \lor x \in C]$$
$$\Leftrightarrow [x \in A \land x \in B] \lor [x \in A \land x \in C]$$
$$\Leftrightarrow [x \in A \cap B] \lor [x \in A \cap C]$$
$$\Leftrightarrow x \in (A \cap B) \cup (A \cap C).$$

This shows that $A \cap (B \cup C)$ and $(A \cap B) \cup (A \cap C)$ have the same members and, therefore, are equal. Notice that the third line of the proof was obtained by using the obvious logical fact that $P \land (Q \lor R)$ is equivalent to $(P \land Q) \lor (P \land R)$, that is, that $P \land (Q \lor R)$ always has the same truth values as $(P \land Q) \lor (P \land R)$. This can be checked by looking at the eight possible ways of assigning truth values to P, Q, and R (see Exercise 1.3d, p. 4).

EXERCISES

1. $\{1, 2, 3,\} \cup \{4, 5, 6\} = ?$

2. $\{1, 2, 3\} \cap \{4, 5, 6\} = ?$

3. $\{0, 2, 4, 6, \ldots\} \cup \{1, 3, 5, \ldots\} = ?$

4. $\{0, 2, 4, 6, \ldots\} \cap \{1, 3, 5, \ldots\} = ?$

5. $\{x : x \text{ is an integer } \land x < 2\} \cup \{x : x \text{ is an integer } \land x > 0\} = ?$

6. Same as Exercise 5, with $\cup$ replaced by $\cap$.

7. Prove: $A \cap B \subseteq A \land A \cap B \subseteq B$.

8. Prove: $A \subseteq A \cup B \land B \subseteq A \cup B$.

9. Prove: $(A \cup B) \cup C = A \cup (B \cup C)$.

10. Prove: $(A \cap B) \cap C = A \cap (B \cap C)$.

11. Prove: $A \subseteq B \Leftrightarrow A \cap B = A$.

12. Prove: $A \subseteq B \Leftrightarrow A \cup B = B$.

13. Prove: $A \cup (B \cap C) = (A \cup B) \cap (A \cup C)$.

14. Prove: $A \cap B = A \cup B \Leftrightarrow A = B$.

15. If $A \cup C = B \cup C$ and $A \cap C = B \cap C$, show that $A = B$.

1.9 DIFFERENCE AND COMPLEMENT

Consider any sets A and B. Let $A - B$ stand for the set of all objects which are in A but not in B. Thus,

$$A - B = \{x : x \in A \wedge x \notin B\}.$$

$A - B$ is called the **difference** of A and B.

EXAMPLES

1. $\{1, 2, 3\} - \{1, 4\} = \{2, 3\}$.
2. $\{1, 2, 3\} - \{4, 5, 6\} = \{1, 2, 3\}$.
3. $\{1, 2\} - \{1, 2, 3\} = \varnothing$.
4. $\{x : x$ is an integer$\} - \{x : x$ is an even integer$\}$
      $= \{x : x$ is an odd integer$\}$.

Notice the following obvious facts;

i. $A - B \subseteq A$.
ii. $A - \varnothing = A$.

We very often restrict our attention to the elements and subsets of some particular set U. For example, in number theory we talk about integers and sets of integers. In that case, U is the set of all integers. In sociology, one talks about people and sets of people. In that case, U is the set of all people. In each case, the set U is temporarily called the **universal set**.

If A is a subset of the universal set U, then by the **complement** A' we mean the set of all objects in U which are not in A. Thus,

$$A' = \{x : x \in U \wedge x \notin A\}.$$

EXAMPLES

5. Let U be the set of all integers. Let A and B be the set of even integers and the set of odd integers, respectively. Then $A' = B$ and $B' = A$. Also,

$$\{x : x \text{ is a positive integer}\}' = \{0\} \cup \{x : x \text{ is a negative integer}\}.$$

6. Let U be the set of all people. Let M and F be the set of all males and the set of all females, respectively. Then $M' = F$ and $F' = M$. In addition let A be the set of all Americans. Then $(M \cap A)' = F \cup A'$, that is, the

complement of the set of all American males is the union of the set of all females and the set of all non-American people.

Some obvious properties of complements are:

iii. $A'' = A$ ("double negation").

iv. $\emptyset' = U$

 v. $U' = \emptyset$.

EXERCISES

1. Prove: $A \subseteq B \Leftrightarrow A - B = \emptyset$.

2. Assume that we are given a universal set U. Prove:

    a. $A - B = A \cap B'$.

    b. $A \subseteq B \Leftrightarrow B' \subseteq A'$.

    c. $(A \cap B)' = A' \cup B'$.

    d. $(A \cup B)' = A' \cap B'$.

    e. $(A \cup B) - (A \cap B) = (A - B) \cup (B - A)$.

3. Let U be the set of all integers. Let E, O, P, N be the sets of all even, odd, positive, negative integers, respectively. Find $E - P, P - E, P \cup N, P', N'$. Which pairs of sets are disjoint?

1.10 POWER SET

Let A be any set. The **power set** $\mathscr{P}(A)$ is defined to be the set of all subsets of A:

$$\mathscr{P}(A) = \{B : B \subseteq A\}.$$

Notice that $\emptyset$ and A are always members of $\mathscr{P}(A)$.

EXAMPLES

1.   Consider the set $A = \{b, c\}$. The subsets of A are $\emptyset$, $\{b\}$, $\{c\}$, and $\{b, c\}$. Thus,

$$\mathscr{P}(\{b, c\}) = \{\emptyset, \{b\}, \{c\}, \{b, c\}\}.$$

2. $\mathscr{P}(\{b\}) = \{\emptyset, \{b\}\}$.

3. $\mathscr{P}(\emptyset) = \{\emptyset\}$.

4. $\mathscr{P}(\{b, c, d\}) = \{\emptyset, \{b\}, \{c\}, \{d\}, \{b, c\}, \{b, d\}, \{c, d\}, \{b, c, d\}\}$.

Observe from these examples that a set with no elements has 1 subset, a set with one element has 2 subsets, a set with two elements has 4 subsets, and a set with three elements has 8 subsets. These are special cases of the general result that, if a set A has n elements, then $\mathscr{P}(A)$ has 2^n elements (see Exercise 5 below).

EXERCISES

1. Prove: $A \subseteq B \Leftrightarrow \mathscr{P}(A) \subseteq \mathscr{P}(B)$.

2. Prove: $\mathscr{P}(A) \cap \mathscr{P}(B) = \mathscr{P}(A \cap B)$.

3. Give a counterexample to show that $\mathscr{P}(A) \cup \mathscr{P}(B) = \mathscr{P}(A \cup B)$ is not always true.

4. Prove: $\mathscr{P}(A) = \mathscr{P}(B) \Rightarrow A = B$.

5. If A is finite and has n elements, prove that the power set $\mathscr{P}(A)$ has 2^n elements. (*Hint*: In choosing a subset of A, there are two possibilities with respect to each of the n elements, that is, it is in the set or not in it.)

1.11 ARBITRARY UNIONS AND INTERSECTIONS

Let $\mathscr{X}$ be a set of sets, that is, the members of $\mathscr{X}$ are themselves sets. By the **union** of $\mathscr{X}$ we mean the set of all objects which are in at least one set in $\mathscr{X}$. We denote the union of $\mathscr{X}$ by $\bigcup_{A \in \mathscr{X}} A$ or $\bigcup_{A \in \mathscr{X}} A$. Thus,

$$\bigcup_{A \in \mathscr{X}} A = \{y : (\exists B)(B \in \mathscr{X} \wedge y \in B)\}.$$

The expression $\bigcup_{A \in \mathscr{X}} A$ is sometimes read: the union of all the sets in $\mathscr{X}$.

EXAMPLES

1. Let $\mathscr{X} = \{\{1\}, \{1, 3\}, \{2, 4, 6\}\}$. Then $\bigcup_{A \in \mathscr{X}} A = \{1, 2, 3, 4, 6\}$.

2. Let $A_n = \{1, 2, 3, \ldots, n\}$, and let $\mathscr{X} = \{A_1, A_2, A_3, \ldots\}$. Then $\bigcup_{A \in \mathscr{X}} A$ is the set of all positive integers.

3. Let $\mathscr{X} = \{B, C\}$. Then $\bigcup_{A \in \mathscr{X}} A = B \cup C$. For, the members of $\bigcup_{A \in \mathscr{X}} A$ are those objects which belong to at least one set in $\{B, C\}$, that is, those objects which belong to at least one of B and C. This example shows that the union operation introduced in Section 1.8 is a special case of the arbitrary union operation.

Let $\mathscr{X}$ be a nonempty set of sets. By the **intersection** of $\mathscr{X}$ we mean the set of all objects which belong to every set in $\mathscr{X}$. It is denoted $\bigcap_{A \in \mathscr{X}} A$ or $\bigcap_{A \in \mathscr{X}} A$. Thus,

$$\bigcap_{A \in \mathscr{X}} A = \{ y : (\forall B)(B \in \mathscr{X} \Rightarrow y \in B) \}.$$

The expression $\bigcap_{A \in \mathscr{X}} A$ is sometimes read: the intersection of all the sets in $\mathscr{X}$.

EXAMPLES

4.   Let $\mathscr{X} = \{\{1, 2\}, \{1, 3, 4\}, \{4, 1\}\}$. Then $\bigcap_{A \in \mathscr{X}} A = \{1\}$.

5.   Let $B_n = W - \{1, 2, \ldots, n\}$, where W is the set of all positive integers. Let $\mathscr{X} = \{B_1, B_2, \ldots\}$. Then $\bigcap_{A \in \mathscr{X}} A = \varnothing$.

6.   Let $\mathscr{X}$ be the set of all sets of integers which contain 1. Then $\bigcap_{A \in \mathscr{X}} A = \{1\}$.

7.   Let $\mathscr{X} = \{B, C\}$. Then $\bigcap_{A \in \mathscr{X}} A = B \cap C$. For, an object belongs to every set in $\mathscr{X}$ if and only if the object belongs to both B and C. This shows that the intersection operation of Section 1.8 is a special case of the arbitrary intersection operation.

EXERCISES

1.   Prove: $B \in \mathscr{X} \Rightarrow B \subseteq \bigcup_{A \in \mathscr{X}} A$.

2.   Prove: $B \in \mathscr{X} \Rightarrow \bigcap_{A \in \mathscr{X}} A \subseteq B$.

3.   Consider a nonempty set W. Let $\mathscr{X}$ be the set of all nonempty finite subsets of W. Find $\bigcup_{A \in \mathscr{X}} A$ and $\bigcap_{A \in \mathscr{X}} A$.

4.   Prove: $\varnothing \in \mathscr{X} \Rightarrow \bigcap_{A \in \mathscr{X}} A = \varnothing$.

5.   Prove: $(\forall B)(B \in \mathscr{X} \Rightarrow C \subseteq B) \Rightarrow C \subseteq \bigcap_{A \in \mathscr{X}} A$.

6.   If $(\forall B)(B \in \mathscr{X} \Rightarrow B \subseteq C)$, show that $\bigcup_{A \in \mathscr{X}} A \subseteq C$.

7.   Let $\mathscr{X}$ be the set of all sets of integers which contain either 1 or 2. Find $\bigcup_{A \in \mathscr{X}} A$ and $\bigcap_{A \in \mathscr{X}} A$.

8.   Let $\mathscr{X}$ be the set of all infinite sets of integers. Find $\bigcup_{A \in \mathscr{X}} A$ and $\bigcap_{A \in \mathscr{X}} A$.

Note   In defining an arbitrary intersection $\bigcap_{A \in \mathscr{X}} A$ we restricted ourselves to the case where $\mathscr{X}$ is nonempty. When $\mathscr{X}$ is empty, $B \in \mathscr{X}$ is always false, and, therefore, $B \in \mathscr{X} \Rightarrow y \in B$ would be trivially true. Hence, when

$\mathscr{X}$ is empty, $(\forall B)(B \in \mathscr{X} \Rightarrow y \in B)$ would be true, no matter what y is. Hence, $\{y : (\forall B)(B \in \mathscr{X} \Rightarrow y \in B)\}$ would be a set containing all objects. But, in most systems of set theory, the existence of such a set would lead to a contradiction (see Appendix E). Therefore, we restrict intersections to nonempty sets.

1.12 ORDERED PAIRS

In mathematics we frequently have need of forming the ordered pair (b, c) of two given objects b and c. For example, in analytic geometry, the points in a plane are represented by ordered pairs (b, c), where b and c are arbitrary real numbers. The order of the objects b and c in the ordered pair (b, c) is important. For example, the point represented by $(1, 2)$ is different from the point represented by $(2, 1)$, see Figure 1.1. For all mathematical

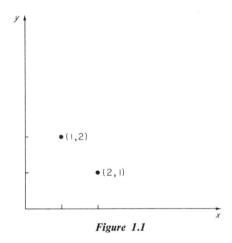

Figure 1.1

applications, the ordered pair (b, c) can be any object associated with b and c, the only condition being that, if we change b or c or both, then we obtain a different ordered pair. This condition can be expressed as

$$(b, c) = (u, v) \Rightarrow [b = u \wedge c = v].$$

In other words, if the ordered pairs (b, c) and (u, v) are the same, then the only case in which this can happen is the case where the first components b and u are the same and the second components c and v are also identical.

It turns out that there is a simple way of defining ordered pairs so that the basic condition is satisfied. (Of course, if such a definition were not possible, we would have to assume the existence of ordered pairs satisfying the basic condition.)

Definition $(b, c) = \{\{b\}, \{b, c\}\}$.

This definition has no intrinsic correctness, that is, it is not the case that this is what we always have meant by the ordered pair (b, c). All that we claim for this rather artificial definition is that, using it, we can prove the basic condition for ordered pairs. This we shall now show.

Theorem 12.1 $(b, c) = (u, v) \Rightarrow [(b = u) \wedge (c = v)]$.

Proof   Assume $(b, c) = (u, v)$. This means:

$$(1) \quad \{\{b\}, \{b, c\}\} = \{\{u\}, \{u, v\}\}.$$

We now have to show that $b = u$ and $c = v$. Since $\{b\}$ belongs to the left side of (1), it also belongs to the right side: $\{b\} \in \{\{u\}, \{u, v\}\}$. Hence, $\{b\} = \{u\} \vee \{b\} = \{u, v\}$. In either case, $b = u$. This is half of what we are trying to prove. We must still show that $c = v$.

Case 1 $u = v$. Then the right side of (1) is $\{\{v\}\}$. But, $\{b, c\}$ belongs to the left side of (1) and, therefore, belongs to the right side of (1): $\{b, c\} \in \{\{v\}\}$. Hence, $\{b, c\} = \{v\}$. This implies $c = v$.

Case 2 $u \neq v$. Now, $\{u, v\}$ belongs to the right side of (1) and, therefore, belongs to the left side: $\{u, v\} \in \{\{b\}, \{b, c\}\}$. Hence,

$$\{u, v\} = \{b\} \vee \{u, v\} = \{b, c\}.$$

But, since $u \neq v$, it follows that $\{u, v\} \neq \{b\}$. Therefore, $\{u, v\} = \{b, c\}$. This implies $\{u, v\} = \{u, c\}$, since $b = u$. Since $v \in \{u, v\}$, it follows that $v \in \{u, c\}$. But, $v \neq u$. Hence, $v = c$.

In both cases, we have shown that $c = v$.   ■

Using ordered pairs, we can define **ordered triples, ordered quadruples,** etc., in the following way.

Definition $(b, c, d) = ((b, c), d)$.

Definition $(b, c, d, e) = ((b, c, d), e)$.

EXERCISES

1. Prove:

$(b, c, d) = (u, v, w) \Rightarrow [b = u \wedge c = v \wedge d = w]$.

2. Prove:

$(b, c, d, e) = (u, v, w, x) \Rightarrow [b = u \wedge c = v \wedge d = w \twoheadrightarrow \wedge e = x]$.

1.13 CARTESIAN PRODUCT

Let A and B be sets. By the **Cartesian product** $A \times B$ we mean the set of all ordered pairs (a, b) such that $a \in A$ and $b \in B$. Thus,

$$A \times B = \{x : (\exists a)(\exists b)(x = (a, b) \wedge a \in A \wedge b \in B)\}.$$

We shall use the customary shorter way of writing:

$$A \times B = \{(a, b) : a \in A \wedge b \in B\}.$$

More generally, $\{(a, b) : P(a, b)\}$ shall represent the set of all ordered pairs (a, b) such that $P(a, b)$ is true.

EXAMPLES

1. Let $A = \{1, 2\}$ and $B = \{2, 3, 4\}$. Then

$$A \times B = \{(1, 2), (1, 3), (1, 4), (2, 2), (2, 3), (2, 4)\}.$$

2. If A and B are both $\{1, 2\}$, then

$$A \times B = \{(1, 1), (1,2), (2, 1), (2, 2)\}.$$

Notation

$$A^2 = A \times A,$$

$$A^3 = A^2 \times A = (A \times A) \times A,$$

$$A^4 = A^3 \times A, \quad \text{etc.}$$

In analytic geometry, if A is the set of real numbers, then A^2 represents the set of points in the plane, A^3 represents the set of points in three-dimensional space, etc.

EXERCISES

1. Let $A = \{a\}$ and $B = \{c, d\}$. Find $A \times B$ and $B \times A$.

2. Let $A = \{a, b\}$ and $B = \{c, d\}$. Find $A \times B$.

3. If A and B are finite sets, A has n elements, and B has k elements, how many elements does $A \times B$ have?

4. a. $A \times \varnothing = ?$
    b. $\varnothing \times A = ?$

5. Show that $A \times B = B \times A \Leftrightarrow [A = \varnothing \lor B = \varnothing \lor A = B]$.

6. Show that $(A \times B) \times C$ is the set of all ordered triples (a, b, c) such that $a \in A$, $b \in B$, and $c \in C$.

7. Let A be the set of real numbers. Describe the geometric meaning of the following subsets of $A \times A$.
    i. $\{(b, c) : b \in A \land c \in A \land b = c\}$.
    ii. $\{(b, c) : b \in A \land c \in A \land b = -c\}$.
    iii. $\{(b, c) : b \in A \land c \in A \land b = 2\}$.

1.14   RELATIONS

By a **relation**† we mean a set of ordered pairs.

EXAMPLES

1.   Let R be the set of all ordered pairs (b, c) such that b is the father of c. R is usually referred to as the fatherhood relation.

2.   Let S be the set of all ordered pairs (b, c) such that b and c are the same integers. S is called the identity relation on the set of integers.

3.   Let L be the set of all ordered pairs (b, c) such that b and c are integers and $b < c$. L is called the less-than relation on the set of integers.

Definitions   Consider a relation R. By the **domain** of R we mean the set $\mathscr{D}(R)$ of all objects x such that $(x, y) \in R$ for some y. Thus,

$$\mathscr{D}(R) = \{x : (\exists y)((x, y) \in R)\}.$$

---

† What we call a relation is actually a binary relation. We shall have no need for a special treatment of relations having more than two arguments.

By the **range** of R we mean the set $\mathscr{R}(R)$ of all objects y such that $(x, y) \in R$ for some x. Thus,

$$\mathscr{R}(R) = \{y : (\exists x)((x, y) \in R)\}.$$

EXAMPLES

4.   Let A be the set of positive integers. Let $R = \{(x, y) : x \in A \wedge y \in A \wedge x < y\}$. Then

$$\mathscr{D}(R) = A \quad \text{and} \quad \mathscr{R}(R) = A - \{1\}.$$

5.   Let B be the set of all real numbers. Let $R = \{(x, y) : x \in B \wedge y \in B \wedge x^2 = y\}$. Then

$$\mathscr{D}(R) = B \quad \text{and} \quad \mathscr{R}(R) = \{x : x \in B \wedge x \geq 0\}.$$

6.   Let P be the set of all people, and let

$$R = \{(x, y) : x \in P \wedge y \in P \wedge x \text{ is the father of } y\}.$$

Then $\mathscr{D}(R)$ is the set of all fathers, and $\mathscr{R}(R)$ is the set of all people.

EXERCISES

1.   Let B be the set of all real numbers. Determine the domain and range of the following relations.
   a. $R = \{(x, y) : x \in B \wedge y \in B \wedge y = |x|\}.$[†]
   b. $R = \{(x, y) : x \in B \wedge y \in B \wedge x + y = 0\}.$
   c. $R = \{(x, y) : x \in B \wedge y \in B \wedge |x| = |y|\}.$
   d. $R = \{(x, y) : x \in B \wedge y \in B \wedge x < y\}.$

2.   If R is a relation, show that $R \subseteq \mathscr{D}(R) \times \mathscr{R}(R)$.

1.15 INVERSE AND COMPOSITION OF RELATIONS

Let R be a relation. By the **inverse** R^{-1} of R we mean the set of all ordered pairs (x, y) for which $(y, x) \in R$. Thus,

$$R^{-1} = \{(x, y) : (y, x) \in R\}.$$

[†] $|x|$ is the absolute value of x:

$$|x| = \begin{cases} x & \text{if } x \geq 0 \\ -x & \text{if } x < 0 \end{cases}$$

R^{-1} is obtained from R by switching the first and second components of every ordered pair in R.

EXAMPLES

1. Let B be the set of real numbers. Let $R = \{(x, y) : x \in B \land y \in B \land x < y\}$. R is the less-than relation. Then, for any x and y in B,

$$(x, y) \in R^{-1} \Leftrightarrow (y, x) \in R$$
$$\Leftrightarrow y < x$$
$$\Leftrightarrow x > y.$$

Thus, R^{-1} is the greater-than relation.

2. Let B be the set of all people. Let $R = \{(x, y) : x \in B \land y \in B \land x \text{ is a parent of } y\}$. Then, for any x and y in B,

$$(x, y) \in R^{-1} \Leftrightarrow (y, x) \in R$$
$$\Leftrightarrow y \text{ is a parent of } x$$
$$\Leftrightarrow x \text{ is a child of } y.$$

3. Let B be the set of all people. Let $R = \{(x, y) : x \in B \land y \in B \land x \text{ is the spouse of } y\}$. Then, for any x and y in B,

$$(x, y) \in R^{-1} \Leftrightarrow (y, x) \in R$$
$$\Leftrightarrow y \text{ is the spouse of } x$$
$$\Leftrightarrow x \text{ is the spouse of } y$$
$$\Leftrightarrow (x, y) \in R.$$

Thus, in this special case, $R^{-1} = R$.

4. Let B be the set of real numbers. Let $R = \{(x, y) : x \in B \land y \in B \land y = x + 1\}$. Then, for any x and y in B,

$$(x, y) \in R^{-1} \Leftrightarrow (y, x) \in R$$
$$\Leftrightarrow x = y + 1$$
$$\Leftrightarrow y = x - 1.$$

Thus, subtraction of 1 is the inverse of addition of 1.

At this point, we should like to introduce the following traditional notation: If R is a relation,

$$xRy \qquad \text{stands for} \qquad (x, y) \in R.$$

For example, one usually writes $x < y$ instead of $(x, y) \in <$, and one writes x is the father of y instead of $(x, y) \in F$, where F is the fatherhood relation.

Consider any relations R and S. Let us define a new relation $S \circ R$, called the **composition** of R and S:

$$S \circ R = \{(x, y) : (\exists z)(xRz \wedge zSy)\}.$$

Thus, $x(S \circ R)y$ means that there is an "intermediate" object z such that xRz and zSy. Notice that R is applied before S.

EXAMPLES

5.   Let A be the set of people. Let F be the fatherhood relation, M the motherhood relation, and P the parenthood relation. Thus, $F = \{(x, y) : x \in A \wedge y \in A \wedge x$ is the father of $y\}$, $M = \{(x, y) : x \in A \wedge y \in A \wedge x$ is the mother of $y\}$, and $P = \{(x, y) : x \in A \wedge y \in A \wedge x$ is a parent of $y\}$. Then,

a. $x(F \circ P)y \Leftrightarrow (\exists z)(x$ is a parent of $z \wedge z$ is the father of $y)$. Thus, $F \circ P$ is the relation of *paternal grandparent*.

b. $x(P \circ F)y \Leftrightarrow (\exists z)(x$ is the father of $z \wedge z$ is a parent of $y)$. Thus, $P \circ F$ is the *grandfather* relation.

Examples a and b show that the relations $R \circ S$ and $S \circ R$ are not necessarily the same (and, in fact, are usually different).

c. $x(F \circ F)y \Leftrightarrow (\exists z)(x$ is the father of $z \wedge z$ is the father of $y)$. Thus, $F \circ F$ is the *paternal grandfather* relation.

d. $x(P \circ P)y \Leftrightarrow (\exists z)(x$ is a parent of $z \wedge z$ is a parent of $y)$. Thus, $P \circ P$ is the *grandparent* relation.

6.   Let B be the set of integers. Let $R = \{(x, y) : x \in B \wedge y \in B \wedge y = x + 1\}$, and $S = \{(x, y) : x \in B \wedge y \in B \wedge y = x^2\}$. Then, for any integers x and y:

$$x(R \circ S)y \Leftrightarrow (\exists z)(xSz \wedge zRy)$$
$$\Leftrightarrow (\exists z)(z = x^2 \wedge y = z + 1)$$
$$\Leftrightarrow y = x^2 + 1.$$

On the other hand,

$$x(S \circ R)y \Leftrightarrow (\exists z)(xRz \wedge zSy)$$
$$\Leftrightarrow (\exists z)(z = x + 1 \wedge y = z^2)$$
$$\Leftrightarrow y = (x + 1)^2$$
$$\Leftrightarrow y = x^2 + 2x + 1.$$

EXERCISES

1. As in Example 5, let A be the set of people, and let F, M, P be the fatherhood, motherhood, parenthood relations, respectively. In addition, let B, S, H, W be the brother, sister, husband, wife relations, respectively.

    a. Verify that $P^{-1} \circ H$ is the son-in-law relation.

    b. Find simple names, if possible, for: H^{-1}, $F \cup M$, $H \cup W$, $M \circ F$, $F \circ M$, $P \circ M$, $M \circ P$, $P \circ B$, $(H \cup W) \circ F$, $(P \circ B) \cup ((P \circ S) \circ H)$.

    c. Express in symbols the following relations: child, sibling, grandchild, daughter-in-law, mother-in-law.

2. Show that, if R is a relation, then

    $$\mathscr{D}(R^{-1}) = \mathscr{R}(R) \quad \text{and} \quad \mathscr{R}(R^{-1}) = \mathscr{D}(R).$$

3. Let B be the set of real numbers.
    Let

    $$S = \{(x, y) : x \in B \wedge y \in B \wedge y = x^2\},$$
    $$W = \{(x, y) : x \in B \wedge y \in B \wedge y = -x\}, \text{ and}$$
    $$R = \{(x, y) : x \in B \wedge y \in B \wedge y = x + 1\}.$$

    a. Describe the following relations: S^{-1}, R^{-1}, W^{-1}, $S \circ S$, $W \circ W$, $R \circ R$, $S \circ W$, $W \circ S$, $S \circ R$, $R \circ S$, $R \circ W$, $W \circ R$.

    b. Determine the domain and range of S, W, R, and of the relations in Part a.

1.16   REFLEXIVITY, SYMMETRY, AND TRANSITIVITY

Consider a relation R and a set A. R is said to be **reflexive** in A if and only if xRx for all x in A. We say that R is **irreflexive** in A if and only if xRx is false for all x in A. Thus,

$$R \text{ is reflexive in } A \Leftrightarrow (\forall x)(x \in A \Rightarrow xRx);$$

$$R \text{ is irreflexive in } A \Leftrightarrow (\forall x)(x \in A \Rightarrow \neg xRx).$$

EXAMPLES

1. Let A be the set of integers. Let

$$R = \{(x, y) : x \in A \wedge y \in A \wedge x \leq y\},$$
$$S = \{(x, y) : x \in A \wedge y \in A \wedge x < y\}$$
$$W = \{(x, y) : x \in A \wedge y \in A \wedge x^2 = y\},$$
$$I_A = \{(x, y) : x \in A \wedge y \in A \wedge x = y\}.$$

Then R is reflexive in A, S is irreflexive in A, and W is neither reflexive nor irreflexive in A. I_A, called the identity relation in A, is reflexive in A.

2. Let A be the set of people. Let

$$L = \{(x, y) : x \in A \wedge y \in A \wedge x \text{ loves } y\},$$
$$T = \{(x, y) : x \in A \wedge y \in A \wedge x \text{ has the same height as } y\}.$$

Then L is neither reflexive nor irreflexive in A, while T is reflexive in A.

We say that a relation R is **symmetric** in A if and only if xRy implies yRx for all x and y in A. Thus,

$$R \text{ is symmetric in } A \Leftrightarrow (\forall x)(\forall y)[(x \in A \wedge y \in A \wedge xRy) \Rightarrow yRx].$$

In Example 1, I_A is symmetric in A, while R, S, and W are not. In Example 2, L is not symmetric in A, while T is symmetric in A.

EXAMPLE

3. Let A be the set of people. Then the relations of *father*, *mother*, *parent*, *brother*, *sister*, *husband*, and *wife* are not symmetric in A. The following relations are symmetric in A:

a. $\{(x, y) : x \in A \wedge y \in A \wedge x \text{ is the spouse of } y\}$;
b. $\{(x, y) : x \in A \wedge y \in A \wedge x \text{ is in the same family as } y\}$;
c. $\{(x, y) : x \in A \wedge y \in A \wedge x \text{ is a sibling of } y\}$;
d. $\{(x, y) : x \in A \wedge y \in A \wedge x \text{ has the same father as } y\}$.

We say that a relation R is **transitive** in A if and only if $xRy \wedge yRz$ implies xRz for all x, y, and z in A. Thus,

R is transitive in $A \Leftrightarrow$
$$(\forall x)(\forall y)(\forall z)([x \in A \wedge y \in A \wedge z \in A \wedge xRy \wedge yRz] \Rightarrow xRz).$$

In Example 1, R, S, and I_A are transitive in A, but W is not. In Example 2, L is not transitive in A, but T is.

EXAMPLE

4.   Let T be the set of triangles in a plane. Let

$$R = \{(x, y) : x \in T \wedge y \in T \wedge x \text{ is congruent to } y\};$$

$$S = \{(x, y) : x \in T \wedge y \in T \wedge x \text{ is similar to } y\};$$

$$W = \{(x, y) : x \in T \wedge y \in T \wedge x \text{ has smaller area than } y\}.$$

Then R, S, and W are all transitive in T. (Notice that R and S also are reflexive and symmetric in T, while W is neither.)

EXERCISES

1.   Determine whether each of the following relations R is reflexive, symmetric, or transitive in the given set A.

a. A is the set of positive integers.

$R = \{(x, y) : x \in A \wedge y \in A \wedge y \text{ is divisible by } x\}.$

b. A is the set of all line segments in a plane.

$R = \{(x, y) : x \in A \wedge y \in A \wedge x \text{ is longer than } y\}.$

c. A is the set of all lines in a plane.

$R = \{(x, y) : x \in A \wedge y \in A \wedge x \text{ is perpendicular to } y\}.$

d. A is the set of all subsets of a given set B.

$R = \{(x, y) : x \in A \wedge y \in A \wedge x \cap y \neq \varnothing\}.$

e. A is the set of all subsets of a given set B.

$R = \{(x, y) : x \in A \wedge y \in A \wedge x \subseteq y\}.$

f. A is the set of all people.

$R = \{(x, y) : x \in A \wedge y \in A \wedge x \text{ is an ancestor of } y\}.$

g. A is any set.

$R = I_A = \{(x, y) : x \in A \wedge y \in A \wedge x = y\}.$

h. A is any set.

$R = \{(x, y) : x \in A \wedge y \in A \wedge x \neq y\}.$

2. A relation R is said to be **asymmetric** in A if and only if xRy implies $\neg(yRx)$ for all x and y in A. A relation R is said to be **antisymmetric** in A if and only if $xRy \wedge yRx$ implies $x = y$ for all x and y in A.

   a. Give examples of relations R and sets A such that

      i. R is asymmetric in A;
      ii. R is not asymmetric in A;
      iii. R is antisymmetric in A;
      iv. R is not antisymmetric in A.

   b. Prove that asymmetry implies antisymmetry.

   c. Let R be the $\leq$ relation on the set of integers. Show that R is antisymmetric but not asymmetric on the set of integers.

   d. If a relation R is irreflexive in a set A and transitive in A, prove that R is asymmetric in A.

3. Show that a relation R is reflexive in a set A if and only if $I_A \subseteq R$, where I_A is the identity relation on A.

4. Show that a relation R is irreflexive in a set A if and only if R and I_A are disjoint.

5. Assume $R \subseteq A \times A$. Show that R is symmetric in A if and only if $R = R^{-1}$.

6. Assume $R \subseteq A \times A$. Show that R is transitive in A if and only if $R \circ R \subseteq R$.

7. Let A be the set of integers. Find a relation R which is

   a. reflexive and symmetric in A but not transitive in A;
   b. reflexive and transitive in A but not symmetric in A;
   c. symmetric and transitive in A but not reflexive in A.

1.17 EQUIVALENCE RELATIONS

A relation R is said to be an **equivalence relation** in a set A if and only if R is reflexive, symmetric, and transitive in A.

EXAMPLES

1. Let A be any set. Then the identity relation $I_A = \{(x, y) : x \in A \wedge y \in A \wedge x = y\}$ is an equivalence relation in A.

2. Let T be the set of triangles in a plane. (a) The congruence relation $R = \{(x, y) : x \in T \wedge y \in T \wedge x \text{ is congruent to } y\}$ is an equivalence relation in T. (b) Likewise, the similarity relation $S = \{(x, y) : x \in T \wedge y \in T \wedge x \text{ is similar to } y\}$ is an equivalence relation in T.

3.   Let P be the set of people. Then the following are equivalence relations in P.

a. $R = \{(x, y) : x \in P \wedge y \in P \wedge x$ is the same height as $y\}$.

b. $S = \{(x, y) : x \in P \wedge y \in P \wedge x$ and y have the same mother and father$\}$.

4.   Let A be the set of integers. Let n be an integer greater than 1. Let $R = \{(x, y) : x \in A \wedge y \in A \wedge x - y$ is divisible by $n\}$. Then R is an equivalence relation in A. (We shall study this relation in more detail in Chapter 3, Section 6.)

Consider an equivalence relation R in a set A. Each element x of A determines the set of all elements z of A such that xRz. This set is denoted $[x]$, and called the **equivalence class** of x (relative to the relation R). Thus,

$$[x] = \{z : z \in A \wedge xRz\}.$$

(We should use the notation $[x]_R$ instead of $[x]$ in order to indicate the equivalence relation relative to which we are forming equivalence classes. However, where we are talking about only one equivalence relation or where it is obvious from the context which equivalence relation is meant, we shall omit the subscript R.)

EXAMPLES

5.   In Example 1, for any x in A,

$$[x] = \{z : z \in A \wedge x = z\}.$$

Hence, $[x] = \{x\}$. Thus, the equivalence classes in this example are the singletons $\{x\}$, where $x \in A$.

6.   In Example 2a, an equivalence class consists of all triangles congruent to a given triangle. With respect to the relation of similarity (Example 2b), an equivalence class consists of all triangles similar to a given triangle.

7.   In Example 3a, an equivalence class consists of all people of some fixed height. In Example 3b, an equivalence class consists of the children of some father and mother.

8.   In Example 4, an equivalence class consists of all integers leaving the same remainder (one of the numbers $0, 1, \ldots, n - 1$) upon division by n. (This will be explained in Chapter 3, Section 6.) Hence, there are precisely n equivalence classes.

Consider an equivalence relation R in a set A. Notice that every element x of A belongs to an equivalence class, namely $[x]$. In addition, two elements x and y determine the same equivalence class if and only if xRy. This can be written:

$$(I) \quad [x] = [y] \Leftrightarrow xRy.$$

Proof If $[x] = [y]$, then, since $y \in [y]$, we have $y \in [x]$, which means that xRy. On the other hand, if xRy, then, for any z,

$$(*) \quad xRz \Leftrightarrow yRz.$$

For, if xRz, then, since xRy implies yRx, we conclude by transitivity of R that yRz. Conversely, if yRz, then, together with xRy, this yields xRz. Thus, $(*)$ has been established. But, from $(*)$, $[x] = \{z : xRz\} = \{z : yRz\} = [y]$. ∎

In addition to (I), we see that, if two equivalence classes are not identical, then they are disjoint, that is,

$$(II) \quad [x] \neq [y] \Rightarrow [x] \cap [y] = \varnothing.$$

Proof Assume $[x] \cap [y] \neq \varnothing$. Then there is some z such that $z \in [x]$ and $z \in [y]$. Hence, xRz and yRz. From yRz we obtain zRy by symmetry. From xRz and zRy we obtain xRy by transitivity. Then, by (I), $[x] = [y]$. ∎

From (II) we see that an equivalence relation on a set A divides A into a collection of nonempty, mutually exclusive subsets (the equivalence classes).

EXERCISES

1. For each relation R and set A, determine whether R is an equivalence relation in A. If it is, describe the equivalence classes.

  a. A is the set of lines in a plane.

  $R = \{(x, y) : x \in A \wedge y \in A \wedge (x = y \text{ or } x \text{ is parallel to } y)\}$.

  b. A is the set of all people.

  $R = \{(x, y) : x \in A \wedge y \in A \wedge x \text{ is related to } y\}$.

c. A is the set of all directed line segments in a plane.

(A directed line segment is a line segment together with a choice of one of the two possible directions on the line.)

$R = \{(x, y) : x \in A \wedge y \in A \wedge x$ and y have the same length and the same direction$\}$.

d. A is the set of people.

$R = \{(x, y) : x \in A \wedge y \in A \wedge x$ and y have at least one parent in common$\}$.

e. A is the set of people.

$R = \{(x, y) : x \in A \wedge y \in A \wedge x$ and y live within one mile of each other$\}$.

f. A is the set of all points on the earth.

$R = \{(x, y) : x \in A \wedge y \in A \wedge x$ and y are at the same distance from the North Pole$\}$.

g. A is the set of all points inside a given circle, excluding the center of the circle.

$R = \{(x, y) : x \in A \wedge y \in A \wedge [x = y$ or the line connecting x and y passes through the center of the circle]$\}$.

h. A is the set of all subsets of the set of integers.

$R = \{(x, y) : x \in A \wedge y \in A \wedge (x - y) \cup (y - x)$ is finite.

(*Hint*: xRy if and only if x and y differ by only a finite number of elements.)

i. A is the set of all integers.

$R = \{(x, y) : x \in A \wedge y \in A \wedge |x| = |y|\}$.

1.18 FUNCTIONS

The notion of **function** is one of the most important mathematical ideas. A function F is customarily explained as a *way of associating* with each object x of some definite collection of objects (called the **domain** of the function) some definite object y (denoted $F(x)$ and called the **value** of F for the argument x). We shall give a more precise definition, eliminating the vague expression *way of associating*.

By a **function** we mean a relation F having the additional property:

$$[(x, y) \in F \wedge (x, z) \in F] \Rightarrow y = z.$$

This means that there cannot be two ordered pairs in F having the same first component but different second components.

A function F, being a relation, has a domain, namely, the set of first components of ordered pairs in F. Notice that, for each element x of the domain of F, there is one and only one object y such that $(x, y) \in F$. This unique object y shall be denoted $F(x)$. It is called the *value of F for the argument x*. (We often also say that F *takes x into* $F(x)$ or that F *maps x into* $F(x)$, or that F *transforms x into* $F(x)$.) This brings us back to the traditional notation for functions. Observe that the range $\mathscr{R}(F)$ of a function F simply consists of the values of the function F. For,

$$y \in \mathscr{R}(F) \Leftrightarrow (\exists x)((x, y) \in F)$$
$$\Leftrightarrow (\exists x)(F(x) = y).$$

EXAMPLES

1. Consider a set A. The identity relation $I_A = \{(x, y) : x \in A \wedge y \in A \wedge x = y\}$ is a function. For, if $(x, y) \in I_A$ and $(x, z) \in I_A$, then $x = y$ and $x = z$, from which it follows that $y = z$. The domain of I_A is A, and, for each x in A, $I_A(x) = x$.

2. Let A be the set of integers. Let $F = \{(x, y) : x \in A \wedge y \in A \wedge y = x + 1\}$. F is a function. For, if $(x, y) \in F$ and $(x, z) \in F$, then $y = x + 1$ and $z = x + 1$, from which $y = z$ follows. The domain of F is A. For each x in A, $F(x) = x + 1$.

3. Let A be the set of real numbers. Let $F = \{(x, y) : x \in A \wedge y \in A \wedge y^2 = x\}$. F is *not* a function. For, $(1, 1) \in F$ and $(1, -1) \in F$.

4. Let A be the set of real numbers. Let $F = \{(x, y) : x \in A \wedge y \in A \wedge y \geq 0 \wedge y^2 = x\}$. Then F is a function. For if $(x, y) \in F$ and $(x, z) \in F$, then $y^2 = x$ and $z^2 = x$, and, therefore, $y^2 = z^2$. Since y and z are nonnegative, it follows that $y = z$. The domain of F is $\{x : (\exists y)(y \geq 0 \wedge y^2 = x)\}$. But this is the set of all nonnegative real numbers. For each nonnegative real number x, $F(x) = \sqrt{x}$.

Very often a function F is defined simply by giving a formula for an arbitrary value $F(x)$. Thus, the function of Example 1 would be defined by saying that $I_A(x) = x$ for all x in A. The function of Example 2 would be introduced by the formula $F(x) = x + 1$ for all integers x. The function of Example 4 would be given by the formula $F(x) = \sqrt{x}$, for all nonnegative real numbers x. Sometimes, the domain of the function is not explicitly stated. If nothing is said about the domain, it usually is intended to be the set of all arguments for which, in that context, the formula makes sense. For example, if the formula $F(x) = \sqrt{x-1}$ is written in a calculus text, the intended domain consists of all real numbers x such that $x \geq 1$. (If $x < 1$, then $x - 1 < 0$, and negative real numbers do not have real square roots.)

EXAMPLES

5. The function defined by

$$F(x) = \text{the father of } x$$

has as its domain the set of all people.

6. The function defined by

$$F(x) = \sqrt[3]{x}$$

has as its domain the set of all real numbers.

7. The function defined by

$$F(x, y) = \sqrt{xy}$$

has as its domain the set of all ordered pairs (x, y) of real numbers x and y such that x and y do not have different signs, that is, if one is positive, the other is not negative.[†]

EXERCISES

1. For each of the following relations F, determine whether it is a function. If it is, find its domain, and, if possible, give an explicit formula for $F(x)$.

    a. Let A be the set of real numbers, and let

    $$F = \{(x, y) : x \in A \land y \in A \land x^2 + y^2 = 1\}.$$

    b. Let A be the set of real numbers, and let

    $$F = \{(x, y) : x \in A \land y \in A \land 2x + 3y = 12\}.$$

    c. Same as b, except that A is the set of positive integers.

[†] Strictly speaking, we should write $F((x, y)) = \sqrt{xy}$ as the formula for F, since the argument is itself an ordered pair. However, in such cases, we shall always drop the extra pair of parentheses.

d.   Let A be the set of integers, and let

$$F = \{(x, y) : x \in A \wedge y \in A \wedge x^2 = y^2\}.$$

e.   Same as d, except that A is the set of positive integers.

f.   Let A be the set of real numbers, and let

$$F = \{(x, y) : x \in A \wedge y \in A \wedge x + 1 > y - 2\}.$$

g.   Let A be the set of real numbers, and let

$$F = \{(x, y) : x \in A \wedge y \in A \wedge x + y = y^2\}.$$

h.   Let A be the set of people, and let

$$F = \{(x, y) : x \in A \wedge y \in A \wedge x \text{ is the mother of } y\}.$$

i.   Let A be the set of people, and let

$$F = \{(x, y) : x \in A \wedge y \in A \wedge x \text{ is a daughter of } y\}.$$

j.   Let A be the set of real numbers, and let

$$F = \{(x, y) : x \in A \wedge y \in A \wedge x = y^2 + 1\}.$$

k.   Let A be the set of real numbers, and let

$$F = \{((x, y), z) : x \in A \wedge y \in A \wedge z \in A \wedge x + y + z = 0).$$

l.   Let A be the set of real numbers, and let

$$F = \{((x, y), z) : x \in A \wedge y \in A \wedge z \in A \wedge x^2 = y^2 + z^2\}.$$

m.   Same as l, except that $z \in A$ is replaced by $z \in A \wedge z \geq 0$.

n. $F = \{(1, 1), (2, 4), (4,16)\}.$

o. $F = \{(0, 3), (1, 4), (2, 6), (3, 3)\}.$

p. $F = \{(0, 3), (1, 4), (2, 6), (0, 5)\}.$

2.   For each of the following formulas determining functions, find the domain which is implicitly assumed.

a. $F(x) = \sqrt{x + 2}.$

b. $F(x) = \sqrt{x^2 - 1}.$

c. $F(x) = $ the employer of x.

d. $F(x) = $ the wife of x.

e. $F(x, y) = \sqrt{x + y} + 2.$

f. $F(x) = \sqrt{4 - x^2} + \sqrt{x^2 - 1}.$

3. Consider any functions F and G. Prove that, if

    i. F and G have the same domain A,

    ii. for any x in A, $F(x) = G(x)$,

then $F = G$. (*Hint*: Show that $(x, y) \in F \Leftrightarrow (x, y) \in G$.) This result provides the method by which we shall, in the course of this book, prove that two functions are equal.

4. Let $\mathscr{X}$ be any set of functions such that any two of them have disjoint domains, that is,

$$(\forall F)(\forall G)(F \in \mathscr{X} \wedge G \in \mathscr{X} \wedge F \neq G \Rightarrow \mathscr{D}(F) \cap \mathscr{D}(G) = \varnothing).$$

Prove that $\bigcup_{F \in \mathscr{X}} F$ is a function and that its domain is the union of the domains of the functions in $\mathscr{X}$.

1.19 FUNCTIONS FROM A INTO (ONTO) B

Let us introduce the following notation for functions.

$$F \colon A \to B$$

shall mean that:

    i. F is a function,

    ii. the domain of F is A $(\mathscr{D}(F) = A)$,

    iii. the range of F is a subset of B $(\mathscr{R}(F) \subseteq B)$.

If $F \colon A \to B$, we shall say that F is a *function from A into B*.

EXAMPLES

1. Let A be the set of real numbers, and let C be the set of nonnegative real numbers. Let $F = \{(x, y) : x \in A \wedge y \in A \wedge y = x^2\}$. Then $F \colon A \to A$, and, in addition, $F \colon A \to C$. (Thus, in general, if $F \colon A \to B$, the range of F may or may not be all of B.)

2. Let A be the set of nonnegative integers. Let $F(x, y) = x^2 + y^2$ for all nonnegative integers x and y. Then, $F \colon A \times A \to A$. Notice that, in this case, it is not obvious how to give a simple characterization of the range of F.

If F is a function from A into B and the range of F is all of B, then we shall say that F is a function from A **onto** B. This will be written $F \colon A \xrightarrow[\text{onto}]{} B$.

Thus,

$$F: A \xrightarrow[\text{onto}]{} B \Leftrightarrow (F: A \to B \wedge \mathscr{R}(F) = B).$$

In Example 1, $F: A \xrightarrow[\text{onto}]{} C$, but F is not a function from A onto A. In Example 2, F turns out *not* to be a function from $A \times A$ onto A. For example, 3 is not in the range of F.

EXAMPLES

3. Let $A \subsetneq B$. Consider the identity function I_A:

$$I_A(x) = x \qquad \text{for all } x \text{ in } A.$$

Then $I_A: A \to B$ but I_A is not a function from A onto B. No element of $B - A$ is in the range of I_A.

4. Let A be the set of nonnegative integers. Let $F(x, y, z, w) = x^2 + y^2 + z^2 + w^2$ for all x, y, z, w in A. Then F: $A^4 \to A$. There is a well-known (but not obvious) result of number theory, namely Lagrange's Theorem, which states that every nonnegative integer is a sum of four squares, that is, $F: A^4 \xrightarrow[\text{onto}]{} A$.

5. Let A be the set of nonnegative integers and let B be the set of positive integers. Consider the function F such that $F(x) = x + 1$ for all x in A. Then $F: A \xrightarrow[\text{onto}]{} B$, but F is not a function from A onto A.

EXERCISE

1. For each of the following functions F and sets A and B, determine whether F is a function from A **onto** b.

   a. A and B are both the set of integers. $F(x) = -x$ for each integer x.

   b. Let C be the set of nonnegative integers. Let $A = C \times C$ and let $B = C$. Let $F(x, y) = 2^x(2y + 1)$ for all x and y in C.

   c. A and B are both equal to the set of real numbers. $F(x) = 3x$ for all x in A.

   d. C and D are arbitrary sets. Let $A = C \times D$ and $B = D \times C$. $F(x, y) = (y, x)$ for all (x, y) in $C \times D$.

   e. A and B are both equal to the set of real numbers. $F(x) = 2x + 1$ for all x in A.

   f. Same as e, except that A and B are both equal to the set of integers.

   g. Same as e, except that A and B are both equal to the set of nonnegative real numbers.

1.20 ONE-ONE FUNCTIONS

We say that a function F is *one-one* (or 1-1) if and only if, for any *distinct* objects x and y of the domain of F, $F(x) \neq F(y)$. In other words, F takes different objects into different objects. Thus,

$$F \text{ is one-one} \Leftrightarrow (\forall x)(\forall y)([x \in \mathscr{D}(F) \wedge y \in \mathscr{D}(F) \wedge x \neq y] \Rightarrow F(x) \neq F(y)).$$

As standard notation, we use

$$F: A \xrightarrow{\text{1-1}} B$$

to signify that $F: A \to B$ and F is one-one.

Notice that the condition for F to be one-one can be reformulated as follows:

$$F(x) = F(y) \Rightarrow x = y.$$

In other words, if F takes x and y into the same object, then $x = y$.

EXAMPLES

1. Let A be any set and let I_A be the identity function on A. Then, $I_A: A \xrightarrow{\text{1-1}} A$. For, if $x \neq y$, then $I_A(x) = x \neq y = I_A(y)$.

2. Let F be the function such that

$$F(x) = x + 1 \text{ for all } x \text{ in the set } A \text{ of integers.}$$

Then, $F: A \xrightarrow{\text{1-1}} A$. For, if $x + 1 = y + 1$, then $x = y$.

3. Let $F(x) = 3x + 1$ for all x in the set A of real numbers. Then, $F: A \xrightarrow{\text{1-1}} A$. For, if $3x + 1 = 3y + 1$, then $x = y$.

4. Let $F(x, y) = 2^x(2y + 1)$ for all x and y in the set A of nonnegative integers. Then $F: A \times A \xrightarrow{\text{1-1}} A$. For, if $2^x(2y + 1) = 2^u(2v + 1)$, then $x = u$ and $y = v$. (Note that x is the number of times 2 divides $2^x(2y + 1)$, and u is the number of times that 2 divides $2^u(2v + 1)$. Hence, if $2^x(2y + 1) = 2^u(2v + 1)$, then $x = u$, and, therefore, $2^x = 2^u$, from which it follows that $2y + 1 = 2v + 1$. Hence, $y = v$.)

5. Consider any sets A and B. Let $F(x, y) = (y, x)$ for all (x, y) in $A \times B$. Then, $F: A \times B \xrightarrow{\text{1-1}} B \times A$. For, if $(x, y) \neq (u, v)$, then $(y, x) \neq (v, u)$.

6. Let A be the set of nonnegative real numbers. Let $F(x) = \sqrt{x}$ for all real numbers x in A. Then $F: A \xrightarrow{1-1} A$. For, if $\sqrt{x} = \sqrt{y}$, then $x = y$.

7. Let A be the set of real numbers and let $F(x) = x^2$ for all x in A. Then F is not one-one, since $F(1) = F(-1)$.

8. Let A be the set of people and let $F(x)$ be the father of x, for all x in A. Then F is not one-one, since different people can have the same father.

Consider a function F. The inverse relation F^{-1} need not be a function. For example, let $F(x) = x^2$ for all real numbers x. Clearly, $(1, 1) \in F$ and $(-1, 1) \in F$. Hence, $(1, 1) \in F^{-1}$ and $(1, -1) \in F^{-1}$. Thus, F^{-1} is not a function.

Theorem 20.1 If F is a one-one function, then F^{-1} must be a function. In fact, F^{-1} is a one-one function.

Proof Assume $(u, v) \in F^{-1}$ and $(u, z) \in F^{-1}$. In order to show that F is a function, we must show that $v = z$. Now, $(v, u) \in F$ and $(z, u) \in F$. Hence, $F(v) = u$ and $F(z) = u$. Since F is one-one, $v = z$. Thus, F^{-1} is a function. To prove that F^{-1} is one-one, we must show that, if $F^{-1}(a) = F^{-1}(b)$, then $a = b$. So, assume $F^{-1}(a) = F^{-1}(b)$. Let $w = F^{-1}(a)$. Then, $(a, w) \in F^{-1}$ and $(b, w) \in F^{-1}$, and, therefore, $(w, a) \in F$ and $(w, b) \in F$. Since F is a function, it follows that $a = b$. ∎

Notice that, if $F: A \xrightarrow{1-1} B$, then $F^{-1}: \mathscr{R}(F) \xrightarrow[\text{onto}]{1-1} A.$[†] For, we know that, if R is any relation, $\mathscr{D}(R^{-1}) = \mathscr{R}(R)$, and $\mathscr{R}(R^{-1}) = \mathscr{D}(R)$. Thus, $\mathscr{D}(F^{-1}) = \mathscr{R}(F)$ and $\mathscr{R}(F^{-1}) = \mathscr{D}(F) = A$. In particular, if $F: A \xrightarrow[\text{onto}]{1-1} B$, then $F^{-1}: B \xrightarrow[\text{onto}]{1-1} A$. For, in this case, $\mathscr{R}(F) = B$.

EXAMPLES

9. In Example 1, $I_A^{-1} = I_A$.

10. In Example 2, $F: A \xrightarrow[\text{onto}]{1-1} A$, $F^{-1}: A \xrightarrow[\text{onto}]{1-1} A$, and $F^{-1}(u) = u - 1$ for all u in A. (In general, to find a formula for F^{-1}, we write $u = F(x)$ and then try to "solve" for x in terms of u.)

[†] We write $G: C \xrightarrow[\text{onto}]{1-1} D$ instead of $G: C \xrightarrow{1-1} D$ and $G: C \xrightarrow[\text{onto}]{} D$.

11.   In Example 3, $F: A \xrightarrow[\text{onto}]{1\text{-}1} A$, and, therefore, $F^{-1}: A \xrightarrow[\text{onto}]{1\text{-}1} A$. The formula for F^{-1} is $F^{-1}(u) = (u - 1)/3$. (For, if $3x + 1 = u$, then $x = (u - 1)/3$.)

12.   In Example 5, $F: A \times B \xrightarrow[\text{onto}]{1\text{-}1} B \times A$, and, therefore, $F^{-1}: B \times A \xrightarrow[\text{onto}]{1\text{-}1} A \times B$. For any (b, a) in $B \times A$, $F^{-1}(b, a) = (a, b)$.

13.   In Example 6, $F: A \xrightarrow[\text{onto}]{1\text{-}1} A$, and, therefore, $F^{-1}: A \xrightarrow[\text{onto}]{1\text{-}1} A$. Notice that $F^{-1}(u) = u^2$ for all u in A. (For, if $\sqrt{x} = u$, then $x = u^2$.)

14.   Let A be the set of all married people in a monogamous society. Let $F(x)$ be the spouse of x, for any x in A. Then $F: A \xrightarrow[\text{onto}]{1\text{-}1} A$. In this case, $F^{-1} = F$.

15.   Let A be the set of all positive real numbers. Let $F(x) = 1/x$ for all x in A. Then $F: A \xrightarrow[\text{onto}]{1\text{-}1} A$, and, therefore, $F^{-1}: A \xrightarrow[\text{onto}]{1\text{-}1} A$. Here again, $F^{-1} = F$. For, if $u = F(x) = 1/x$, then $F^{-1}(u) = x = 1/u$.

Observe that F^{-1} merely undoes the effect produced by F. If F takes x into y, then F^{-1} takes y back into x.

When $F: A \xrightarrow[\text{onto}]{1\text{-}1} B$, we say that F is a *one-one correspondence* between A and B. If there is a one-one correspondence between A and B, then A and B are said to be **equinumerous** because, in a certain sense, each has as many elements as the other.

EXAMPLES

16.   Let C and D be arbitrary sets. Then $C \times D$ and $D \times C$ are equinumerous. For, if $F(c, d) = (d, c)$, then $F: C \times D \xrightarrow[\text{onto}]{1\text{-}1} D \times C$.

17.   Let A be the set of nonnegative integers and let B be the set of positive integers. Then A and B are equinumerous. Let $F(x) = x + 1$ for all x in A. Then, $F: A \xrightarrow[\text{onto}]{1\text{-}1} B$. Notice that B is a proper subset of A. Thus, it is possible for a set to be equinumerous with one of its proper subsets. (However, this is not possible for finite sets.)

18.   Let A be the set of nonnegative integers. Let $F(x, y) = 2^x(2y + 1) - 1$ for all x and y in A. Then, $F: A \times A \xrightarrow[\text{onto}]{1\text{-}1} A$. Hence, $A \times A$ is equinumerous with A.

19.   Let A be the set of nonnegative integers and let Z be the set of all integers. We can *enumerate* Z as follows: $0, 1, -1, 2, -2, 3, -3, \ldots$. More precisely, we define

$$F(n) = \begin{cases} (n+1)/2 & \text{if } n \text{ is odd,} \\ -(n/2) & \text{if } n \text{ is even.} \end{cases}$$

Thus, $F(0) = 0$, $F(1) = 1$, $F(2) = -1$, $F(3) = 2$, $F(4) = -2$, $F(5) = 3$, $F(6) = -3$, etc. Then, $F: A \xrightarrow[\text{onto}]{1\text{-}1} Z$, and, therefore, A is equinumerous with Z.

EXERCISES

1.   Determine whether each of the following functions F is one-one. If it is one-one, describe F^{-1} (by a formula or otherwise), and find the domain of F^{-1}.

a. $F(x) = |x|$ for all real numbers x.

b. $F(x) = x^2 - 1$ for all real numbers x.

c. $F(x) = 7x + 3$ for all real numbers x.

d. $F(x) = 7x + 3$ for all integers x.

e. $F(x)$ is the father of x, for all x such that x is an only child.

f. $F(x, y) = \sqrt{x^2 + y^2}$ for all real numbers x and y.

g. $F(x)$ is the height of x in inches, for all people x.

h. $F(x, y) = x + y$ for all real numbers x and y.

i.   Consider an equivalence relation R in a set A. Let $F(x) = [x]$ for all x in A. (Remember that $[x]$ is the equivalence class of x, relative to R.)

2.   In each of the following cases, show that A and B are equinumerous.

a. A is the set of all even integers and B is the set of all odd integers.

b. A is the set of all even integers and B is the set of all integers.

c. A is the set of all real numbers between 0 and 1 (including 0 and 1), and B is the set of all real numbers between 1 and 3 (including 1 and 3).

1.21   COMPOSITION OF FUNCTIONS

We already have defined the composition $R \circ S$ of two relations S and R. Since a function is a special kind of relation, the composition of any two functions G and F is a well-defined relation $F \circ G$. However, more can be said about $F \circ G$.

Theorem 21.1   If F and G are functions, then $F \circ G$ is a function.

Proof Assume $(x, y) \in F \circ G$ and $(x, u) \in F \circ G$. We must show that $y = u$. Now, $(\exists w)((x, w) \in G \wedge (w, y) \in F)$ and $(\exists v)((x, v) \in G \wedge (v, u) \in F$.) Take such elements w and v. Then, $G(x) = w$, $F(w) = y$, $G(x) = v$, and $F(v) = u$. Hence, $y = F(w) = F(G(x)) = F(v) = u$. ∎

Observe also that $x \in \mathscr{D}(F \circ G)$ if and only if $x \in \mathscr{D}(G)$ and $G(x) \in \mathscr{D}(F)$. In addition, if $x \in \mathscr{D}(F \circ G)$, then

$$(F \circ G)(x) = F(G(x)).$$

For, let $y = (F \circ G)(x)$. Then, $(\exists w)((x, w) \in G \wedge (w, y) \in F)$. Hence, $(\exists w)(G(x) = w \wedge F(w) = y)$. Therefore, $y = F(G(x))$.

Thus, the value of $(F \circ G)(x)$ is obtained by first applying G to x and then applying F to the result $G(x)$.

EXAMPLES

1.   Let $F(x) = x + 1$ for all real numbers x, and let $G(x) = \sqrt{x}$ for all nonnegative real numbers x. Then,

$$(F \circ G)(x) = F(G(x)) = F(\sqrt{x}) = \sqrt{x} + 1$$

for all nonnegative real numbers x. On the other hand,

$$(G \circ F)(x) = G(F(x)) = G(x + 1) = \sqrt{x + 1}$$

for all real numbers x such that $x + 1 \geq 0$, that is, for all real numbers x such that $x \geq -1$. In this case, the domain of $G \circ F$ is a proper subset of the domain of F.

2.   Let $F(x) = \sin x$ for all real numbers x, and $G(x) = x^2$ for all real numbers x. Then

$$(F \circ G)(x) = F(G(x)) = F(x^2) = \sin(x^2)$$

for all real numbers x, and

$$(G \circ F)(x) = G(F(x)) = G(\sin x) = (\sin x)^2$$

for all real numbers x. In addition, $(G \circ G)(x) = G(G(x)) = G(x^2) = x^4$, and $(F \circ F)(x) = F(F(x)) = F(\sin x) = \sin(\sin x)$.

3.   For each person x, let $H(x)$ be the father of x, and let $G(x)$ be the mother of x. Then, $(H \circ G)(x) = H(G(x)) = H(\text{the mother of } x) = \text{the father}$

of the mother of x = the maternal grandfather of x. On the other hand, $(G \circ H)(x) = G(H(x)) = G(\text{the father of } x) = $ the mother of the father of $x = $ the paternal grandmother of x. (Notice that $H = \{(x, y) : y$ is the father of $x\}$. This is not the same thing as the father relation: $F = \{(x, y) : x$ is the father of $y\}$. F is not even a function.)

Theorem 21.2 Consider functions $G: A \to B$ and $F: B \to C$. Then, $F \circ G: A \to C$.

Proof We already know that $F \circ G$ is a function. The domain of $F \circ G$ is $\{x : x \in A \wedge G(x) \in B\}$. But this set is precisely A itself. Finally, $\mathscr{R}(F \circ G) \subseteq \mathscr{R}(F)$, and, therefore, $\mathscr{R}(F \circ G) \subseteq C$. ∎

We have two important extensions of this result.

Theorem 21.3 If $G: A \xrightarrow{\text{1-1}} B$ and $F: B \xrightarrow{\text{1-1}} C$, then $F \circ G: A \xrightarrow{\text{1-1}} C$.

Proof Assume $(F \circ G)(x) = (F \circ G)(y)$. We must show that $x = y$. We are given $F(G(x)) = F(G(y))$. Since F is one-one, we conclude that $G(x) = G(y)$. Since G is also one-one, we obtain $x = y$. ∎

Theorem 21.4 If $G: A \xrightarrow[\text{onto}]{} B$ and $F: B \xrightarrow[\text{onto}]{} C$, then $F \circ G: A \xrightarrow[\text{onto}]{} C$.

Proof Assume $c \in C$. We must find some a in A such that $(F \circ G)(a) = c$. Now, since $F: B \xrightarrow[\text{onto}]{} C$, there is some b in B with $F(b) = c$. Since $G: A \xrightarrow[\text{onto}]{} B$, there is some a in A such that $G(a) = b$. Hence,

$$(F \circ G)(a) = F(G(a)) = F(b) = c. \quad ∎$$

Putting Theorems 21.3 and 21.4 together, we obtain:

Corollary 21.5 If $G: A \xrightarrow[\text{onto}]{\text{1-1}} B$ and $F: B \xrightarrow[\text{onto}]{\text{1-1}} C$, then $F \circ G: A \xrightarrow[\text{onto}]{\text{1-1}} C$.

EXERCISES

1. Assume $F: A \to B, G: B \to C$, and $H: C \to D$. Show that $H \circ (G \circ F) = (H \circ G) \circ F$. (This is the associative law for composition of functions.)

2. Assume $F: A \xrightarrow[\text{onto}]{\text{1-1}} B$. Prove:
   a. $F^{-1} \circ F = I_A$.
   b. $F \circ F^{-1} = I_B$.
   c. $(F^{-1})^{-1} = F$.

3. For each of the following functions F and G, determine the compositions $F \circ G$ and $G \circ F$, and find their domains.

   a. $F(x) = x + 5$ for all integers x.
       $G(x) = x^3$ for all integers x.

   b. $F(x) = x^2$ for all real numbers x.
       $G(x) = \sqrt{x}$ for all nonnegative real numbers x.

   c. $F(x) = -x$ for all positive integers x.
       $G(x) = \sqrt{x}$ for all positive integers x.

   d. $F(x) = x/2$ for all integers x.
       $G(x) = 6x$ for all integers x.

   e. $F(x) = -x$ for all real numbers x.
       $G(x) = |x|$ for all real numbers x.

   f. $F(x) = x/(x^2 + 1)$ for all real numbers x.
       $G(x) = 1/x$ for all nonzero real numbers x.

4. Assume $G: A \xrightarrow{1\text{-}1} B$ and $F: B \xrightarrow{1\text{-}1} C$. Prove: $(F \circ G)^{-1} = G^{-1} \circ F^{-1}$.

5. Assume $G: A \twoheadrightarrow B$ and $F: B \twoheadrightarrow C$.

   a. If $F \circ G$ is one-one, prove that G is one-one.

   b. If $F \circ G: A \xrightarrow[\text{onto}]{} C$, prove that $F: B \xrightarrow[\text{onto}]{} C$.

1.22 OPERATIONS

Let A be a set. By a **singulary operation on A** we mean a function F from A into A. Thus, F is a singulary operation on A if and only if $F: A \rightarrow A$. This means that the values of F as well as its arguments must belong to A.

EXAMPLES

1. Let $F(x) = 2x$ for all integers x. F is a singulary operation on the set of integers, since, for every integer x, $2x$ is also an integer.

2. Let $F(x) = x/2$ for all integers x. F is *not* a singulary operation on the set of integers. For example, $F(3) = 3/2$, which is not an integer.

3. Let $F(x) = x/2$ for all rational numbers x. Then F is a singulary operation on the set of rational numbers.

4. Let $F(x) = x + 1$ for all positive integers x. Then F is a singulary operation on the set of positive integers.

5. Let $F(x) = x - 1$ for all positive integers x. Then F is *not* a singulary operation on the set of positive integers. $F(1) = 0$, which is not a positive integer.

By a **binary operation on** A we mean a function from $A \times A$ into A. Thus, F is a binary operation on A if and only if $F: A \times A \to A$.

EXAMPLES

6.   Let $F(x, y) = x + y$ for all positive integers x and y. F is a binary operation on the set of positive integers.

7.   Let $F(x, y) = x - y$ for all positive integers x and y. Then, F is *not* a binary operation on the set of positive integers. For example, $F(1, 1) = 0$, which is not a positive integer.

8.   Let $F(x, y) = x/y$ for all nonzero integers x and y. F is *not* a binary operation on the set of nonzero integers. For example, $F(2, 3) = 2/3$, which is not an integer.

9.   Let $F(x, y) = (x + y)/2$ for any real numbers x and y. This is a binary operation on the set of real numbers.

10.   Let $F(x, y) = x + y$ for all integers x and y such that $x > 2$ and $y > 2$. This is a binary operation on the set of all integers greater than 2. For, if $x > 2$ and $y > 2$, then $x + y > 2$.

11.   Let $F(x, y) = x \times y$ for all real numbers x and y such that $0 \le x \le 1$ and $0 \le y \le 1$. F is a binary operation on the set of all real numbers x such that $0 \le x \le 1$.

We could go on to define **ternary, quaternary,** etc. operations. For example, F is a ternary operation on A if and only if $F: A^3 \to A$. However such operations will be of little importance in the rest of this book.

EXERCISES

1.   If F is a binary operation on A, and $C \subseteq A$, we say that C is **closed** under F if and only if $F(x, y) \in C$ whenever x and y are in C. In each of the following cases, determine whether C is closed under F.

    a. $F(x, y) = x - y$ for all integers x and y.
        i. C is the set of all integers.
        ii. C is the set of all nonnegative integers.
        iii. C is the set of all even integers.
        iv. C is the set of all odd integers.
        v. $C = \{-1, 0, 1\}$.

b. $F(x, y) = x \times y$ for all real numbers x and y.

  i. $C = \{0, 1\}$.
  ii. C is the set of positive real numbers.
  iii. C is the set of all nonnegative real numbers.
  iv. C is the set of all negative real numbers.
  v. C is the set of all real numbers x such that $-1 \le x \le 1$.
  vi. C is the set of all integers.
  vii. C is the set of all rational numbers.
  viii. C is the set of all rational numbers of the form $k/2^n$, where k and n are integers.
  ix. C is the set of all real numbers of the form $a + b\sqrt{2}$, where a and b are integers.

c. $F(x, y) = x/y$ for all nonzero real numbers x and y.

  i. C is the set of all nonzero rational numbers.
  ii. C is the set of all nonzero integers.
  iii. C is the set of all positive rational numbers.
  iv. C is the set of all rational numbers of the form 2^n, where n is any integer.

2.   In each of the following cases, determine whether the given function F is a singulary or binary operation on the given set A.

a. $F(x)$ is the spouse of x, for all x in the set A of married people.

b. $F(x)$ is the father of x, for all x in the set A of people.

c. $F(x) = 1/x$ for all x in the set A of nonzero integers.

d. $F(x, y) = x + y - 1$ for all x and y in the set A of positive integers.

e. $F(x, y) = (x \times y) - (x + y)$ for all integers x and y such that $x \ge 2$ and $y \ge 2$. $A = \{x : x \text{ is an integer} \wedge x \ge 2\}$.

f. $F(x, y) = (x \times y) - (x + y)$ for all x and y in the set A of integers.

g. $F(x, y) = x \times y$ for all x and y in the set $A = \{a + b\sqrt{3} : a \text{ and } b \text{ are rational numbers}\}$.

h. $F(x, y) = x/y$ for all x and y in the set $A = \{a + b\sqrt{2} : a \text{ and } b \text{ are integers which are not both } 0\}$.

i.   Same as h, except that a and b are rational numbers instead of integers.

3.   Consider a binary operation F on a set A:

$$F: A \times A \to A.$$

Assume $C \subseteq A$. By the **restriction of the operation F to C**, we mean the function G such that $\mathscr{D}(G) = C \times C$ and

$$G(x, y) = F(x, y) \qquad \text{for all } x \text{ and } y \text{ in } C.$$

a.   Prove that the restriction G of F to C is an operation on C if and only if C is closed under F.

b.   Let $F(x, y) = (x \times y) - |x| \, |y|$ for all real numbers x and y. F is a binary operation on the set of real numbers. What is the restriction of F to the set of all nonnegative real numbers?

THE NATURAL NUMBERS

Among mathematical objects the natural numbers[†] 1, 2, 3, ... are the most familiar. Their origin in human thought goes back well beyond the earliest known civilizations. In this chapter we shall present an axiomatic development of the theory of natural numbers. Remarks of a philosophical nature concerning the significance and justification of our axiomatic approach and of other approaches will be postponed until the end of the chapter.

2.1 PEANO SYSTEMS

By a **Peano system** we mean a set P, a particular element 1 in P, and a singulary operation S on P such that the following axioms are satisfied.

(P1) 1 is not the *successor* $S(x)$ of any object x in P. In symbols, $(\forall x)(S(x) \neq 1)$.

(P2) Different objects in P have different successors. This can be formulated as follows:

$$(\forall x)(\forall y)(x \neq y \Rightarrow S(x) \neq S(y)).$$

(P3) Principle of Mathematical Induction: Any subset of P containing 1 and closed under the successor operation must be identical with P. This can be symbolically rendered as follows:

$$(\forall B)([B \subseteq P \wedge 1 \in B \wedge (\forall x)(x \in B \Rightarrow S(x) \in B] \Rightarrow P = B).$$

[†] Synonym: positive integers.

Such a Peano system will be denoted by the ordered triple $(P, S, 1)$. P is called the **underlying set**, S the **successor operation**, and 1 the **distinguished element**. The distinguished element 1 need not have anything to do with the ordinary integer 1.

We now shall mention a few examples of Peano systems. In these examples and in others used later, the reader is assumed to be familiar with the basic ideas and results of secondary-school mathematics. Or course, some of these ideas themselves form the subject matter of this book so that, to avoid circular reasoning, these examples will not be used in the rigorous mathematical development of the number systems.

EXAMPLES OF PEANO SYSTEMS

1. Let P be the set N of natural numbers. "1" is to denote the ordinary integer 1. S is the operation of adding $1: S(x) = x + 1$ for all natural numbers x. This example will be called the **standard** Peano system. The axioms are obviously true. (1) holds, for, if x is a natural number, then $x + 1 \neq 1$. (2) holds, since, if x and y are natural numbers and $x \neq y$, it follows that $x + 1 \neq y + 1$. To see that (3) holds, assume that B is a subset of P containing 1 and closed under the operation of adding 1, that is, if $x \in B$, then $x + 1 \in B$. Then, since $1 \in B$, it follows that $1 + 1 \in B$, that is, $2 \in B$. Hence, $2 + 1 \in B$, that is, $3 \in B$, etc. Thus, it is obvious[†] that all natural numbers belong to P.

2. Let P be the set NN of nonnegative integers. "1" is to denote the integer 0. S is the operation of adding 1 (the ordinary integer 1). Thus, $S(0) = 1$, $S(1) = 2$, $S(2) = 3, \ldots$.

3. Let P be the set of all integers greater than or equal to 1000. "1" is to denote the integer 1000. S is the operation of adding the ordinary integer 1. Thus, $S(1000) = 1001$, $S(1001) = 1002$, etc.

4. Let P be the set of negative integers. "1" is to denote the integer -1. S is the operation of subtracting 1. Thus, $S(-1) = -2$, $S(-2) = -3$, etc.

5. Let P be the set of even positive integers. Let "1" denote the integer 2. S is to be the operation of adding 2. Thus, $S(2) = 4$, $S(4) = 6$, $S(6) = 8$, etc.

[†] Remember that in examples we rely upon our intuitive mathematical ideas. The obviousness of what we have asserted is undeniable. Its justification is another matter.

The reader should give the arguments showing that Examples 2–5 are Peano systems.

It will have been observed that all of these examples consist simply of

$$1, \quad S(1), \quad S(S(1)), \quad S(S(S(1))), \quad \ldots, \quad \text{etc.}$$

However, the use of dots (...) or *etc.* is not mathematically legitimate, especially in this context, where the notion lying in back of the use of dots or *etc.* is connected with what we are studying. Hence, the statement that a Peano system consists of a sequence of pairwise-distinct objects 1, $S(1)$, $S(S(1))$, etc., is not a rigorous mathematical sentence, although it gives us a good intuitive picture of the situation.

Now a few simple properties of Peano systems will be derived. These properties and almost all of the other results in this chapter will not be new or surprising to the reader. Novelty will be found only in the definitions and proofs, as well as in the discovery that *all of our previous knowledge concerning the positive integers can be derived from the simple axioms for a Peano system.* Throughout this chapter, if nothing is said to the contrary, $(P, S, 1)$ will stand for an arbitrary Peano system.

Theorem 1.1 Every element different from 1 is a successor. This can be formulated as follows:

$$(\forall x)(x = 1 \lor (\exists y)(x = S(y))).$$

Proof We shall use the Principle of Mathematical Induction. Let B be the set of all elements of P which are either equal to 1 or a successor. Thus,

$$B = \{x : x \in P \land (x = 1 \lor (\exists y)(x = S(y))).$$

We must show that $P = B$. First, it is obvious that $1 \in B$. Second, assume $x \in B$. Then $x \in P$. Hence, $S(x) \in P$, and $(\exists y)(S(x) = S(y))$, namely, take $y = x$. Thus, $S(x) \in B$. We have shown that $x \in B \Rightarrow S(x) \in B$. Therefore, by Axiom P3, $P = B$. ■

Theorem 1.2 No object is its own successor, that is, $(\forall x)(S(x) \neq x)$.

Proof Again we shall use mathematical induction. Let B be the set of all elements of P that are not equal to their own successor. Thus, $B = \{x : x \in P \land S(x) \neq x\}$. We must prove that $B = P$. First, $1 \in B$, by Axiom P1. Second, assume $x \in B$. Then, $x \in P \land S(x) \neq x$. Since

$S(x) \neq x$, it follows by Axiom P2, that $S(S(x)) \neq S(x)$. Thus, $S(x)$ is not equal to its own successor. Hence, $S(x) \in B$. We have shown that $x \in B \Rightarrow S(x) \in B$. By Axiom P3, $P = B$.   ■

EXERCISES

1. Prove that every object different from 1 is the successor of a *unique* object, that is,

$$(\forall x)(x = 1 \vee (\exists! y)(x = S(y))).$$

2. Prove: $(\forall x)(x = 1 \vee x = S(1) \vee (\exists y)(x = S(S(y))))$.

3. Determine whether or not the following structures $(P, S, 1)$ are Peano systems.

    a. P is the set of all integers greater than 9. "1" stands for the integer 10. $S(u) = u + 1$ for any u in P, that is S is the operation of adding the ordinary integer 1.

    b. P is the set of all integers. "1" stands for the ordinary integer 1. $S(u) = u + 1$ for all u in P.

    c. P is the set of all integers. "1" stands for the ordinary integer 0.

    $$S(u) = \begin{cases} 1 & \text{if} \quad u = 0 \\ -u & \text{if} \quad u > 0 \\ -(u - 1) & \text{if} \quad u < 0. \end{cases}$$

    d. P is the set of all rational numbers of the form $1/2^n$, where n is a nonnegative integer. "1" stands for the integer 1. $S(u) = \frac{1}{2}u$ for all u in P.

    e. $P = \{1, 2, 3, 4\}$. "1" stands for the integer 1, and $S(1) = 2$, $S(2) = 3$, $S(3) = 4$, $S(4) = 1$.

    f. $P = \{1, 2, 3, \ldots\} \cup \{1\#, 2\#, 3\#, \ldots\}$, where $1\#, 2\#, 3\#, \ldots$ are pairwise-distinct new objects different from the natural numbers. Let "1" denote the integer 1. Let $S(n) = n + 1$ for any natural number n, and let $S(n\#) = (n + 1)\#$ for any natural number n. Thus, $S(1) = 2$, $S(2) = 3$, $\ldots$, and $S(1\#) = 2\#$, $S(2\#) = 3\#, \ldots$.

2.2 THE ITERATION THEOREM

The reader will have noticed that many common arithmetical notions such as $+$ and $\times$ are not mentioned in the definition of a Peano system. These notions will be *defined by induction*. The idea of definition by induction is different from that of proof by mathematical induction, which is embodied in Axiom P3. When we define a function f by induction, we first specify $f(1)$ and then indicate a rule for obtaining $f(S(x))$ from the previous value $f(x)$. This determines a unique function f defined on P. The method of definition by induction requires an independent justification, which is given by the following theorem.

Theorem 2.1 *Iteration Theorem* Consider any Peano system $(P, S, 1)$. Let W be an arbitrary set, let c be a fixed element of W, and let g be a singulary operation on W (that is, $g: W \to W$). Then, there is a unique function $F: P \to W$ such that

a. $F(1) = c$.
b. $F(S(x)) = g(F(x))$ for all x in P.

Before we present the rather intricate proof of the Iteration Theorem, let us give two examples. (These examples are relatively trivial. More important applications will be developed after the proof of the theorem.) In both examples, we take $(P, S, 1)$ to be the standard Peano system, that is, P is the set of natural numbers, $S(x) = x + 1$, and "1" denotes 1.

EXAMPLES

1. Let W be the set of real numbers, let $c = \sqrt{2}$, and let $g(x) = x + \sqrt{2}$ for all real numbers x. Then

$$F(1) = \sqrt{2}.$$
$$F(2) = F(S(1)) = g(F(1)) = g(\sqrt{2}) = \sqrt{2} + \sqrt{2} = 2\sqrt{2}.$$
$$F(3) = F(S(2)) = g(F(2)) = g(2\sqrt{2}) = 2\sqrt{2} + \sqrt{2} = 3\sqrt{2}.$$

In general, for any natural number n, $F(n) = n\sqrt{2}$.

2. Let W be the set of rational numbers, let $c = \frac{1}{2}$, and let $g(x) = x/2$ for all rational numbers x. Then,

$$F(1) = \tfrac{1}{2},$$
$$F(2) = F(S(1)) = g(F(1)) = g(\tfrac{1}{2}) = \tfrac{1}{4}.$$
$$F(3) = F(S(2)) = g(F(2)) = g(\tfrac{1}{4}) = \tfrac{1}{8}.$$

In general, for any natural number n, $F(n) = 1/2^n$.

Proof of the Iteration Theorem First, we shall prove the *existence* of a suitable function F.[†] Take any n in P. A function $f: A \to W$ is said

---

[†] The idea of the proof is to define F piecemeal: for each n in P, we show by induction that we can define a suitable restriction of F for all numbers from 1 up to n. The complexity of the proof results from the fact that we are not yet able to define the set of elements of P *from* 1 *up to* n. An alternative proof is sketched in Exercise 2 at the end of the chapter. The proof of the Iteration Theorem is the hardest proof in this chapter. Although newcomers to abstract mathematics should do their best to follow the proof, they should concentrate on understanding the meaning and applications of the theorem.

to be **n-admissible** if and only if:

   i. $A \subseteq P$

   ii. $1 \in A$

   iii. $n \in A$ conditions on A

   iv. $(\forall u)(S(u) \in A \Rightarrow u \in A)$

   v. $f(1) = c$ conditions on f

   vi. $(\forall u)(S(u) \in A \Rightarrow f(S(u)) = g(f(u)))$

(I) If f is $S(n)$-admissible, then f is n-admissible. Conditions (i), (ii), (iv)–(vi) automatically go over from $S(n)$ to n. It only remains to check (iii), that is, to prove that $n \in A$. But, by the assumption of $S(n)$-admissibility of f, $S(n) \in A$. But, then, by (iv), $n \in A$.

(II) For any n in P, there exists at least one n-admissible function.

We shall prove this by mathematical induction. Let B be the set of all members n of P for which there exists at least one n-admissible function. We must show that $B = P$. First, $1 \in B$. Simply define the function f with domain $\{1\}$ such that $f(1) = c$. Clearly, f is 1-admissible. Second, assume $n \in B$. Then there is some n-admissible function f. Let A be the domain of f. We define another function f^* as follows: For any x in A, let $f^*(x) = f(x)$, and, if $S(n) \notin A$, let $f^*(S(n)) = g(f(n))$. Note that, if $S(n) \in A$, then, by (vi), $f(S(n)) = g(f(n))$. Hence, in this case also, $f^*(S(n)) = g(f(n))$. The domain A^* of f^* is $A \cup \{S(n)\}$, which is equal to A in the case where $S(n) \in A$. The verification that f^* is $S(n)$-admissible is straightforward and is left as an exercise for the reader. Thus, $S(n) \in B$. We have shown that $n \in B \Rightarrow S(n) \in B$. By mathematical induction, $B = P$.

(III) If f is n-admissible and h is n-admissible, then $f(n) = h(n)$.

We shall prove this by mathematical induction. Let $Y = \{n : n \in P \wedge (\forall f)(\forall h)(f \text{ is } n\text{-admissible} \wedge h \text{ is } n\text{-admissible} \Rightarrow f(n) = h(n))$. We must show that $Y = P$. First $1 \in Y$ since $f(1) = c = h(1)$. Second, assume $n \in Y$, and let f and h be $S(n)$-admissible. By (I), f and h are n-admissible. Hence, since $n \in Y$, it follows that $f(n) = h(n)$. Then, by (vi), $f(S(n)) = g(f(n)) = g(h(n)) = h(S(n))$. Thus, $S(n) \in Y$. We have shown that $n \in Y \Rightarrow S(n) \in Y$. Hence, by mathematical induction, $P = Y$.

(IV) For any n in P, by virtue of (II), we may take some n-admissible function f. Let us define $F(n) = f(n)$. By (III), the value of $F(n)$ does not depend upon the particular choice of the n-admissible function f.

It is obvious that the domain of F is P. Also, $F(1) = f(1)$ for some 1-admissible function f. But, $f(1) = c$. Hence, we have condition (a): $F(1) = c$.

In addition, $F(S(n)) = f(S(n))$ for some $S(n)$-admissible function f. Since f is also n-admissible, by virtue of (I), $F(n) = f(n)$. But, since f is $S(n)$-admissible, $f(S(n)) = g(f(n))$. This yields (b): $F(S(n)) = g(F(n))$. Thus, conditions (a), (b) on the function F are satisfied. We must still show that the range of F is included in W. For any n in P, $F(n) = f(n)$ for some n-admissible function f. Since $\mathscr{R}(f) \subseteq W$, it follows that $F(n) \in W$.

Uniqueness. Assume F_1 and F_2 are functions from P into W satisfying conditions (a), (b). Let $K = \{n : n \in P \wedge F_1(n) = F_2(n)\}$. Now, $1 \in K$, since $F_1(1) = c = F_2(1)$. Moreover, if $n \in K$, then $F_1(n) = F_2(n)$. Hence, $F_1(S(n)) = g(F_1(n)) = g(F_2(n)) = F_2(S(n))$, that is, $S(n) \in K$. We have shown that $n \in K \Rightarrow S(n) \in K$. Hence, by mathematical induction, $P = K$. Thus, $F_1(n) = F_2(n)$ for all n in P. Since P is the domain of both F_1 and F_2, it follows that $F_1 = F_2$. ∎

2.3 APPLICATION OF THE ITERATION THEOREM: ADDITION

Theorem 3.1 (*Addition*) Consider any Peano system $(P, S, 1)$. Then there is a unique binary operation $+$ on P such that

(α) $x + 1 = S(x)$ for all x in P;

(β) $x + S(y) = S(x + y)$ for all x and y in P.

(We adopt the standard convention of writing $x + y$ instead of $+(x, y)$.)

Proof Take any x in P. Let us apply the Iteration Theorem, with $W = P$, $c = S(x)$, and $g = S$. Then there is a unique function $F : P \to P$ such that $F(1) = c = S(x)$ and $F(S(y)) = g(F(y)) = S(F(y))$ for any y in P. We denote this unique function F by f_x, since it is determined by x. Thus, $f_x(1) = S(x)$ and $f_x(S(y)) = S(f_x(y))$ for any y in P. Now, let $x + y = f_x(y)$ for any x and y in P. Then, $x + 1 = f_x(1) = S(x)$, and $x + S(y) = f_x(S(y))$ $= S(f_x(y)) = S(x + y)$.

To prove the uniqueness of $+$, assume that there is another binary operation h such that

$$h(x, 1) = S(x) \qquad \text{for all } x \text{ in } P;$$

$$h(x, S(y)) = S(h(x, y)) \qquad \text{for all } x \text{ and } y \text{ in } P.$$

Take any x in P. Let us prove that $h(x, y) = x + y$, for all y in P. Let $B = \{y : y \in P \wedge h(x, y) = x + y\}$. First, $1 \in B$, since $h(x, 1) = S(x)$

$= x + 1$. Second, assume $y \in B$. Then, $h(x, y) = x + y$. But, $h(x, S(y))$ $= S(h(x, y)) = S(x + y) = x + S(y)$. Thus, $S(y) \in B$, and we have shown that $y \in B \Rightarrow S(y) \in B$. Hence, by mathematical induction, $P = B$. Thus, $h(x, y) = x + y$ for all x and y in P. This means that $h = +$.   ■

Terminology $+$ is called the **addition** operation on P. $x + y$ is called the **sum** of x and y.

Observe that, by virtue of Equation (α), Equation (β) can be rewritten as

(β') $x + (y + 1) = (x + y) + 1$.

Notice also that Equations (α) and (β) enable us to compute any particular sum. We use the standard definitions:

$$2 = S(1), \; 3 = S(2), \; 4 = S(3), \; 5 = S(4), \quad \text{etc.}$$

EXAMPLES

1. $2 + 2 = 2 + S(1)$ (Definition of "2")

            $= S(2 + 1)$ (Equation (β))

            $= S(3)$ (Equation (α) and definition of "3")

            $= 4$ (Definition of "4").

2. $2 + 3 = 2 + S(2)$ (Definition of "3")

            $= S(2 + 2)$ (Equation (β))

            $= S(4)$ (Example 1)

            $= 5$ (Definition of "5").

3. $3 + 3 = 3 + S(2)$ (Definition of "3")

            $= S(3 + 2)$ (Equation (β))

            $= S(3 + S(1))$ (Definition of "2")

            $= S(S(3 + 1))$ (Equation (β))

            $= S(S(4))$ (Equation (α) and definition of "4")

            $= S(5)$ (Definition of "5")

            $= 6$ (Definition of "6").

EXERCISE

1. Compute $4 + 2$ and $2 + 4$. (We do not know yet that these two values must be the same.)

The following theorems bring together some of the most important properties of addition. We would suggest to the reader that, before he looks at the proof of any theorem, he should try to construct a proof on his own. If he does not succeed, then, after studying the proof in the text, he should try again to write out the proof on his own.

Theorem 3.2 (*Associativity of Addition*)

$$x + (y + z) = (x + y) + z \qquad \text{for all} \quad x, y, \text{ and } z \text{ in } P.$$

Proof Consider any x and y in P. Let

$$A = \{z : z \in P \land x + (y + z) = (x + y) + z\}.$$

We must show that $A = P$. First, $1 \in A$, since $x + (y + 1) = (x + y) + 1$ has already been established (see Equation (β'), page 60). Second, assume $z \in A$. Then, $z \in P \land x + (y + z) = (x + y) + z$. Now,

$$
\begin{aligned}
x + (y + S(z)) &= x + (S(y + z)) & &\text{(Equation } (\beta)) \\
&= S(x + (y + z)) & &\text{(Equation } (\beta)) \\
&= S((x + y) + z) & &\text{(Inductive hypothesis}^\dagger\text{:} \\
& & & \quad x + (y + z) = (x + y) + z) \\
&= (x + y) + S(z) & &\text{(Equation } (\beta)).
\end{aligned}
$$

Thus, $S(z) \in A$. We have shown that $z \in A \Rightarrow S(z) \in A$. Hence, by mathematical induction, $P = A$. ∎

From the associative law it follows that it does not matter how parentheses are inserted in a sum of three or more terms, and that, as a consequence, parentheses may be omitted. For example, let us write down the various ways of putting parentheses in $a + b + c + d$, and let us show that they all yield the same result.

1. $((a + b) + c) + d$.

2. $(a + b) + (c + d)$. This is equal to (1), by taking $x = a + b$, $y = c$, $z = d$ in Theorem 3.2.

---

† According to traditional usage, when attempting to show $z \in A \Rightarrow S(z) \in A$ in a proof by mathematical induction that $A = P$, the hypothesis $z \in A$ is called the **inductive hypothesis**.

3. $(a + (b + c)) + d$. This is equal to (1), since $(a + b) + c = a + (b + c)$ by virtue of Theorem 3.2.

4. $a + ((b + c) + d)$. This is equal to (3), by taking $x = a$, $y = b + c$, $z = d$ in Theorem 3.2.

5. $a + (b + (c + d))$. This is equal to (2), by taking $x = a$, $y = b$, $z = c + d$ in Theorem 3.2.

The reader should verify that these five expressions represent all possible ways of inserting parentheses in $a + b + c + d$. Therefore, there is no ambiguity in writing $a + b + c + d$, since, no matter which of the expressions (1)–(5) is meant, we obtain the same result.

We still do not know that addition is commutative: $x + y = y + x$. To show this, we need the following preliminary facts.

Lemma 3.3

a. $x + 1 = 1 + x$ for all x in P. (Hence, by Equation (α), $1 + x = S(x)$.)

b. $S(y) + x = S(y + x)$ for all x and y in P.

Proof (a) Let $A = \{x : x \in P \wedge x + 1 = 1 + x\}$. we must show that $A = P$. First, $1 \in A$, since $1 + 1 = 1 + 1$. Second, assume $x \in A$. Then $x \in P \wedge x + 1 = 1 + x$. Hence,

$$
\begin{aligned}
S(x) + 1 &= (x + 1) + 1 && \text{(Equation } (\alpha)) \\
&= (1 + x) + 1 && \text{(Inductive hypothesis: } x + 1 = 1 + x) \\
&= 1 + (x + 1) && \text{(Theorem 3.2: associativity of } +) \\
&= 1 + S(x) && \text{(Equation } (\alpha)).
\end{aligned}
$$

Thus, $S(x) \in A$. We have shown that $x \in A \Rightarrow S(x) \in A$. By mathematical induction, $A = P$.

b. $S(y) + x = (y + 1) + x = y + (1 + x) = y + (x + 1) = (y + x) + 1$ $= S(y + x)$. ∎

Theorem 3.4 (*Commutativity of Addition*)

$$x + y = y + x \qquad \text{for all} \quad x \text{ and } y \text{ in } P.$$

Proof Consider any x in P. Let $A = \{y : y \in P \wedge x + y = y + x\}$. We must show that $A = P$. First, $1 \in A$, since $x + 1 = 1 + x$ by Lemma

3.3a. Second, assume $y \in A$. Then, $y \in P$ and $x + y = y + x$. Hence,

$$x + S(y) = S(x + y)^\dagger$$
$$= S(y + x)$$
$$= S(y) + x \qquad \text{(Lemma 3.3b)}.$$

Thus, $S(y) \in A$. Hence, $y \in A \Rightarrow S(y) \in A$. Therefore, by mathematical induction, $A = P$. ∎

Theorem 3.5 (*Cancellation Law for Addition*)

$$x + z = y + z \Rightarrow x = y \qquad \text{for all} \quad x, y, \text{ and } z \text{ in } P.^\ddagger$$

Proof Consider any x and y in P. Let $A = \{z : z \in P \wedge [x + z = y + z \Rightarrow x = y]\}$. We must prove that $A = P$. First, $1 \in A$, since $x + 1 = y + 1 \Rightarrow x = y$ is merely a restatement of Axiom P2, using Equation (α). Second, assume $z \in A$. Then, $z \in P \wedge (x + z = y + z \Rightarrow x = y)$. Now, assume $x + S(z) = y + S(z)$. Then, by Equation (β), $S(x + z) = S(y + z)$. Therefore, by Axiom P2, $x + z = y + z$. Hence, by inductive hypothesis, $x = y$. Thus, we have shown that $x + S(z) = y + S(z) \Rightarrow x = y$, that is, $S(z) \in A$. Therefore, $z \in A \Rightarrow S(z) \in A$. By mathematical induction, $A = P$. ∎

Theorem 3.6 $y \neq x + y$ for all x and y in P.

Proof Consider any x in P. Let $A = \{y : y \in P \wedge y \neq x + y\}$. We must prove that $A = P$. Clearly, $1 \in A$, since $1 \neq x + 1$ by Equation (α) and Axiom P1. Assume now that $y \in A$. Then, $y \in P \wedge y \neq x + y$. Hence, by Axiom P2, $S(y) \neq S(x + y)$. But, $S(x + y) = x + S(y)$ by Equation (β). Hence, $S(y) \neq x + S(y)$, that is, $S(y) \in A$. Thus, we have shown that $y \in A \Rightarrow S(y) \in A$. By mathematical induction, $A = P$. ∎

EXERCISE

2. Prove: $(x + y) + (u + v) = (x + u) + (y + v)$.

---

† From this point on, we usually will not justify routine steps, such as applications of Equations (α) or (β) or of the inductive hypothesis.

‡ Of course, the converse $x = y \Rightarrow x + z = y + z$ also holds, as a consequence of the substitutivity of equality (see Appendix A).

2.4 THE ORDER RELATION

Once the usual properties of addition are available it is possible to introduce an order relation in a Peano system.

Definition $x < y$ for $(\exists z)(x + z = y)$.[†]

Notice that this is a purely abbreviational definition. The expression $x < y$ is simply shorthand for $(\exists z)(x + z = y)$.

Definition $x \not< y$ for $\neg(x < y)$.

Theorem 4.1

a. $x \not< x$ for all x in P (irreflexivity of $<$).
b. $[x < y \wedge y < z] \Rightarrow x < z$ for all x, y, and z in P (transitivity of $<$).
c.   For any x, y in P, exactly one of the following conditions holds: $x < y$, $x = y$, $y < x$ (trichotomy).

Proof (a) $x < x \Rightarrow (\exists z)(x + z = x)$. But $x + z \neq x$ by Theorems 3.6 and 3.4. Hence, $x \not< x$.

(b)   Assume $x < y \wedge y < z$. Then $x + u = y$ and $y + v = z$ for some u and v in P. Then, $x + (u + v) = (x + u) + v = y + v = z$. Thus, $(\exists w)(x + w = z)$, that is, $x < z$.

(c)   That at most one of $x < y$, $x = y$, $y < x$ holds is obvious. (If $x < y \wedge x = y$, then $x < x$, contradicting (a). If $x = y$ and $y < x$, then $x < x$, again contradicting (a). If $x < y \wedge y < x$, then, by (b), $x < x$, contradicting (a).) To prove that at least one of $x < y$, $x = y$, $y < x$ holds, take any y in P, and let $A = \{x : x \in P \wedge [x < y \vee x = y \vee y < x]\}$. We must prove that $A - P$. First, $1 \in A$. For, by Theorem 1.1, we have $x - 1 \vee (\exists z)(x = S(z))$, and, hence, by Equation (α) and Lemma 3.3a, $x = 1 \vee (\exists z)(1 + z = x)$, that is, $x = 1 \vee 1 < x$. Second, assume $x \in A$. Then, $x < y \vee x = y \vee y < x$.

Case i $x < y$. Then $x + z = y$ for some z. By Theorem 1.1, $z = 1 \vee (\exists u)(z = S(u))$. If $z = 1$, then $x + 1 = y$, that is, $S(x) = y$. If $z = S(u)$, then $S(x) + u = x + S(u) = x + z = y$. Hence, $S(x) < y$. In either event, $S(x) < y \vee S(x) = y$.

Case ii $x = y$. Then, $S(x) = S(y) = y + 1$. Hence, $y < S(x)$.

[†] The word "for" in the definition means "is an abbreviation for."

Case iii $y < x$. Then $x = y + u$ for some u. Hence, $S(x) = S(y + u)$ $= y + S(u)$. Thus, $y < S(x)$.

In all cases, $S(x) < y \lor S(x) = y \lor y < S(x)$. Hence, $S(x) \in A$. We have shown that $x \in A \Rightarrow S(x) \in A$. By mathematical induction, $A = P$. ∎

Theorem 4.2

a. $x < S(x)$.
b. $\neg(\exists y)(x < y < S(x))$.

Proof (a) $x + 1 = S(x)$. Therefore, $x < S(x)$.

(b) Assume $x < y < S(x)$. From $x < y$, we obtain $x + z = y$ for some z. By Theorem 1.1, $z = 1 \lor (\exists u)(z = S(u))$. If $z = 1$, $S(x) = x + 1$ $= x + z = y$, contradicting Theorem 4.1c, since $y < S(x)$. If $z = S(u)$ for some u, then $S(x) + u = x + S(u) = x + z = y$. Thus, $S(x) < y$, contradicting Theorem 4.1c, since $y < S(x)$. In either case, our assumption has led to a contradiction. ∎

Definitions

$$x \leq y \quad \text{for} \quad x < y \lor x = y.$$
$$x \nleq y \quad \text{for} \quad \neg(x \leq y).$$
$$x > y \quad \text{for} \quad y < x.$$
$$x \geq y \quad \text{for} \quad y \leq x, \text{ etc.}$$

EXERCISE

1. Prove: $x < y \Leftrightarrow (x \leq y \land x \neq y)$.

Theorem 4.3 For any x, y, and z in P:

a. $x \leq x$.   (Reflexivity of $\leq$)
b. $[x < y \land y \leq z] \Rightarrow x < z$.
c. $[x \leq y \land y < z] \Rightarrow x < z$.
d. $[x \leq y \land y \leq z] \Rightarrow x \leq z$.   (Transitivity of $\leq$)
e. $x \leq y \lor y \leq x$.
f. $[x \leq y \land y \leq x] \Rightarrow x = y$.

Proof (a) $x = x$.

(b) Assume $x < y \land y \leq z$. Since $y \leq z$, we have $y = z \lor y < z$. If $y = z$, then $x < z$, by substitutivity of equality. If $y < z$, then $x < z$ by Theorem 4.1b.

(c)   Similar, to (b) and left to the reader.

(d)   Follows from (b), (c), and the transitivity of equality.

(e)   Follows from Theorem 4.1c.

(f)   Assume $x \leq y \wedge y \leq x$. It follows by definition that $x = y \vee x < y$ and $x = y \vee y < x$. If $x \neq y$, then $x < y \wedge y < x$, contradicting Theorem 4.1c.   ∎

The next theorem establishes some of the order properties of 1 and $+$.

Theorem 4.4   For any x, y, and z in P:

a. $x \neq 1 \Rightarrow 1 < x$.

b. $x < x + y$.

c. $x < y \Leftrightarrow x + z < y + z$.

d. $x \leq y \Leftrightarrow x + z \leq y + z$.

e. $[x < y \wedge u < v] \Rightarrow x + u < y + v$.

f. $[x < y \wedge u \leq v] \Rightarrow x + u < y + v$.

g. $[x \leq y \wedge u < v] \Rightarrow x + u < y + v$.

h. $[x \leq y \wedge u \leq v] \Rightarrow x + u \leq y + v$.

Proof   (a)   Assume $x \neq 1$. Then, $(\exists u)(x = S(u))$, by Theorem 1.1. Hence, $x = u + 1 = 1 + u$ for some u. Therefore, $1 < x$.

(b)   There is some u such that $x + u = x + y$, namely, $u = y$. Hence, by definition, $x < x + y$.

(c)   Let us first prove that

$$(*) \quad x < y \Rightarrow x + z < y + z \text{ for all } x, y, \text{ and } z \text{ in } P.$$

Assume $x < y$. Then $x + w = y$ for some w. Hence, $(x + z) + w = (x + w) + z = y + z$. Thus, $x + z < y + z$. To prove the converse of $(*)$, assume $x + z < y + z$. We shall prove $x < y$ by reductio ad absurdum. Assume that $x \not< y$. By Theorem 4.1c, $x = y \vee y < x$. But, if $x = y$, then $x + z = y + z$, contradicting $x + z < y + z$. Likewise, if $y < x$, then, by $(*)$, $y + z < x + z$, contradicting $x + z < y + z$.

(d)   Follows easily from (c) and Theorem 3.5, and is left to the reader.

(e)   Assume $x < y \wedge u < v$. By (c), $x + u < y + u$ and $y + u < y + v$. Hence, by the transitivity of $<$ (Theorem 4.1b), $x + u < y + v$.

(f)–(h)   Follow easily from (e) and are left as exercises.   ∎

EXERCISES

Prove:

2. $1 \leq x$.

3. $x \leq 1 \Leftrightarrow x = 1$.

4. $x < S(y) \Leftrightarrow x \leq y$.

5. $x < y \Leftrightarrow S(x) \leq y$.

6. $x + u < y + v \Rightarrow (x < y \lor u < v)$.

7. $1 < 2 < 3 < 4$, etc.

8. If $g : P \rightarrow P$ and $(\forall x)(\forall y)(x < y \Rightarrow g(x) < g(y))$, prove that $(\forall u)(g(u) \geq u)$.

Definition Let A be a subset of P. By a **least element** of A we mean an object z of A such that z is less than or equal to every element of A. Thus z is a least element of A if and only if

i. $z \in A$, and

ii. $(\forall u)(u \in A \Rightarrow z \leq u)$.

Observe that, by Theorem 4.3f, A has at most one least element. Also notice that 1 is the least element of any set to which it belongs.

The following theorem provides a very useful fact about the order relation.

Theorem 4.5 (*Least Number Principle*) Any nonempty subset of P has a least element.

Proof We are given $\varnothing \neq A \subseteq P$. Assume A has no least element, and let us show that this leads to a contradiction. Let B be the set of all elements x of P such that x and all smaller elements are not in A. In symbols,

$$B = \{x : x \in P \land (\forall u)(u \leq x \Rightarrow u \notin A)\}.$$

Observe that no element of B can be an element of A. Let us show now by mathematical induction that $B = P$. First, $1 \in B$. For, assume not. Then $(\exists u)(u \leq 1 \land u \in A)$. If $u \leq 1$, then $u = 1$ by Theorem 4.4a. Thus, $1 \in A$. But then 1 would be a least element of A, contradicting our assumption. Hence, $1 \in B$. Now, assume $x \in B$. Then,

$$(**) \quad (\forall u)(u \leq x \Rightarrow u \notin A).$$

Therefore, $S(x) \notin A$. (For, if $S(x)$ were in A, then, by $(**)$ and Theorem

4.2b, $S(x)$ would be a least element of A.) Hence, by (**) and Theorem 4.2b, $(\forall u)(u \leq S(x) \Rightarrow u \notin A)$. Then, $S(x) \in B$. We have shown that $x \in B$ $\Rightarrow S(x) \in B$. By mathematical induction, $B = P$. But, $B \cap A = \emptyset$. Hence, $A = \emptyset$, contradicting our assumption that A is nonempty. ∎

EXERCISES

8. Prove the Principle of Complete Induction:

$$(\forall B)([B \subseteq P \wedge (\forall x)((\forall y)(y < x \Rightarrow y \in B) \Rightarrow x \in B] \Rightarrow P = B).$$

(*Hint*: Assume $P \neq B$. Then, $P - B \neq \emptyset$. Apply the Least Number Principle. Notice that, when $x = 1$, $(\forall y)(y < x \Rightarrow y \in B)$ is trivially true, since $y < 1$ is always false.)

9. By a **greatest** element of a set $B \subseteq P$ we mean an object z such that $z \in B \wedge (\forall u)(u \in B \Rightarrow u \leq z)$. Prove:

    a.   A set has at most one greatest element.
    b.   A set need not have a greatest element.
    c.   If $\emptyset \neq A \subseteq P$ and A is bounded above (that is, $(\exists w)(\forall u)(u \in A \Rightarrow u \leq w)$), then A has a greatest element. (*Hint*: Let $B = \{w : (\forall u)(u \in A \Rightarrow u \leq w)\}$. By hypothesis, $B \neq \emptyset$. Apply the Least Number Principle.)

2.5 MULTIPLICATION

The existence of the familiar operation of multiplication can be derived from the Iteration Theorem.

Theorem 5.1   Consider any Peano system $(P, S, 1)$. There is a unique binary operation $\times$ on P such that

(γ) $x \times 1 = x$ for all x in P.
(δ) $x \times S(y) = (x \times y) + x$ for all x and y in P.

(Here again we use the traditional notation $x \times y$ instead of the functional notation $\times(x, y)$.)

Proof   Consider any x in P. In the Iteration Theorem, take $W = P$, $c = x$, and the function $g: P \to P$ such that $g(u) = u + x$ for any u in P. Then, there must be a unique function $F: P \to P$ such that $F(1) = c = x$ and, for any y in P, $F(S(y)) = g(F(y)) = F(y) + x$. We denote this unique function F by ψ_x, since it is determined by x. Thus, $\psi_x(1) = x$ and $\psi_x(S(y))$

$= \psi_x(y) + x$. Now define $x \times y = \psi_x(y)$. We then have

$$x \times 1 = \psi_x(1) = x \qquad \text{(Equation } (\gamma)),$$

and

$$x \times S(y) = \psi_x(S(y)) = \psi_x(y) + x = (x \times y) + x \qquad \text{(Equation } (\delta)).$$

The uniqueness of $\times$ is proved easily. Let $\theta(x, y)$ be any other binary operation on P satisfying (γ) and (δ):

$$\theta(x, 1) = x \text{ and } \theta(x, S(y)) = \theta(x, y) + x \text{ for any } x \text{ and } y \text{ in } P.$$

Take any x in P. Let $B = \{y : y \in P \wedge \theta(x, y) = x \times y\}$. $1 \in B$, since $\theta(x, 1) = x = x \times 1$. Next, assume $y \in B$. Then, $y \in P \wedge \theta(x, y) = x \times y$. Hence, $\theta(x, S(y)) = \theta(x, y) + x = (x \times y) + x = x \times S(y)$. Thus, $S(y) \in B$. We have shown that $y \in B \Rightarrow S(y) \in B$. By mathematical induction, $B = P$, that is, $\theta(x, y) = x \times y$ for all x and y in P. This means that $\times$ is the unique operation satisfying (γ) and (δ). ∎

$\times$ is called the **multiplication** operation on P, and $x \times y$ is called the **product** of x and y.

Notice that, by virtue of Equation (α), $S(y) = y + 1$, Equation (δ) takes the form

(δ') $x \times (y + 1) = (x \times y) + x$.

Also observe that Equations (γ), (δ), along with Equations (α), (β) for addition, enable us to compute any particular product.

EXAMPLES

1. $\begin{aligned}
2 \times 2 &= 2 \times S(1) && \text{(Definition of “2”)} \\
&= (2 \times 1) + 2 && \text{(Equation } (\delta)) \\
&= 2 + 2 && \text{(Equation } (\gamma)) \\
&= 4 && \text{(Example 3.1, p. 60)}
\end{aligned}$

2. $\begin{aligned}
2 \times 3 &= 2 \times S(2) && \text{(Definition of “3”)} \\
&= (2 \times 2) + 2 && \text{(Equation } (\delta)) \\
&= 4 + 2 && \text{(Computation 1 above)} \\
&= 4 + S(1) && \text{(Definition of “2”)} \\
&= S(4 + 1) && \text{(Equation } (\beta)) \\
&= S(5) && \text{(Equation } (\alpha) \text{ and definition of “5”)} \\
&= 6 && \text{(Definition of “6”).}
\end{aligned}$

EXERCISES

1. Compute 3×2. (Observe that at this point we have not established the commutativity of multiplication: $x \times y = y \times x$. Hence, we have no right to use the computation for 2×3 given in Example 2.)

2. Compute 3×3.

3. Compute 1×3.

It is important now to show that the familiar properties of multiplication are derivable. First we establish a basic relation between addition and multiplication.

Theorem 5.2 (*Distributive Laws*)

a. $x \times (y + z) = (x \times y) + (x \times z)$.
b. $(y + z) \times x = (y \times x) + (z \times x)$.

Proof (a) Consider any x and y in P. Let

$$B = \{z : z \in P \wedge x \times (y + z) = (x \times y) + (x \times z)\}.$$

We must prove that $B = P$. First, $1 \in B$, since

$$
\begin{aligned}
x \times (y + 1) &= (x \times y) + x && \text{(Equation (δ'), p. 69)} \\
&= (x \times y) + (x \times 1) && \text{(Equation (γ))}.
\end{aligned}
$$

Second, assume $z \in B$. Then, $z \in P \wedge x \times (y + z) = (x \times y) + (x \times z)$. Therefore,

$$
\begin{aligned}
x \times (y + S(z)) &= x \times S(y + z) && \text{(Equation (β))} \\
&= (x \times (y + z)) + x && \text{(Equation (δ))} \\
&= [(x \times y) + (x \times z)] + x && \text{(Inductive hypothesis)} \\
&= (x \times y) + [(x \times z) + x] && \text{(Associativity of $+$)} \\
&= (x \times y) + (x \times S(z)) && \text{(Equation (δ))}.
\end{aligned}
$$

Thus, $S(z) \in B$. We have shown that $z \in B \Rightarrow S(z) \in B$. Hence, by mathematical induction, $B = P$.

(b) Consider any y, z in P. Let $A = \{x : x \in P \wedge (y + z) \times x = (y \times x) + (z \times x)\}$. We must prove $A = P$. First, $1 \in A$, since

$$
\begin{aligned}
(y + z) \times 1 &= y + z && \text{(Equation (γ))} \\
&= (y \times 1) + (z \times 1) && \text{(Equation (γ))}.
\end{aligned}
$$

Second, assume $x \in A$. Then $x \in P \wedge (y + z) \times x = (y \times x) + (z \times x)$. Hence,

$$(y + z) \times S(x) = ((y + z) \times x) + (y + z) \qquad \text{(Equation } (\delta))$$
$$= ((y \times x) + (z \times x)) + (y + z) \qquad \text{(Inductive hypothesis)}$$
$$= ((y \times x) + y) + ((z \times x) + z) \qquad \text{(Associativity and}$$
$$\text{commutativity of } +)$$
$$= (y \times S(x)) + (z \times S(x)) \qquad \text{(Equation } (\delta)).$$

Thus, $S(x) \in A$. We have shown that $x \in A \Rightarrow S(x) \in A$. By mathematical induction, $A = P$. ∎

Remark If we already knew that multiplication is commutative, Theorem 5.2b would be a consequence of Theorem 5.2a. However, we still have not proved the commutativity of multiplication. This will be done in Theorem 5.4, and the proof will depend upon Theorem 5.2b.

EXERCISES

4. Prove: $2 \times x = x + x$.
5. Prove: $3 \times x = (x + x) + x$.

Lemma 5.3 $1 \times x = x \times 1$ for all x in P.

Proof Let $A = \{x : x \in P \wedge 1 \times x = x \times 1\}$. We must prove that $A = P$. Clearly, $1 \in A$, since $1 \times 1 = 1 \times 1$. Assume now that $x \in A$. Then, $x \in P \wedge 1 \times x = x \times 1$. Hence,

$$1 \times S(x) = (1 \times x) + 1 \qquad \text{(Equation } (\delta))$$
$$= (x \times 1) + 1 \qquad \text{(Inductive hypothesis)}$$
$$= x + 1 \qquad \text{(Equation } (\gamma))$$
$$= S(x) \qquad \text{(Equation } (\alpha))$$
$$= S(x) \times 1 \qquad \text{(Equation } (\gamma)).$$

Thus, $S(x) \in A$. We have shown that $x \in A \Rightarrow S(x) \in A$. By mathematical induction, $A = P$. ∎

Theorem 5.4 (*Commutativity of Multiplication*)

$$x \times y = y \times x \qquad \text{for all} \quad x \text{ and } y \text{ in } P.$$

Proof Consider any x in P. Let $A = \{y : y \in P \wedge x \times y = y \times x\}$. We must prove that $A = P$. By Lemma 5.3, $1 \in A$. Assume now that $y \in A$. Thus, $y \in P \wedge x \times y = y \times x$. Hence,

$$
\begin{aligned}
x \times S(y) &= (x \times y) + x && \text{(Equation (δ))}\\
&= (y \times x) + x && \text{(Inductive hypothesis)}\\
&= (y \times x) + (x \times 1) && \text{(Equation (γ))}\\
&= (y \times x) + (1 \times x) && \text{(Lemma 5.3)}\\
&= (y + 1) \times x && \text{(Theorem 5.2b)}\\
&= S(y) \times x && \text{(Equation (α)).}
\end{aligned}
$$

Thus, $S(y) \in A$. We have shown that $y \in A \Rightarrow S(y) \in A$. By mathematical induction, $A = P$. ∎

Theorem 5.5 *(Associativity of Multiplication)*

$$x \times (y \times z) = (x \times y) \times z \quad \text{for all} \quad x, y, \text{ and } z \text{ in } P.$$

Proof Consider any x and y in P. Let

$$B = \{z : z \in P \wedge x \times (y \times z) = (x \times y) \times z\}.$$

$1 \in B$, since

$$
\begin{aligned}
x \times (y \times 1) &= x \times y && \text{(Equation (γ))}\\
&= (x \times y) \times 1 && \text{(Equation (γ)).}
\end{aligned}
$$

Assume $z \in B$. Thus, $z \in P \wedge x \times (y \times z) = (x \times y) \times z$. Hence,

$$
\begin{aligned}
x \times (y \times S(z)) &= x \times ((y \times z) + y) && \text{(Equation (δ))}\\
&= (x \times (y \times z)) + (x \times y) && \text{(Distributivity: Theorem 5.2a)}\\
&= ((x \times y) \times z) + (x \times y) && \text{(Inductive hypothesis)}\\
&= (x \times y) \times S(z) && \text{(Equation (δ)).}
\end{aligned}
$$

Thus, $S(z) \in B$. We have shown that $z \in B \Rightarrow S(z) \in B$. By mathematical induction, $B = P$. ∎

EXERCISE

6. Prove: $(x \times y) \times (u \times v) = (x \times u) \times (y \times v)$.

Theorem 5.6 $x < y \Leftrightarrow x \times z < y \times z$ for all x, y, and z in P.

Proof Let us first prove

$$(*) \quad x < y \Rightarrow x \times z < y \times z \qquad \text{for all } x, y, \text{ and } z \text{ in } P.$$

Assume $x < y$. Then $y = x + u$ for some u in P. Hence, we have $y \times z = (x + u) \times z = (x \times z) + (u \times z)$ by Theorem 5.2b. Thus, $(\exists v)(y \times z = (x \times z) + v)$, namely, let $v = u \times z$. Hence, $x \times z < y \times z$.

To prove the converse of $(*)$, assume $x \times z < y \times z$. We shall show $x < y$ by reductio ad absurdum. Assume $x \nless y$. Then, by the Trichotomy Law (Theorem 4.1c), $x = y \lor y < x$. If $x = y$, then $x \times z = y \times z$, contradicting the assumption that $x \times z < y \times z$. If $y < x$, then, by $(*)$, $y \times z < x \times z$, contradicting the assumption that $x \times z < y \times z$. In both cases, we have obtained a contradiction. ∎

Corollary 5.7 (*Cancellation Law for Multiplication*)

$$x \times z = y \times z \Rightarrow x = y \qquad \text{for all} \quad x, y, \text{ and } z \text{ in } P.$$

Proof Assume $x \times z = y \times z$. We shall prove $x = y$ by reductio ad absurdum. Assume $x \neq y$. Then, by the Trichotomy Law (Theorem 4.1c), $x < y \lor y < x$. If $x < y$, then, by Theorem 5.6, $x \times z < y \times z$, contradicting our assumption that $x \times z = y \times z$. If $y < x$, then, again by Theorem 5.6, $y \times z < x \times z$, contradicting our assumption that $x \times z = y \times z$. In both cases, we have obtained a contradiction. ∎

Corollary 5.8
a. $x \leq y \Leftrightarrow x \times z \leq y \times z.$
b. $z \leq y \times z.$
c. $1 < y \Leftrightarrow z < y \times z.$
d. $[x < u \land y < v] \Rightarrow x \times y < u \times v.$
e. $[x < u \land y \leq v] \Rightarrow x \times y < u \times v.$
f. $[x \leq u \land y < v] \Rightarrow x \times y < u \times v.$
g. $[x \leq u \land y \leq v] \Rightarrow x \times y \leq u \times v.$

Proof (a) Follows immediately from Theorem 5.6 and Corollary 5.7.
(b) Let $x = 1$ in Part (a).
(c) Let $x = 1$ in Theorem 5.6.
(d) Assume $x < u \land y < v$. By Theorem 5.6, we may write $x \times y < u \times y \land u \times y < u \times v$. By transitivity of $<$, $x \times y < u \times v$.
(e)–(g) are easy consequences of (d) and properties of equality. ∎

EXERCISES

7. We use the standard notation: x^2 for $x \times x$, and x^3 for $x^2 \times x$,

    a. Prove $1^2 = 1$.
    b. Prove $(x + y)^2 = x^2 + (2 \times x \times y) + y^2$.

8. a. Prove the Division Theorem:

    $$(\forall x)(\forall y)(\exists! q)(y = q \times x \vee (\exists! r)(y = (q \times x) + r \wedge r < x)).$$

    (In words, if y is "divided" by x, there is a unique quotient q and remainder r.)
    Hint: Take x in P, and use mathematical induction with respect to y.

    b. Define: x is *even* for $(\exists u)(x = 2 \times u)$.
    $$x \text{ is } odd \text{ for } x = 1 \vee (\exists u)(x = (2 \times u) + 1).$$
    Prove:

    i.   Every element of P is either even or odd.
    ii. No element of P is both even and odd.
    iii. x^2 is even $\Leftrightarrow$ x is even.
    iv. x^2 is odd $\Leftrightarrow$ x is odd.
    v.   The sum and product of even elements is even.
    vi. The product of odd elements is odd.
    vii. The sum of two odd elements is even.
    viii. If x is odd and y is even, then $x + y$ is odd and $x \times y$ is even.

2.6 EXPONENTIATION

Theorem 6.1   There is a unique binary operation $\tau(x, y)$ on P such that:

(ϱ) $\tau(x, 1) = x$ \qquad\qquad for all x in P.
(σ) $\tau(x, S(y)) = \tau(x, y) \times x$ \qquad for all x and y in P.

(*Notation*:   We shall let x^y stand for $\tau(x, y)$. Then (ϱ), (σ) become the familiar equations:

(ϱ') $x^1 = x$.
(σ') $x^{y+1} = x^y \times x$.)

Proof   Take any x in P. In the Iteration Theorem, take $W = P$, $c = x$, and $g(u) = u \times x$ for all u in P. We obtain a unique function $F: P \to P$ such that $F(1) = c = x$ and $F(S(y)) = g(F(y)) = F(y) \times x$ for all x and y. Since F is determined by x, we denote F by h_x. Then, $h_x(1) = x$ and $h_x(S(y)) = h_x(y) \times x$ for all x and y in P. Define $\tau(x, y) = h_x(y)$. This yields:

$$\tau(x, 1) = h_x(1) = x,$$

and

$$\tau(x, S(y)) = h_x(S(y)) = h_x(y) \times x = \tau(x, y) \times x.$$

To prove the uniqueness of τ, assume that there is a function $\mu(x, y)$ also satisfying (ϱ) and (σ): $\mu(x, 1) = x$ and $\mu(x, S(y)) = \mu(x, y) \times x$. Take any x in P. Let $B = \{y : y \in P \land \tau(x, y) = \mu(x, y)\}$. We must prove $B = P$. First, $1 \in B$, since $\tau(x, 1) = x = \mu(x, 1)$. Assume that $y \in B$. Thus, $y \in P \land \tau(x, y) = \mu(x, y)$. Hence,

$$\tau(x, S(y)) = \tau(x, y) \times x = \mu(x, y) \times x = \mu(x, S(y)).$$

This means that $S(y) \in B$. We have shown that $y \in B \Rightarrow S(y) \in B$. Hence, by mathematical induction, $B = P$.   ∎

Remark Our notation for exponentiation does not conflict with our previous agreement that x^2 stands for $x \times x$, and x^3 for $x^2 \times x$. For, according to Equations (ϱ'), (σ'), $x^2 = x^{1+1} = x^1 \times x = x \times x$, and $x^3 = x^{2+1} = x^2 \times x$.

EXERCISE

1. Compute the values of 2^2, 2^3, 3^2.

Lemma 6.2 $1^x = 1$ for all x in P.

Proof Let $A = \{x : x \in P \land 1^x = 1\}$. We must prove $A = P$. First, $1 \in A$, since $1^1 = 1$. Assume $x \in A$. Thus, $x \in P \land 1^x = 1$. Then $1^{S(x)} = 1^{x+1} = 1^x \times 1 = 1 \times 1 = 1$. Thus, $S(x) \in A$. We have shown that $x \in A \Rightarrow S(x) \in A$. Hence, by mathematical induction, $A = P$.   ∎

Theorem 6.3 (*Basic Properties of Exponentiation*)
a. $x^{y+z} = x^y \times x^z$.
b. $(x^y)^z = x^{y \times z}$.
c. $(x \times y)^z = x^z \times y^z$.

Proof (a) Consider any x and y in P. Let

$$A = \{z : z \in P \land x^{y+z} = x^y \times x^z\}.$$

We must prove $A = P$. Now, $1 \in A$, since

$$
\begin{aligned}
x^{y+1} &= x^y \times x && \text{(Equation (σ'))} \\
&= x^y \times x^1 && \text{(Equation (ϱ')).}
\end{aligned}
$$

Assume that $z \in A$. Thus, $z \in P \wedge x^{y+z} = x^y \times x^z$. Hence,

$$
\begin{aligned}
x^{y+S(z)} &= x^{S(y+z)} & &\text{(Equation }(\beta)) \\
&= x^{(y+z)+1} & &\text{(Equation }(\alpha)) \\
&= x^{y+z} \times x & &\text{(Equation }(\sigma')) \\
&= (x^y \times x^z) \times x & &\text{(Inductive hypothesis)} \\
&= x^y \times (x^z \times x) & &\text{(Associativity of } \times : \text{ Theorem 5.5)} \\
&= x^y \times x^{S(z)} & &\text{(Equation }(\sigma')).
\end{aligned}
$$

Thus, $S(z) \in A$. We have shown that $z \in A \Rightarrow S(z) \in A$. Therefore, by mathematical induction, $A = P$.

(b) Consider any x and y in P. Let $B = \{z : z \in P \wedge (x^y)^z = x^{y \times z}\}$. We must show that $B = P$. First, $1 \in B$, since

$$
\begin{aligned}
(x^y)^1 &= x^y & &\text{(Equation }(\varrho')) \\
&= x^{y \times 1} & &\text{(Equation }(\gamma)).
\end{aligned}
$$

Assume now that $z \in B$. Thus, $z \in P \wedge (x^y)^z = x^{y \times z}$. Then,

$$
\begin{aligned}
(x^y)^{S(z)} &= (x^y)^z \times x^y & &\text{(Equation }(\sigma')) \\
&= x^{y \times z} \times x^y & &\text{(Inductive hypothesis)} \\
&= x^{(y \times z)+y} & &\text{(Part (a))} \\
&= x^{y \times S(z)} & &\text{(Equation }(\delta)).
\end{aligned}
$$

Thus, $S(z) \in B$. We have shown that $z \in B \Rightarrow S(z) \in B$. By mathematical induction, $B = P$.

(c) Consider any x and y in P. Let $C = \{z : z \in P \wedge (x \times y)^z = x^z \times y^z\}$. We must show that $C = P$. First, $1 \in C$, since $(x \times y)^1 = x \times y = x^1 \times y^1$. Assume now that $z \in C$. Thus, $z \in P \wedge (x \times y)^z = x^z \times y^z$. Then,

$$
\begin{aligned}
(x \times y)^{S(z)} &= (x \times y)^{z+1} & &\text{(Equation }(\alpha)) \\
&= (x \times y)^z \times (x \times y) & &\text{(Equation }(\sigma')) \\
&= (x^z \times y^z) \times (x \times y) & &\text{(Inductive hypothesis)} \\
&= (x^z \times x) \times (y^z \times y) & &\text{(Associativity and Commutativity of } \times) \\
&= (x^{z+1}) \times (y^{z \times 1}) & &\text{(Equation }(\sigma')) \\
&= x^{S(z)} \times y^{S(z)} & &\text{(Equation }(\alpha)).
\end{aligned}
$$

Thus, $S(z) \in C$. We have proved: $z \in C \Rightarrow S(z) \in C$. Hence, by mathematical induction, $C = P$. ■

Theorem 6.4 (*Exponentiation and the Order Relation*) For any x, y, and z in P:

a. $x < y \Leftrightarrow x^z < y^z$.

b. $(1 < z \wedge x < y) \Leftrightarrow z^x < z^y$.

c. $1 < x \Leftrightarrow z < x^z$.

d. $(1 < z \wedge 1 < y) \Leftrightarrow z < z^y$.

Proof (a) Let us first prove:

$$(\$) \quad x < y \Rightarrow x^z < y^z \text{ for any } x, y, \text{ and } z \text{ in } P.$$

Assume $x < y$. Let $A = \{z : z \in P \wedge x^z < y^z\}$. We must show that $A = P$. $1 \in A$, since $x^1 = x < y = y^1$. Assume $z \in A$. Thus, $z \in P \wedge x^z < y^z$, Then,

$$
\begin{aligned}
x^{S(z)} = x^{z+1} = x^z \times x \quad &\text{(Equation } (\sigma')) \\
< y^z \times y \quad &\text{(Corollary 5.8d)} \\
= y^{z+1} = y^{S(z)}.
\end{aligned}
$$

Thus, $S(z) \in A$. We have shown that $z \in A \Rightarrow S(z) \in A$. By mathematical induction, $A = P$.

To prove the converse of ($\$$), assume $x^z < y^z$. We shall show $x < y$ by reductio ad absurdum. So, assume $x \not< y$. Then $x = y \vee y < x$, by the Trichotomy Law (Theorem 4.1c). If $x = y$, then $x^z = y^z$, contradicting $x^z < y^z$. If $y < x$, then, by ($\$$), $y^z < x^z$, contradicting $x^z < y^z$. In both cases, we have obtained a contradiction.

(b) First, we shall prove:

$$(\#) \quad (1 < z \wedge x < y) \Rightarrow z^x < z^y.$$

Assume $1 < z \wedge x < y$. Then $y = x + u$ for some u, by definition of $<$. Since $1 < z$, we obtain $1^u < z^u$ (by Part (a) of this theorem). Hence, $1 < z^u$, by Lemma 6.2. Therefore,

$$
\begin{aligned}
z^x = z^x \times 1 \quad &\text{(Equation } (\gamma)) \\
< z^x \times z^u \quad &\text{(Corollary 5.8f)} \\
= z^{x+u} \quad &\text{(Theorem 6.3a)} \\
= z^y.
\end{aligned}
$$

Thus, $z^x < z^y$.

Now, to prove the converse of ($\#$), assume $z^x < z^y$. Clearly, $z \neq 1$, by Lemma 6.2. Hence, $1 < z$, by Theorem 4.4a. We shall prove $x < y$ by reductio ad absurdum. Assume $x \not< y$. By the Trichotomy Law (Theorem 4.1c), $x = y \lor y < x$. If $x = y$, then $z^x = z^y$, contradicting $z^x < z^y$. If $y < x$, then, by ($\#$), $z^y < z^x$, contradicting $z^x < z^y$. In both cases, we have obtained a contradiction.

(c) Let us first prove:

$$(\mathfrak{c}) \quad 1 < x \Rightarrow z < x^z.$$

Assume $1 < x$. Let $A = \{z : z \in P \land z < x^z\}$. First, $1 \in A$, since $1 < x = x^1$. Assume $z \in A$. Thus, $z \in P \land z < x^z$. Then,

$$
\begin{aligned}
x^{S(z)} &= x^z \times x \quad &&\text{(Equation (σ'))} \\
&> z \times x \quad &&\text{(Inductive hypothesis and Corollary 5.8d)} \\
&\geq z \quad &&\text{(Corollary 5.8b).}
\end{aligned}
$$

Thus, $S(z) \in A$. We have shown that $z \in A \Rightarrow S(z) \in A$. By mathematical induction, $A = P$.

To prove the converse of ($\mathfrak{c}$), assume $1 \not< x$. Then $x = 1$, by Theorem 4.4a. Hence, $x^z = 1^z = 1 \leq z$. Thus, $z \not< x^z$ by the Trichotomy Law (Theorem 4.1c).

(d) Substitute 1 for x in (b).   ∎

EXERCISES

2. Prove: $1 < x \Leftrightarrow 1 < x^z$.

3. Prove: $x \leq x^z$.

4. Prove: $x \leq y \Leftrightarrow x^z \leq y^z$.

5. Prove: $x \leq y \Rightarrow z^x \leq z^y$.

6. Prove: $(1 < z \land z^x \leq z^y) \Rightarrow x \leq y$.

7. Prove: $(1 + z)^y \geq 1 + (y \times z)$.

2.7 ISOMORPHISM. CATEGORICITY

Do the axioms for Peano systems completely characterize the natural numbers? The meaning of this question is not quite clear, but we shall give a positive answer in the following sense. Any Peano system is *essentially* the same as the natural numbers. Of course, a precise definition is required for the notion that two Peano systems are *essentially the same*. This leads to the important idea of *isomorphic* Peano systems.

Definition   Consider any two Peano systems $\mathscr{P} = (P, S, 1)$ and $\mathscr{P}^* = (P^* \, S^*, 1^*)$. We say that $\mathscr{P}$ and $\mathscr{P}^*$ are **isomorphic** if and only if there is a one-one correspondence H between P and P^*

$$H: P \xrightarrow[\text{onto}]{1\text{-}1} P^*$$

such that:

i. $H(1) = 1^*$, and
ii. $H(S(x)) = S^*(H(x))$ for all x in P.

Such a function H is called an **isomorphism** between $\mathscr{P}$ and $\mathscr{P}^*$.

EXAMPLES

1.   Let $\mathscr{P} = (P, S, 1)$ be the standard Peano system, where P is the set of natural numbers, "1" stands for the ordinary integer 1, and $S(x) = x + 1$ for all x in P. Let $\mathscr{P}^* = (P^*, S^*, 1^*)$ be the Peano system in which P^* is the set of even natural numbers, "1^*" denotes the integer 2, and $S^*(x) = x + 2$ for all x in P^*. Then the function $H: P \to P^*$ such that $H(x) = 2 \times x$ for all x in P is an isomorphism between $\mathscr{P}$ and $\mathscr{P}^*$. Note first that $H(1) = 2 = 1^*$, verifying condition (i). Moreover,

$$H(S(x)) = H(x + 1) = 2 \times (x + 1) = (2 \times x) + 2 = H(x) + 2 = S^*(H(x)),$$

verifying (ii). It is clear that H is one-one (since $x \neq y \Rightarrow 2 \times x \neq 2 \times y$), and that the range of H is P^*.

2.   Again let $\mathscr{P} = (P, S, 1)$ be the standard Peano system and let $\mathscr{P}^\# = (P^\#, S^\#, 1^\#)$, where $P^\#$ is the set of negative integers, $1^\# = -1$, and $S^\#(u) = u - 1$ for all u in $P^\#$. Then the function $G: P \to P^\#$ such that $G(x) = -x$ for all x in P is an isomorphism between $\mathscr{P}$ and $\mathscr{P}^\#$. First, $G(1) = -1 = 1^\#$. In addition,

$$G(S(x)) = G(x + 1) = -(x + 1) = (-x) - 1 = G(x) - 1 = S^\#(G(x)).$$

Finally, it is obvious that G is one-one (since $x \neq y \Rightarrow -x \neq -y$), and that the range of G is $P^\#$.

An isomorphism H between two Peano systems $\mathscr{P}$ and $\mathscr{P}*$ shows that the two structures are essentially the same. Condition (i) asserts that the distinguished object 1 of $\mathscr{P}$ corresponds to the distinguished object $1*$ of $\mathscr{P}*$. Condition (ii) says that the successor operations correspond under the function H; this is pictured in the following diagram:

$$
\begin{array}{ccc}
x & \xrightarrow{\;H\;} & H(x) \\
{\scriptstyle S}\downarrow & & \downarrow{\scriptstyle S*} \\
S(x) & \xrightarrow{\;H\;} & H(S(x)) = S*(H(x)).
\end{array}
$$

Theorem 7.1　Any two Peano systems are isomorphic.

Proof　Let $\mathscr{P} = (P, S, 1)$ and $\mathscr{P}* = (P*, S*, 1*)$ be two Peano systems. In the Iteration Theorem, let $W = P*$, $c = 1*$, and $g = S*$. We obtain a unique function $F: P \to P*$ such that $F(1) = c = 1*$ and $F(S(x)) = g(F(x)) = S*(F(x))$ for any x in P. To complete the proof that F is an isomorphism, it must be verified that F is one-one and that its range is $P*$. To show that the range of F is $P*$, let B be the range of F. Since $F(1) = 1*$, it follows that $1* \in B$. Now assume $z \in B$. Then $z = F(x)$ for some x in P. Hence, $S*(z) = S*(F(x)) = F(S(x))$. Thus, $S*(z) \in B$. We have shown that $z \in B \Rightarrow S*(z) \in B$. Hence, by mathematical induction (Axiom P3 for $\mathscr{P}*$), $B = P*$.

We now turn to the proof that F is one-one. Let

$$A = \{x : x \in P \wedge (\forall y)[(y \in P \wedge y \neq x] \Rightarrow F(x) \neq F(y)).$$

We must show that $A = P$. Let us show first that $1 \in A$. To this end, assume $y \in P \wedge y \neq 1$. Then, by Theorem 1.1, $y = S(u)$ for some u in P. Hence, $F(y) = F(S(u)) = S*(F(u)) \neq 1*$ by Axiom P1 for $\mathscr{P}*$. Since $F(1) = 1*$, it follows that $F(y) \neq F(1)$. Thus, $1 \in A$. Now assume $x \in A$. Thus, $x \in P \wedge (\forall y)([y \in P \wedge y \neq x] \Rightarrow F(x) \neq F(y))$. Assume $y \in P$ and $y \neq S(x)$.

Case (i)　$y = 1$. Then $F(S(x)) = S*(F(x)) \neq 1* = F(1)$. Thus, in this case, $F(S(x)) \neq F(y)$.

Case (ii)　$y \neq 1$. Then, by Theorem 1.1, $y = S(u)$ for some u in P. Since $y \neq S(x)$, it follows that $x \neq u$. Hence, by inductive hypothesis, $F(x) \neq F(u)$. Then, by Axiom P2, $S*(F(x)) \neq S*(F(u))$. Therefore,

$$F(S(x)) = S*(F(x)) \neq S*(F(u)) = F(S(u)) = F(y).$$

In both cases, $F(S(x)) \neq F(y)$. This shows that $(\forall y)([y \in P \land y \neq S(x)]$ $\Rightarrow F(S(x)) \neq F(y))$, and, therefore, that $S(x) \in A$. We have proved: $x \in A$ $\Rightarrow S(x) \in A$. Hence, by mathematical induction, $A = P$. ∎

Remark 1 The proof of Theorem 7.1 established an additional fact: There is precisely one isomorphism between any two Peano systems. (The function F was unique).

Remark 2 The isomorphism F between $\mathscr{P}$ and $\mathscr{P}^*$ preserves all defined operations and relations. For example, if $+$, $\times$, $<$ are the addition, multiplication, and order in $\mathscr{P}$, while $+^*$, $\times^*$, $<^*$ are the corresponding notions in $\mathscr{P}^*$, then, for any x and y in P:

  i. $F(x + y) = F(x) +^* F(y)$.
 ii. $F(x \times y) = F(x) \times^* F(y)$.
iii. $x < y \Leftrightarrow F(x) <^* F(y)$.

Let us see how, for example, (i) is established. Take any x in P. Let $A = \{y : y \in P \land F(x + y) = F(x) +^* F(y)\}$. We must prove $A = P$. First $1 \in A$, since

$$F(x + 1) = F(S(x)) = S^*(F(x)) = F(x) +^* 1^* = F(x) +^* F(1).$$

Now assume $y \in A$. Thus, $y \in P \land F(x + y) = F(x) +^* F(y)$. Then

$$F(x + S(y)) = F(S(x + y)) = S^*(F(x + y)) = S^*(F(x) +^* F(y))$$
$$= F(x) +^* S^*(F(y)) = F(x) +^* F(S(y)).$$

Thus, $S(y) \in A$. We have shown that $y \in A \Rightarrow S(y) \in A$. By mathematical induction, $A = P$.

The reader should verify (ii) and (iii) on his own.

Theorem 7.1 is often stated in another way. A set of axioms is said to be **categorical** if and only if any two models for those axioms are isomorphic. Thus, Theorem 7.1 asserts that Axioms P1–P3 for Peano systems are categorical.[†] It seems plausible, by Theorem 7.1, that any property concerning

---

[†] The notion of Peano system really has been only partially axiomatized by means of Axioms P1–P3. The logical notions and the concept of *set* have been left informal. If we add logical and set-theoretical axioms to Axioms P1–P3 in such a way that all proofs are based upon explicitly stated axioms and rules of inference, then the enlarged axiom system is no longer categorical. This noncategoricity of the completely formalized axioms for Peano systems is a special case of a more general logical phenomenon (see Mendelson [1964], p. 69 and p. 116).

a Peano system which holds for one Peano system must hold for any other Peano system. This is indeed the case, since an isomorphism "carries over" any property of a Peano system onto an isomorphic system.[†]

2.8 A BASIC EXISTENCE ASSUMPTION

It would seem strange to ask at this point whether any Peano systems exist. In fact, in Section 2.1, we listed several examples of Peano systems, the primary example being the standard system. However, if we look carefully at this example, some doubts may arise.

What is our intuitive understanding of the natural numbers? Surely this, being the firmest of all our mathematical ideas, should have a definite, transparent meaning. Let us examine a few attempts to make this meaning clear.

(1)   The natural numbers may be thought of as symbolic expressions: 1 is |, 2 is ||, 3 is |||, 4 is ||||, etc. Thus, we start with a vertical stroke | and obtain new expressions by appending additional vertical strokes. There are some obvious objections to this approach. First, we cannot be talking about particular physical marks on paper, since a vertical stroke for the number 1 may be repeated in different physical locations. The number 1 cannot be a class of all congruent strokes, since the length of the stroke may vary; we would even acknowledge as a 1 a somewhat wiggly stroke written by a very nervous person. Even if we should succeed in giving a sufficiently general geometric characterization of the curves which would be recognized as 1's, there is still another objection. Different people and different civilizations may use different symbols for the basic unit, for example, a circle or a square instead of a stroke. Yet, we could not give priority to one symbolism over any of the others. Nevertheless, in all cases, we would have to admit that, regardless of the difference in symbols, we are all talking about the same things.

(2)   The natural numbers may be conceived to be set-theoretic objects. In one very appealing version of this approach, the number 1 is defined as the set of all singletons $\{x\}$; the number 2 is the set of all unordered pairs $\{x, y\}$, where $x \neq y$; the number 3 is the set of all sets $\{x, y, z\}$, where $x \neq y$, $x \neq z$, $y \neq z$; and so on. Within a suitable axiomatic presentation of set theory, clear rigorous definitions can be given along these lines for the general notion of natural number and for the familiar operations and relations

---

[†] A precise statement and proof of this fact (in the case of so-called elementary theories) may be found in Mendelson [1964], p. 90.

involving natural numbers. Indeed, the axioms for a Peano system are easy consequences of the definitions and simple theorems of set theory. Nevertheless, there are strong deficiencies in this approach as well. First, there are many competing forms of axiomatic set theory. In some of them, the approach sketched above cannot be carried through, and a completely different definition is necessary. For example, one can define the natural numbers as follows: $1 = \{\varnothing\}$, $2 = \{\varnothing, 1\}$, $3 = \{\varnothing, 1, 2\}$, etc. Alternatively, one could use: $1 = \{\varnothing\}$, $2 = \{1\}$, $3 = \{2\}$, etc. Thus, even in set theory, there is no single way to handle the natural numbers. However, even if a set-theoretic definition is agreed upon, it can be argued that the clear mathematical idea of the natural numbers should not be defined in set-theoretic terms. The paradoxes (that is, arguments leading to a contradiction) arising in set theory have cast doubt upon the clarity and meaningfulness of the general notions of set theory. It would be inadvisable then to define our basic mathematical concepts in terms of set-theoretic ideas.

This discussion leads us to the conjecture that the natural numbers are not *particular* mathematical objects. Different people, different languages, and different set theories may have different systems of natural numbers. However, they all satisfy the axioms for Peano systems and, therefore, by Theorem 7.1, are isomorphic. There is no one system which has priority in any sense over all the others. For Peano systems, as for all the mathematical systems we shall encounter later, it is the form (or structure) which is important, not the "content."

Since the natural numbers are necessary in the further development of mathematics, we shall make one simple assumption.

Basic Axiom   There exists a Peano system.[†]

SUPPLEMENTARY EXERCISES

1. By a *Dedekind system* we mean any structure (P, S), where P is a set, S is a singulary operation on P, and the following axioms are satisfied.

   (1')   The range of S is a proper subset of P: $\mathscr{R}(S) \neq P$.

   (2') S is one-one: $(\forall x)(\forall y)([x \in P \land y \in P \land x \neq y] \Rightarrow S(x) \neq S(y))$.

   (3')   Principle of Mathematical Induction:

   $$(\forall z)(\forall B)([B \subseteq P \land z \in P - \mathscr{R}(S) \land z \in B \land (\forall x)(x \in B \Rightarrow S(x) \in B)] \Rightarrow B = P).$$

[†] In Appendix E, there is a proof of the Basic Axiom within a system of axiomatic set theory. As the discussion above has indicated, this is not a clear-cut justification for the Basic Axiom.

a. Prove: $(\exists!u)(u \in P - \mathscr{R}(S))$.

   Hint: Let $z \in P - \mathscr{R}(S)$, and apply the Principle of Mathematical Induction (Axiom (3′)) to the set $B = \mathscr{R}(S) \cup \{z\}$.

b. Let the unique u in $P - \mathscr{R}(S)$ be denoted 1. Prove that $(P, S, 1)$ is a Peano system.

2. A Second Proof of the Iteration Theorem (Theorem 2.1). Consider a set W, an element c in W and a singulary operation g on W. Let $(P, S, 1)$ be a Peano system. We must show that there is a unique function $F: P \to W$ such that $F(1) = c$ and $F(S(x)) = g(F(x))$ for all x in P.
   Let $\mathscr{C} = \{H : H \subseteq P \times W \wedge (1, c) \in H \wedge (\forall x)(\forall w)((x, w) \in H \Rightarrow (S(x), g(w)) \in H\}$. Observe that $P \times W \in \mathscr{C}$. Let $F = \bigcap_{H \in \mathscr{C}} H$.

   Prove:

   a. $F \in \mathscr{C}$.

   b. $F: P \to W$. (*Hint*: Let $B = \{x : (\exists!w)(w \in W \wedge (x, w) \in F)\}$. By mathematical induction, prove $B = P$.)

   c. $F(1) = c$.

   d. $F(S(x)) = g(F(x))$ for all x in P.

3. Why is the following not a correct proof that any two Peano systems $\mathscr{P} = (P, S, 1)$ and $\mathscr{P}^* = (P^*, S^*, 1^*)$ are isomorphic?
   P consists of

   $$1, \quad S(1), \quad S(S(1)), \quad S(S(S(1))), \ldots,$$

   while P^* consists of

   $$1^*, \quad S^*(1^*), \quad S^*(S^*(1^*)), \quad S^*(S^*(S^*(1^*))), \quad \ldots.$$

   Simply let 1 correspond to 1^*, $S(1)$ to $S^*(1^*)$, $S(S(1))$ to $S^*(S^*(1^*))$, and so on (each object in the list for P being associated with the corresponding object in the list for P^*).

4. Prove that the relation "isomorphic with" is reflexive, symmetric, and transitive, that is, for any Peano systems $\mathscr{P} = (P, S, 1)$, $\mathscr{P}^* = (P^*, S^*, 1^*)$, and $\mathscr{P}^\# = (P^\#, S^\#, 1^\#)$:

   a. $\mathscr{P}$ is isomorphic with $\mathscr{P}$. (Reflexivity)

   b. If $\mathscr{P}$ is isomorphic with $\mathscr{P}^*$, then $\mathscr{P}^*$ is isomorphic with $\mathscr{P}$. (Symmetry)

   c. If $\mathscr{P}$ is isomorphic with $\mathscr{P}^*$, and $\mathscr{P}^*$ is isomorphic with $\mathscr{P}^\#$, then $\mathscr{P}$ is isomorphic with $\mathscr{P}^\#$. (Transitivity) (*Hints*: (a) Use the identity function on P. (b) If F is an isomorphism of $\mathscr{P}$ with $\mathscr{P}^*$, use the inverse F^{-1}. (c) If F is an isomorphism of $\mathscr{P}$ with $\mathscr{P}^*$, and G is an isomorphism of $\mathscr{P}^*$ with $\mathscr{P}^\#$, consider the composition $G \circ F$.)

5.[D] Consider a structure $(P, S, 1)$, where $1 \in P$ and $S: P \to P$. Show that $(P, S, 1)$ is a Peano system if and only if it satisfies the Iteration Theorem, that is, for any set W, any object c in W, and any singulary operation g on W, there is a unique function $F: P \to W$ such that $F(1) = c$ and $F(S(x)) = g(F(x))$ for all x in P. (see Maclane and Birkhoff [1967], pp. 67–70).

6. Show that each of Axioms P1–P3 is independent of the other axioms. (For example, to show the independence of Axiom P1, give an example of a structure $(P, S, 1)$ which satisfies Axioms P2 and P3 but does not satisfy Axiom P1.)

7. (a) *Recursion Theorem.* Let Y be any set, $d \in Y$, and $h: P \times Y \to Y$. (Here, $(P, S, 1)$ is a Peano system.) Then there is a unique function $\Phi: P \to Y$ such that $\Phi(1) = d$ and $\Phi(n + 1) = h(n, \Phi(n))$ for all n in P.

The reader should fill in the details of the following sketch of a proof. Let $W = P \times Y$ and define $g: P \times Y \to P \times Y$ by $g(n, y) = (n + 1, h(n, y))$. By the Iteration Theorem, with $c = (1, d)$, there is a unique function $F: P \to P \times Y$ such that $F(1) = (1, d)$ and $F(n + 1) = g(F(n))$ for all n in P. By mathematical induction, it follows easily that the first component of $F(n)$ is n. Let $\Phi(n)$ be the second component of $F(n)$. Thus, $\Phi(1) = d$, and

$$(n + 1, \Phi(n + 1)) = F(n + 1) = g(F(n)) = g(n, \Phi(n)) = (n + 1, h(n, \Phi(n))).$$

Hence, $\Phi(n + 1) = h(n, \Phi(n))$.

(b) Show that there is a **factorial** function $n!$ from P into P such that $1! = 1$ and $(n + 1)! = n \times (n!)$ for all n in P.

SUGGESTIONS FOR FURTHER READING

Barker [1964], pp. 67–104.

Benecerraf [1965].

Dedekind [1901]. (Read the second essay, *The Nature and Meaning of Numbers*, a translation of the 1887 paper, *Was sind und was sollen die Zahlen?*. Part of the *Preface to the First Edition* (pp. 31–36) is a lucid explanation and justification of the study of number systems.)

Mendelson [1964].

Parsons [1965].

Peano [1891].

Wang [1957]. (Historical material on the contributions of Dedekind, Peano, and Frege to the foundations of arithmetic.)

CHAPTER 3

THE INTEGERS

3.1 DEFINITION OF THE INTEGERS

Zero and the negative integers were introduced into mathematics relatively late in recorded history. Of course, their universal usefulness in such diverse areas as finance and physics is well known. We shall show in this chapter how to construct the integers from the natural numbers (that is, from a Peano system), and we shall derive the basic and familiar properties of the system of integers.

By our Basic Axiom, we may assume the existence of a Peano system $(P, S, 1)$. The elements of P will be called **natural numbers.**

We could construct the integers by adjoining a new object 0, together with a new object $-x$ for each natural number x. We would then have to define the usual arithmetic operations and relations and derive the fundamental properties of the integers However, this straightforward and natural approach turns out to be long and tedious because many definitions and proofs require subdivision into a large number of cases. (See Exercise 1 on p. 153 for an outline of such a development.) For this reason, we shall employ a more sophisticated, but easier, method.

From our intuitive knowledge of the integers we know that every integer r can be represented as the difference $n - k$ of some natural numbers n and k. For example,

$$3 = 4 - 1,$$
$$0 = 1 - 1,$$
$$-7 = 1 - 8, \quad \text{etc.}$$

Of course, the same integer can be represented as the difference of many different pairs:

$$3 = 4 - 1 = 5 - 2 = 6 - 3 = 7 - 4 = \cdots,$$
$$0 = 1 - 1 = 2 - 2 = 3 - 3 = \cdots,$$
$$-7 = 1 - 8 = 2 - 9 = 3 - 10 = \cdots, \qquad \text{etc.}$$

It then appears plausible to define an integer r to be the set of all pairs of natural numbers (n, k) such that $r = n - k$. For example, -7 will be the set of all ordered pairs (n, k) of natural numbers such that $-7 = n - k$. This is roughly what we shall do. However, we cannot proceed directly as indicated because:

a.   Not having defined -7 as yet, we cannot ascribe any meaning to $-7 = n - k$;

b.   The subtraction operation has not yet been defined, so that, for example, $1 - 8$ has no meaning at this point.

We shall avoid these difficulties in the following way.

Definition   For any natural numbers n, k, j, i, let

$$(n, j) \sim (k, i)$$

mean that

$$n + i = k + j.$$

This defines a relation $\sim$ between ordered pairs of natural numbers.

EXAMPLES

1. $(1, 3) \sim (5, 7)$.

2. $(4, 2) \sim (13, 11)$.

3. $(2, 2) \sim (3, 3)$.

4.   For any natural numbers n, j:

$$(n, j) \sim (1, 1) \Leftrightarrow n = j.$$

For, $(n, j) \sim (1, 1) \Leftrightarrow n + 1 = 1 + j$. By the Cancellation Law,

$$n + 1 = 1 + j \Leftrightarrow n = j.$$

5.   For any natural numbers n, j:

$$(n, j) \sim (2, 1) \Leftrightarrow n = j + 1.$$

For, $(n, j) \sim (2, 1) \Leftrightarrow n + 1 = 2 + j$. But, $2 + j = 1 + 1 + j$. Hence, by the Cancellation Law, $n + 1 = 2 + j \Leftrightarrow n = j + 1$.

From our intuitive knowledge of the integers we see that $(n, j) \sim (k, i)$ corresponds to the equation

$$n - j = k - i.$$

Instead of $n - j = k - i$, we have used the equivalent equation $n + i = k + j$, which involves only natural numbers and the addition operation on natural numbers. (Of course, this remark and similar ones later that are based upon our intuitive understanding are meant only as informal guides for the reader. They have no official standing, since it is our intuitive knowledge about numbers that we are trying to make precise and to justify.)

Theorem 1.1 $\sim$ is an equivalence relation in the set $P \times P$ of ordered pairs of natural numbers, that is, for all natural numbers h, i, j, k, m, n:

a. $(h, i) \sim (h, i)$ (Reflexivity)
b. $(h, i) \sim (j, k) \Rightarrow (j, k) \sim (h, i)$ (Symmetry)
c. $[(h, i) \sim (j, k) \wedge (j, k) \sim (m, n)] \Rightarrow (h, i) \sim (m, n)$ (Transitivity)

Proof   (a) $h + i = h + i$.

(b)   Assume $(h, i) \sim (j, k)$. Thus, $h + k = j + i$. Then $j + i = h + k$, that is, $(j, k) \sim (h, i)$.

(c)   Assume $(h, i) \sim (j, k)$ and $(j, k) \sim (m, n)$. Thus,

$$h + k = j + i \quad \text{and} \quad j + n = m + k.$$

By adding these equations, we obtain

$$h + k + j + n = j + i + m + k.$$

Hence, by the Associative, Commutative, and Cancellation Laws for addition, $h + n = m + i$. Thus, $(h, i) \sim (m, n)$.   ∎

Since $\sim$ is an equivalence relation in $P \times P$, we obtain a division of $P \times P$ into equivalence classes with respect to $\sim$. Given (n, j) in $P \times P$, we use the standard notation $[(n, j)]$ for the equivalence class of (n, j):

$$[(n, j)] = \{(k, i) : (k, i) \in P \times P \wedge (n, j) \sim (k, i)\}.$$

EXAMPLES

6.
$$[(2, 3)] = \{(k, i) : (k, i) \in P \times P \wedge (2, 3) \sim (k, i)\}$$
$$= \{(k, i) : (k, i) \in P \times P \wedge 2 + i = k + 3\}$$
$$= \{(1, 2), (2, 3), (3, 4), (4, 5), \ldots\}.$$

7.
$$[(4, 4)] = \{(k, i) : (k, i) \in P \times P \wedge (4, 4) \sim (k, i)\}$$
$$= \{(k, i) : (k, i) \in P \times P \wedge 4 + i = k + 4\}$$
$$= \{(k, i) : (k, i) \in P \times P \wedge i = k\}$$
$$= \{(1, 1), (2, 2), (3, 3), (4, 4), \ldots\}.$$

We let Z denote the set of all equivalence classes with respect to $\sim$. The elements of Z will be called **integers**. We shall use lowercase Greek letters $\alpha, \beta, \gamma, \delta, \ldots$ as variables for integers.

Let us introduce special notation for two specific integers.

Definitions $0_Z = [(1, 1)].$
$$1_Z = [(2, 1)].$$

By Example 4, $(n, j) \in 0_Z \Leftrightarrow n = j$. Thus,

$$0_Z = \{(1, 1), (2, 2), (3, 3), \ldots\}.$$

By Example 5, $(n, j) \in 1_Z \Leftrightarrow n = j + 1$. Thus,

$$1_Z = \{(2, 1), (3, 2), (4, 3), \ldots\}.$$

Notice that $0_Z \neq 1_Z$, since $(1, 1) \in 0_Z$ but $(1, 1) \notin 1_Z$.

3.2   ADDITION AND MULTIPLICATION OF INTEGERS

We can be led to suitable definitions of addition and multiplication of integers if we return to our intuitive picture of the integers. For example, since

$$3 = 4 - 1$$

and

$$-7 = 1 - 8,$$

we obtain

$$3 + (-7) = (4 + 1) - (1 + 8) = 5 - 9 = -4.$$

More generally, if

$$r = n - j$$

and

$$s = k - i,$$

then

$$r + s = (n + k) - (j + i).$$

This suggests a definition of addition of integers, but first we must confirm the following fact.

Lemma 2.1   Let $n, j, k, i, n_1, j_1, k_1, i_1$ be natural numbers. If $(n, j) \sim (n_1, j_1)$ and $(k, i) \sim (k_1, i_1)$, then

$$(n + k, j + i) \sim (n_1 + k_1, j_1 + i_1).$$

Proof   By assumption, $n + j_1 = n_1 + j$ and $k + i_1 = k_1 + i$. Adding, we obtain $(n + j_1) + (k + i_1) = (n_1 + j) + (k_1 + i)$. By associativity and commutativity of $+$,

$$(n + k) + (j_1 + i_1) = (n_1 + k_1) + (j + i),$$

that is,

$$(n + k, j + i) \sim (n_1 + k_1, j_1 + i_1). \quad \blacksquare$$

EXAMPLE

1.   We know that $(1, 2) \sim (3, 4)$ (since $1 + 4 = 3 + 2$) and $(2, 2) \sim (3, 3)$ (since $2 + 3 = 3 + 2$). Hence, by the Lemma, we have $(1 + 2, 2 + 2) \sim (3 + 3, 4 + 3)$. Thus, $(3, 4) \sim (6, 7)$. This is correct, since $3 + 7 = 6 + 4$.

Lemma 2.1 permits us to make the following definition.

Definition   Let α and β be integers. Choose any element (n, j) of α, and choose any element (k, i) of β. Then

$$\alpha +_Z \beta \qquad \text{stands for} \qquad [(n + k, j + i)].$$

The reason that Lemma 2.1 is required for this definition to make sense is that we must be sure that our particular choice of pairs (n, j) in α and (k, i) in β does not affect the outcome $\alpha +_Z \beta$. If we had chosen different pairs (n_1, j_1) in α and (k_1, i_1) in β, then Lemma 2.1 tells us that

$$(n + k, j + i) \sim (n_1 + k_1, j_1 + i_1),$$

and, therefore,

$$[(n + k, j + i)] = [(n_1 + k_1, j_1 + i_1)].$$

Theorem 2.2 (*Properties of Addition*) Consider any α, β, γ in Z.

a. $\alpha +_Z \beta = \beta +_Z \alpha$ (Commutativity).

b. $\alpha +_Z (\beta +_Z \gamma) = (\alpha +_Z \beta) +_Z \gamma$ (Associativity).

c. $\alpha +_Z 0_Z = \alpha$.

d. $(\exists! \delta)(\alpha +_Z \delta = 0_Z)$.   (This unique δ is denoted $-_Z\alpha$. If $(h, i) \in \alpha$, then $(i, h) \in -_Z\alpha$.)

Proof (a) Take any (n, j) in α and any (k, i) in β. By definition,

$$\alpha +_Z \beta = [(n + k, j + i)] \quad \text{and} \quad \beta +_Z \alpha = [(k + n, i + j)].$$

Hence, $\alpha +_Z \beta = \beta +_Z \alpha$, by virtue of the commutativity of $+$.[†]

(b) Take (n, j) in α, (k, i) in β, and (m, h) in γ. Then $\beta +_Z \gamma = [(k + m, i + h)]$. Thus, $(k + m, i + h) \in \beta +_Z \gamma$. Hence, $\alpha +_Z (\beta +_Z \gamma) = [(n + (k + m), j + (i + h))]$. Similarly, $\alpha +_Z \beta = [(n + k, j + i)]$, and $(\alpha +_Z \beta) +_Z \gamma = [((n + k) + m, (j + i) + h)]$. Therefore, $\alpha +_Z (\beta +_Z \gamma) = (\alpha +_Z \beta) +_Z \gamma$, by the associativity of $+$.

(c) Take (n, j) in α. Remember that $(1, 1) \in 0_Z$. Then, $\alpha +_Z 0_Z = [(n + 1, j + 1)]$. However, $(n + 1, j + 1) \sim (n, j)$, since $(n + 1) + j = n + (j + 1)$. Hence, $[(n + 1, j + 1)] = [(n, j)]$. But, since $(n, j) \in \alpha$, $[(n, j)] = \alpha$. Therefore, $\alpha +_Z 0_Z = \alpha$.

(d) Take (n, j) in α. Let $\delta = [(j, n)]$. Then, $\alpha +_Z \delta = [(n + j, j + n)] = [(n + j, n + j)] = 0_Z$. To prove the uniqueness of δ, assume that $\alpha +_Z \mu = 0_Z$. We must show that $\mu = \delta$. But,

$$\mu = \mu +_Z 0_Z = \mu +_Z (\alpha +_Z \delta) = (\mu +_Z \alpha) +_Z \delta$$
$$= (\alpha +_Z \mu) +_Z \delta = 0_Z +_Z \delta = \delta +_Z 0_Z = \delta.$$

[†] Remember that $+$ indicates addition in the Peano system $(P, S, 1)$.

In addition, for any natural numbers h and i,

$$(h, i) \in \alpha \Leftrightarrow (h, i) \sim (n, j)$$
$$\Leftrightarrow h + j = n + i$$
$$\Leftrightarrow i + n = j + h$$
$$\Leftrightarrow (i, h) \sim (j, n)$$
$$\Leftrightarrow (i, h) \in \delta.$$

(Example: $(i, h) \in {}_{-z}1_Z \Leftrightarrow (h, i) \in 1_Z \Leftrightarrow h = i + 1.$)   ∎

Multiplication is handled in much the same way as addition. First we look at the intuitive picture for hints as to how to proceed. If $r = n - j$ and $s = k - i$, then

$$r \times s = (n - j) \times (k - i) = ((n \times k) + (j \times i)) - ((j \times k) + (n \times i)).$$

This gives us the idea for the multiplication rule. But, as before, a preliminary lemma is needed.

Lemma 2.3   Let $n, j, k, i, n_1, j_1, k_1, i_1$ be   natural   numbers.   If $(n, j) \sim (n_1, j_1)$ and $(k, i) \sim (k_1, i_1)$, then

$$((n \times k) + (j \times i), (j \times k) + (n \times i))$$
$$\sim ((n_1 \times k_1) + (j_1 \times i_1), (j_1 \times k_1) + (n_1 \times i_1)).$$

Proof   Assume $(n, j) \sim (n_1, j_1)$ and $(k, i) \sim (k_1, i_1)$. Thus,

(i) $n + j_1 = n_1 + j.$

(ii) $k + i_1 = i + k_1.$

We must prove:

$$\underbrace{(n \times k) + (j \times i) + (j_1 \times k_1) + (n_1 \times i_1)}_{(A)} = \underbrace{(n_1 \times k_1) + (j_1 \times i_1) + (j \times k) + (n \times i)}_{(B)}.$$

Multiplying (i) by k,

(iii) $(n \times k) + (j_1 \times k) = (n_1 \times k) + (j \times k).$

Multiplying (ii) by j_1 and reversing the two sides of the equality,

(iv) $(j_1 \times k_1) + (j_1 \times i) = (j_1 \times k) + (j_1 \times i_1).$

Multiplying (i) by i and reversing the two sides of the equality

(v) $(i \times j) + (i \times n_1) = (i \times n) + (i \times j_1).$

Multiplying (ii) by n_1,

(vi) $\quad (n_1 \times k) + (n_1 \times i_1) = (n_1 \times i) + (n_1 \times k_1).$

Adding (iii)–(vi), we obtain

$$(A) + (C) = (B) + (C),$$

where $(C) = (j_1 \times k) + (j_1 \times i) + (i \times n_1) + (n_1 \times k)$. Hence, by the Cancellation Law for $+$, $(A) = (B)$. ∎

EXAMPLE

2. We know that $(1, 2) \sim (2, 3)$ and $(2, 4) \sim (1, 3)$. Hence, by the Lemma,

$$((1 \times 2) + (2 \times 4), (1 \times 4) + (2 \times 2)) \sim ((2 \times 1) + (3 \times 3), (2 \times 3) + (3 \times 1)),$$

that is, $(10, 8) \sim (11, 9)$. This holds, since $10 + 9 = 11 + 8$.

Lemma 2.3 permits us to make the following definition.

Definition Given α and β in Z, take any (n, j) in α and any (k, i) in β. Let $\alpha \times_z \beta$ stand for $[(n \times k) + (j \times i), (j \times k) + (n \times i))$.

Lemma 2.3 tells us that the definition gives us the same result for $\alpha \times_z \beta$ no matter what ordered pairs (n, j) in α and (k, i) in β we may happen to choose.

Theorem 2.4 (*Properties of Multiplication*) For any α, β, γ in Z:

a. $\alpha \times_z \beta = \beta \times_z \alpha$ (Commutativity).
b. $\alpha \times_z (\beta \times_z \gamma) = (\alpha \times_z \beta) \times_z \gamma$ (Associativity).
c. $\alpha \times_z (\beta +_z \gamma) = (\alpha \times_z \beta) +_z (\alpha \times_z \gamma)$ (Distributivity).
d. $\alpha \times_z 1_z = \alpha$.
e. $[\alpha \neq 0_z \wedge \beta \neq 0_z] \Rightarrow \alpha \times_z \beta \neq 0_z$.

Proof In all the parts below, choose any (n, j) in α, any (k, i) in β, and any (m, h) in γ.

(a) $\qquad \alpha \times_z \beta = [(n \times k) + (j \times i), (j \times k) + (n \times i)],$

and

$$\beta \times_z \alpha = [(k \times n) + (i \times j), (i \times n) + (k \times j)].$$

The result follows from the commutativity of $+$ and $\times$.

(b) $\qquad \beta \times_Z \gamma = [(k \times m) + (i \times h), (i \times m) + (k \times h)].$

Hence,

$$
\begin{aligned}
\alpha \times_Z (\beta \times_Z \gamma) &= [(n \times ((k \times m) + (i \times h)) + (j \times ((i \times m) + (k \times h))), \\
&\qquad (j \times ((k \times m) + (i \times h)) + (n \times ((i \times m) + (k \times h)))] \\
&= [(n \times (k \times m)) + (n \times (i \times h)) + (j \times (i \times m)) + (j \times (k \times h)), \\
&\qquad (j \times (k \times m)) + (j \times (i \times h)) + (n \times (i \times m)) \\
&\qquad + (n \times (k \times h))]
\end{aligned}
$$
$\qquad\qquad\qquad$ (by the Distributive Law for $\times$ and $+$).

On the other hand, $\alpha \times_Z \beta = [(n \times k) + (j \times i), (j \times k) + (n \times i)]$. Hence,

$$
\begin{aligned}
(\alpha \times_Z \beta) \times_Z \gamma &= [((n \times k) + (j \times i)) \times m) + (((j \times k) + (n \times i)) \times h), \\
&\qquad (((j \times k) + (n \times i)) \times m) + (((n \times k) + (j \times i)) \times h)] \\
&= [((n \times k) \times m) + ((j \times i) \times m) + ((j \times k) \times h) + ((n \times i) \times h), \\
&\qquad ((j \times k) \times m) + ((n \times i) \times m) + ((n \times k) \times h) \\
&\qquad + ((j \times i) \times h)]
\end{aligned}
$$
$\qquad\qquad\qquad$ (by the Distributive Law for $\times$ and $+$).

The result $\alpha \times_Z (\beta \times_Z \gamma) = (\alpha \times_Z \beta) \times_Z \gamma$ follows by the associativity of $\times$ and the commutativity of $+$.

(c) $\qquad$
$$
\begin{aligned}
\alpha \times_Z (\beta +_Z \gamma) &= [(n, j)] \times_Z [(k + m, i + h)] \\
&= [((n \times (k + m)) + (j \times (i + h)), (j \times (k + m)) \\
&\qquad + (n \times (i + h)))] \\
&= [((n \times k) + (n \times m) + (j \times i) + (j \times h), \\
&\qquad (j \times k) + (j \times m) + (n \times i) + (n \times h))]
\end{aligned}
$$
$\qquad\qquad\qquad$ (by the Distributive Law for $\times$ and $+$)
$$
\begin{aligned}
&= [((n \times k) + (j \times i), (j \times k) + (n \times i))] \\
&\quad +_Z [((n \times m) + (j \times h), (j \times m) + (n \times h))] \\
&= (\alpha \times_Z \beta) +_Z (\alpha \times_Z \gamma).
\end{aligned}
$$

(d) $\qquad$
$$
\begin{aligned}
\alpha \times_Z 1_Z &= [(n, j)] \times_Z [(2, 1)] \\
&= [((n \times 2) + (j \times 1), (j \times 2) + (n \times 1))] \\
&= [((2 \times n) + j, (2 \times j) + n)] = [(n, j)] = \alpha.
\end{aligned}
$$

(The penultimate equation is obtained as follows: By virtue of the equation $2 \times i = i + i$, we obtain $((2 \times n) + j) + j = n + ((2 \times j) + n))$, which means that $((2 \times n) + j, (2 \times j) + n) \sim (n, j)$.)

(e)   Assume $\alpha \neq 0_Z$ and $\beta \neq 0_Z$. Since $(n, j) \in \alpha$ and $(k, i) \in \beta$, it follows that $n \neq j$ and $k \neq i$. Since $k \neq i$, then $k < i \vee i < k$, by the Trichotomy Law (Theorem II.4.1c).

Case i $i < k$.   Then $k = i + u$ for some u in P. Since $n \neq j$, it follows by Corollary II.5.7 that $n \times u \neq j \times u$. Hence, $(n \times k) + (j \times i) = (n \times (i + u)) + (j \times i) = (n \times i) + (n \times u) + (j \times i) \neq (n \times i) + (j \times u) + (j \times i)$ by the Cancellation Law for $+$ and the fact that $n \times u \neq j \times u$. But,

$$(n \times i) + (j \times u) + (j \times i) = (j \times (i + u)) + (n \times i) = (j \times k) + (n \times i).$$

Thus, $(n \times k) + (j \times i) \neq (j \times k) + (n \times i)$. This means that

$$\alpha \times_Z \beta = [(n \times k) + (j \times i), (j \times k) + (n \times i)] \neq 0_Z.$$

(Remember that $[(v, w)] = 0_Z \Leftrightarrow v = w$.)

Case ii $k < i$.   The proof here is similar to that for Case (i) and is left as an exercise for the reader.   ∎

3.3   RINGS AND INTEGRAL DOMAINS

We have proved a few elementary facts about the integers (Theorems 2.2, 2.4), and we could go on to derive a great many more results. However, it is more convenient to proceed in a more general way. We shall study a class of algebraic structures, of which the integers form just one example, and we shall prove a large number of theorems about such structures. All of these theorems will, in particular, hold true for the integers, but they will also be applicable to many other mathematical structures. In this way, instead of proving essentially the same thing over and over again for each particular structure, we do it once and for all. The algebraic structures introduced here are used not only in the development of number systems, but also are of importance in the study of many other parts of contemporary mathematics.

Definition   By a **ring** we mean a triple $(R, +, \times)$ such that R is a set and $+$ and $\times$ are binary operations in R satisfying:

1. $x + (y + z) = (x + y) + z$ (Associativity of $+$).
2. $x + y = y + x$ (Commutativity of $+$).

3. There is an element 0 in R such that

3a. $(\forall x)(x + 0 = x)$;

3b. $(\forall x)(\exists y)(x + y = 0)$.

Note i   Such an element 0 is unique. For, if there were another such element $\square$, then, by (3a) and (2),

$$\square = \square + 0 = 0 + \square = 0.$$

The unique element 0 is called the **zero element** of the ring.

Note ii   (3b) implies $(\forall x)(\exists! y)(x + y = 0)$. For, assume $x + y = 0 = x + z$. Then,

$$y = 0 + y = (x + z) + y = (x + y) + z = 0 + z = z.$$

The unique y such that $x + y = 0$ is denoted $-x$ and is called the **additive inverse** of x. Thus,

$$u + v = 0 \Leftrightarrow v = -u.$$

4. $x \times (y \times z) = (x \times y) \times z$ (Associativity of $\times$).

5. a. $x \times (y + z) = (x \times y) + (x \times z)$ ⎱ (Distributivity of $\times$ with respect
    b. $(y + z) \times x = (y \times x) + (z \times x)$ ⎰   to $+$).

Definition   A **ring with unit element** is a ring $(R, +, \times)$ such that:

6. There is an element 1 such that

$$x = x \times 1 = 1 \times x \qquad \text{for all } x \text{ in } R.$$

Note   Such an element 1 is unique. For, if there were another such element 1^* such that $x \times 1^* = x = 1^* \times x$ for all x in R, then $1^* = 1^* \times 1 = 1$. The unique element 1 is called the **unit element** of R.

Definition   A ring $(R, +, \times)$ is **commutative** if and only if

7. $x \times y = y \times x$ for all x and y in R.

EXAMPLES†

1. The system of integers $(Z, +_Z, \times_Z)$ is a commutative ring with unit element (Theorems 2.2, 2.4).

† As mentioned before, in examples we rely upon the intuitive knowledge of the reader.

2.   The rational numbers under ordinary addition and multiplication form a commutative ring with unit element.

3.   The real numbers under ordinary addition and multiplication form a commutative ring with unit element.

4.   The complex numbers under ordinary addition and multiplication

$$((a + bi) \oplus (c + di) = (a + c) + (b + d)i$$

and

$$(a + bi) \otimes (c + di) = (ac - bd) + (ad + bc)i$$

form a commutative ring with unit element.

5.   The even integers under ordinary addition and multiplication form a commutative ring without a unit element. (Observe that there is no even integer u such that $x \times u = x$ for all even integers x.)

6.   Let R be the set of all 2×2 matrices $\left(\begin{smallmatrix} a & b \\ c & d \end{smallmatrix}\right)$ with integral entries, that is, a, b, c, d are integers. Let

$$\begin{pmatrix} a & b \\ c & d \end{pmatrix} + \begin{pmatrix} a^* & b^* \\ c^* & d^* \end{pmatrix} = \begin{pmatrix} a + a^* & b + b^* \\ c + c^* & d + d^* \end{pmatrix},$$

and

$$\begin{pmatrix} a & b \\ c & d \end{pmatrix} \times \begin{pmatrix} a^* & b^* \\ c^* & d^* \end{pmatrix} = \begin{pmatrix} aa^* + bc^* & ab^* + bd^* \\ ca^* + dc^* & cb^* + dd^* \end{pmatrix}.$$

$(R, +, \times)$ is a noncommutative ring with unit element. The unit element is $\left(\begin{smallmatrix} 1 & 0 \\ 0 & 1 \end{smallmatrix}\right)$. To show noncommutativity, consider $\left(\begin{smallmatrix} 1 & 0 \\ 0 & 0 \end{smallmatrix}\right)$ and $\left(\begin{smallmatrix} 0 & 1 \\ 0 & 0 \end{smallmatrix}\right)$.

Theorem 3.1   (*Properties of Rings*)

a. $x + y = x + z \Rightarrow y = z$ (Cancellation Law for $+$).
b. $-0 = 0$.
c. $-(-x) = x$.
d. $-x = -y \Rightarrow x = y$.
e. $-(x + y) = (-x) + (-y)$.
f. $x + y = x \Rightarrow y = 0$.
g. $x \times 0 = 0 = 0 \times x$.
h. $-(x \times y) = (-x) \times y = x \times (-y)$.
i. $(-x) \times (-y) = x \times y$.

Proof (a)   Assume $x + y = x + z$. Adding $-x$ to both sides, we obtain $(-x) + x + y = (-x) + x + z$. Hence, $0 + y = 0 + z$. Therefore, $y = z$.

(b) $0 + 0 = 0$ by Axiom (3a). Since -0 is the unique element u such that $0 + u = 0$, it follows that $-0 = 0$.

(c) $(-x) + x = x + (-x) = 0$. Since $-(-x)$ is the unique element u such that $(-x) + u = 0$, it follows that $-(-x) = x$.

(d)   Assume $-x = -y$. Then $-(-x) = -(-y)$. Now apply (c).

(e)   Observe that

$$(x + y) + ((-x) + (-y)) = (x + (-x)) + (y + (-y)) = 0 + 0 = 0.$$

Since $-(x + y)$ is the unique element u such that $(x + y) + u = 0$, it follows that $-(x + y) = (-x) + (-y)$.

(f)   This follows from (a) by taking $z = 0$.

(g) $x \times 0 = x \times (0 + 0) = (x \times 0) + (x \times 0)$. Now apply (f), substituting $(x \times 0)$ for both x and y. Similarly,

$$0 \times x = (0 + 0) \times x = (0 \times x) + (0 \times x),$$

and we again use (f).

(h) $(x \times y) + ((-x) \times y) = (x + (-x)) \times y = 0 \times y = 0$. Since $-(x \times y)$ is the unique u such that $(x \times y) + u = 0$, it follows that $-(x \times y) = (-x) \times y$. Similarly,

$$(x \times y) + (x \times (-y)) = x \times (y + (-y)) = x \times 0 = 0.$$

Hence, $x \times (-y) = -(x \times y)$.

(i) $(-x) \times (-y) = -(x \times (-y))$ by (h)
$$= -(-(x \times y)) \text{ by (h)}$$
$$= x \times y \qquad \text{ by (c).} \qquad \blacksquare$$

Definition   In any ring, let

$$x - y = x + (-y).$$

The operation $-$ is called *subtraction*.

Theorem 3.2   (*Properties of Subtraction*)

a. $x - x = 0$.

b. $-(x - y) = y - x$.

c. $x - y = 0 \Leftrightarrow x = y$.

d. $x \times (y - z) = (x \times y) - (x \times z)$ $\left.\right\}$ (Distributivity of $\times$ over $-$)
    $(y - z) \times x = (y \times x) - (z \times x)$

e. In a ring with unit element,

$$(-1) \times x = -x = x \times (-1) \qquad \text{and} \qquad (-1) \times (-1) = 1.$$

Proof (a) $x - x = x + (-x) = 0.$

(b) $-(x - y) = -(x + (-y))$
$$\begin{aligned} &= (-x) + (-(-y)) &&\text{by Theorem 3.1e,} \\ &= (-x) + y &&\text{by Theorem 3.1c,} \\ &= y + (-x) \\ &= y - x. \end{aligned}$$

(c) $x = y \Rightarrow x - y = 0$ follows from (a). Conversely, assume $x - y = 0$. Then, $x + (-y) = 0$. Hence, $-y = -x$, since $-x$ is the unique element u such that $x + u = 0$. Then $x = y$ by Theorem 3.1d.

(d) $x \times (y - z) = x \times (y + (-z)) = (x \times y) + (x \times (-z))$
$$= (x \times y) + (-(x \times z)) = (x \times y) - (x \times z).$$

Similarly,

$$\begin{aligned} (y - z) \times x = (y + (-z)) \times x &= (y \times x) + ((-z) \times x) \\ &= (y \times x) + (-(z \times x)) = (y \times x) - (z \times x). \end{aligned}$$

(e) $(-1) \times x = -(1 \times x) = -x.$
$x \times (-1) = -(x \times 1) = -x.$

Also,

$$\begin{aligned} (-1) \times (-1) &= 1 \times 1 &&\text{by Theorem 3.1i,} \\ &= 1. \quad \blacksquare \end{aligned}$$

EXERCISES

1. In any ring, prove $(\forall x)(\forall y)(\exists! z)(x + z = y)$.

2. In a ring, is $+$ distributive over $\times$, that is does $x + (y \times z) = (x + y) \times (x + z)$ hold for all x, y, z? (*Hint*: Consider the ring of integers.)

3. Prove: $(x - y) + (y - z) = x - z.$

4. Prove: $0 - x = -x.$

5. Prove: $x + (y - x) = y.$

6. Prove: $(y + z) - (x + z) = y - x.$

Definition   By an **integral domain** we mean a commutative ring with unit element $(R, +, \times)$ satisfying the additional axiom:

8. $[x \neq 0 \wedge y \neq 0] \Rightarrow x \times y \neq 0$ for any x and y in R.

(An equivalent form of (8) is: $x \times y = 0 \Rightarrow [x = 0 \vee y = 0]$.)

Terminology   By a **zero divisor** we mean a nonzero element x for which there exists some nonzero element y such that $x \times y = 0$. Thus, the property (8) is equivalent to saying that R contains no zero divisors.

Theorem 3.3   For a commutative ring with unit element, property (8) is equivalent to:

$8'$.   For any x, y, z in R,

$$[x \times y = x \times z \wedge x \neq 0] \Rightarrow y = z. \qquad \text{(Cancellation Law for } \times)$$

(Thus, $(8')$ could be used to characterize integral domains instead of (8).)

Proof   Assume (8). Assume $x \times y = x \times z \wedge x \neq 0$. Then $(x \times y) - (x \times z) = 0$ by Theorem 3.2c. Hence, $x \times (y - z) = 0$ by Theorem 3.2d. Since $x \neq 0$, it follows by (8) that $y - z = 0$. Then $y = z$, by Theorem 3.2c. Conversely, assume $(8')$. Assume that $x \times y = 0$. We wish to prove that $x = 0$ or $y = 0$. So, assume $x \neq 0$. It remains to prove that $y = 0$. Now, $x \times 0 = 0$ by Theorem 3.1g. Hence, $x \times y = 0 = x \times 0$. Therefore, by $(8')$, $y = 0$.   ∎

Since integral domains are rings, everything proved about rings automatically holds for integral domains. Observe that all the results proved in this section hold for the system of integers, since $(Z, +_Z, \times_Z)$ is an integral domain (see Theorems 2.2 and 2.4).

EXAMPLES

Of the examples of rings on pp. 96–97, 1–4 are integral domains, 5 is not an integral domain because it lacks a unit element, and 6 is not an integral domain because it is noncommutative.

7.   Let R be the set of all functions from Z into Z. For any f, g in R, let $f + g$ and $f \times g$ be defined as follow:

$$(f + g)(x) = f(x) + g(x) \qquad \text{for all } x \text{ in } Z;$$
$$(f \times g)(x) = f(x) \times g(x) \qquad \text{for all } x \text{ in } Z.$$

Then $(R, +, \times)$ is a commutative ring with unit element, but it is not an integral domain. The zero element is the constant function 0. The unit element is the constant function 1. A counterexample to (8) is constructed as follows:

Let

$$f(x) = \begin{cases} 1 & \text{if } x = 0, \\ 0 & \text{if } x \neq 0; \end{cases}$$

$$g(x) = \begin{cases} 0 & \text{if } x = 0, \\ 1 & \text{if } x \neq 0. \end{cases}$$

Then neither f nor g is the zero element, but $f \times g = 0$. We leave the verification of the ring axioms to the reader.

Theorem 3.4 In a ring with unit element $(R, +, \times)$,

$$0 = 1 \Leftrightarrow R \text{ consists of a single element.}$$

Proof If R consists of a single element, then $0 = 1$. Conversely, if $0 = 1$, then, for any x in R,

$$x = x \times 1 = x \times 0 = 0.$$

Hence, R consists of only the single element 0.   ■

Remark $(\{0\}, +, \times)$ is an integral domain, where $0 + 0 = 0 \times 0 = 0$.

EXERCISES

7. In each of the following examples, R is a set, and $+$ and $\times$ are binary operations in R. Determine in each case whether $(R, +, \times)$ is a ring. Also find out which of the rings are commutative, which have unit elements, and which are integral domains. (For those commutative rings with unit elements which are not integral domains, give examples of zero divisors.)

a. R is the set of all functions from Z into Z. For any f, g in R, define:

$$(f + g)(x) = f(x) + g(x) \qquad \text{for each } x \text{ in } Z;$$
$$(f \times g)(x) = f(g(x)) \qquad \text{for each } x \text{ in } Z.$$

Thus, $\times$ is the composition operation.

b. $R = Z$, $+$ is ordinary addition $+_Z$, and $x \times y = 0_Z$ for all x, y in Z.

c.   Let A be some nonempty set. Let R be the set of all subsets of A. Let $B + C = (B \cup C) - (B \cap C)$ and $B \times C = B \cap C$ for all B and C in R.

d.   Same as (c), except that R is the set of all finite subsets of A.

e.   Let α be a fixed element of Z. Let R be the set of all multiples of α, that is, R $= \{\beta \times_Z \alpha : \beta \in Z\}$. Let $+$ and $\times$ be the ordinary operations $+_Z$ and $\times_Z$ restricted to elements of R.

f. R is the set of all polynomials $a_n x^n + a_{n-1} x^{n-1} + \cdots + a_1 x + a_0$, with integral coefficients $a_0, \ldots, a_n$. $+$ and $\times$ are the usual operations of addition and multiplication of polynomials (see Appendix C).

g.   Same as (f), except that R consists of all polynomials of the form $a_n x^n + a_{n-1} x^{n-1}$ $+ \cdots + a_1 x$, that is, all polynomials with constant term 0.

h. $R = \{\alpha + \beta \sqrt{2} : \alpha \in Z \wedge \beta \in Z\}$, and $+$ and $\times$ are ordinary addition and multiplication of real numbers.

i. $R = \{\alpha + \beta i : \alpha \in Z \wedge \beta \in Z\}$, and $+$ and $\times$ are ordinary addition and multiplication of complex numbers. (As usual, $i = \sqrt{-1}$.)

j. R is the set of all continuous functions from the interval $0 \leq x \leq 1$ of the real numbers into the real numbers. For f, g in R, let $(f + g)(x) = f(x) + g(x)$ and $(f \times g)(x) = f(x) \times g(x)$ for all x in the interval. (Readers not familiar with the notion of continuous function should skip this problem.)

k. $R = \{\text{even, odd}\}$.

$$\begin{cases} \text{even} + \text{even} - \text{odd} + \text{odd} \ - \text{even}; \\ \text{even} + \text{odd} \ = \text{odd} + \text{even} = \text{odd}; \end{cases}$$
$$\begin{cases} \text{even} \times \text{even} = \text{even} \times \text{odd} = \text{odd} \times \text{even} = \text{even}; \\ \text{odd} \ \times \text{odd} \ = \text{odd}. \end{cases}$$

l.   Let A be a fixed nonempty set. Let R be the set of all subsets of A. For B, C in R, let $B + C = B \cup C$ and $B \times C = B \cap C$.

m.   Let $R = \{0, 1, 2\}$.

+	0	1	2
0	0	1	2
1	1	2	0
2	2	0	1

In this table for addition, the entry in the intersection of the row headed by x and the column headed by y is $x + y$. Thus, $1 + 2 = 0$, $2 + 2 = 1$, etc.

×	0	1	2
0	0	0	0
1	0	1	2
2	0	2	1

n.   Let $R = \{0, 1, 2, 3\}$.

+	0	1	2	3
0	0	1	2	3
1	1	2	3	0
2	2	3	0	1
3	3	0	1	2

×	0	1	2	3
0	0	0	0	0
1	0	1	2	3
2	0	2	0	2
3	0	3	2	1

o. R is the set $Z \times Z$ of all ordered pairs of integers. Let

$$(\alpha, \beta) + (\gamma, \delta) = (\alpha +_z \gamma, \beta +_z \delta)$$

and

$$(\alpha, \beta) \times (\gamma, \delta) = (\alpha \times_z \gamma, \beta \times_z \delta) \qquad \text{for all } \alpha, \beta, \gamma, \delta \text{ in } Z.$$

p.   Let R be the set of natural numbers. Let $+$ and $\times$ be the ordinary addition and multiplication of natural numbers.

8.   Give two examples of:

a.   A noncommutative ring with a unit element.

b.   A noncommutative ring without a unit element.

c.   A commutative ring without a unit element.

d.   A commutative ring with a unit element which is not an integral domain.

9.   A ring for which $x^2 = x$ holds for all x is called a **Boolean ring.**

a.   Give an example of a Boolean ring.

b.   Show that a Boolean ring satisfies $x + x = 0$ for all x, and then prove that it is a commutative ring.

3.4 ORDERED INTEGRAL DOMAINS

Definition   By an **ordered integral domain** we mean a quadruple $(R, +, \times, <)$ consisting of an integral domain $(R, +, \times)$ together with a binary relation $<$ in R satisfying the following axioms.

(O1) $x \not< x.$[†]   (Irreflexivity)

(O2) $[x < y \wedge y < z] \Rightarrow x < z.$ (Transitivity)

(O3) $x < y \vee x = y \vee y < x.$ (Trichotomy)

(O4) $x < y \Rightarrow x + z < y + z.$

(O5) $[x < y \wedge 0 < z] \Rightarrow x \times z < y \times z.$

In addition, we stipulate that $0 \neq 1$ must hold.

Definition $x \leq y$ for $x < y \vee x = y.$

$x > y$ for $y < x.$

$x \geq y$ for $y \leq x.$

[†] As usual, we write $x \not< y$ instead of $\neg(x < y).$

Theorem 4.1 In any ordered integral domain:

a. Exactly one of the condition:

$$x < y, \qquad x = y, \qquad y < x$$

holds.

b. $[x < y \wedge u < v] \Rightarrow x + u < y + v.$

c. $[x < y \wedge u \leq v] \Rightarrow x + u < y + v.$

d. $[x \leq y \wedge u < v] \Rightarrow x + u < y + v.$

e. $[x \leq y \wedge u \leq v] \Rightarrow x + u \leq y + v.$

f. $[0 < z \wedge x \times z < y \times z] \Rightarrow x < y.$

Proof (a) From Axiom O3 we have $x < y \vee x = y \vee y < x$. We must show that no two of the conditions hold simultaneously. If $x < y$ and $x = y$, then $x < x$, contradicting Axiom O1. If $x = y$ and $y < x$, then $x < x$, contradicting Axiom O1. If $x < y$ and $y < x$, then, by Axiom O2, $x < x$, again contradicting Axiom O1.

　(b) Assume $x < y$ and $u < v$. From $x < y$, we obtain $x + u < y + u$ by Axiom O4. From $u < v$, we obtain $y + u < y + v$ by Axiom O4. Hence, by Axiom O2, $x + u < y + v$.

　(c)–(e) follow easily from (b) and are left as exercises for the reader.

　(f) Assume $0 < z \wedge x \times z < y \times z$. Assume $x \not< y$. Then $y < x \vee y = x$. Hence, either $y \times z < x \times z \vee y \times z = x \times z$, contradicting $x \times z < y \times z$. ∎

Definitions　x is *positive* $\Leftrightarrow 0 < x$.

　　　　　　　x is *negative* $\Leftrightarrow x < 0$.

Theorem 4.2 In any ordered integral domain:

a. $x < y \Leftrightarrow y - x$ is positive.

b. $x < y \Leftrightarrow x - y$ is negative.

c. y is positive $\Leftrightarrow -y$ is negative.

d. x is negative $\Leftrightarrow -x$ is positive.

e. $x < y \Leftrightarrow -y < -x.$

f. $[x < y \wedge z < 0] \Rightarrow y \times z < x \times z.$

g. The sum and product of two positive elements are positive.

h. The product of two negative elements is positive.

i.   The product of a positive and a negative element is negative.

j. $x \neq 0 \Rightarrow x^2 > 0$.

k. $x^2 \geq 0$.

l. $-1 < 0 < 1.$[†]

Proof   (a)   Assume $x < y$. By O4, add $-x$ to both sides, obtaining $x + (-x) < y + (-x)$, and, therefore, $0 < y - x$. Conversely, assume $0 < y - x$. By O4, add x to both sides, obtaining $x + 0 < x + (y - x)$, and, therefore, $x < y$.

(b)   Assume $x < y$. By O4, add $-y$ to both sides, obtaining $x + (-y) < y + (-y)$, and, therefore, $x - y < 0$. Conversely, assume $x - y < 0$. By O4, add y to both sides, obtaining $y + (x - y) < y$, and, therefore, $x < y$.

(c)   Substitute 0 for x in (b).

(d)   Substitute 0 for y in (a).

(e)   By (a), $x < y \Leftrightarrow 0 < y - x$. Also, by (a),

$$-y < -x \Leftrightarrow 0 < (-x) - (-y).$$

But, $(-x) - (-y) = (-x) + (-(-y)) = (-x) + y = y - x$.

(f)   Assume $x < y \wedge z < 0$. By (d), $0 < -z$. Hence, by Axiom O5, $x \times (-z) < y \times (-z)$. But, $x \times (-z) = -(x \times z)$ and $y \times (-z) = -(y \times z)$, by Theorem 3.1h. Hence, $-(x \times z) < -(y \times z)$. Therefore, by (e),

$$y \times z < x \times z.$$

(g)   Assume $0 < x \wedge 0 < y$. By Theorem 4.1b, $0 < x + y$. In addition, by Axiom O5, $0 \times y < x \times y$. But, $0 \times y = 0$.

(h)   If $x < 0$ and $y < 0$, then $0 < -x$ and $0 < -y$, by (d). Hence, $0 < (-x) \times (-y)$, by (g). But, $(-x) \times (-y) = x \times y$, by Theorem 3.1i.

(i)   Assume $0 < x \wedge y < 0$. Then $0 < -y$ by (d). Hence, $0 < x \times (-y)$ by (g). But $x \times (-y) = -(x \times y)$ by Theorem 3.1h. Hence, $0 < -(x \times y)$. Then $x \times y < 0$, by (d).

(j)   If $x \neq 0$, then $x < 0$ or $0 < x$. If $0 < x$, apply (g), and, if $x < 0$, apply (h).

(k)   Use (j) and the fact that $0 \times 0 = 0$.

(l)   By (j), $0 < 1^2$. But $1^2 = 1$. Hence, $0 < 1$. By (c), $-1 < 0$. ■

[†] As usual, $u < v < w$ stands for $u < v \wedge v < w$.

EXERCISES

1. The complex numbers $C = \{a + bi : a$ and b are real numbers$\}$ form an integral domain under ordinary addition and multiplication. Show that a relation $<$ cannot be defined on C to obtain an ordered integral domain. (*Hint:* Use Theorem 4.2j, l.)

2. In an ordered integral domain, prove that $1 < 2 < 3 < \ldots .$[†]

3. Prove that every ordered integral domain is infinite.

4. In an ordered integral domain, prove that $x + x = 0$ implies $x = 0$.

5. Prove that $x < y \Rightarrow x^3 < y^3$ holds in an ordered integral domain.

6. Show that $x^2 - (x \times y) + y^2 \geq 0$ holds in an ordered integral domain.

7. In the definition of ordered integral domains, what structures were excluded by the stipulation that $0 \neq 1$?

8. Prove that Axiom (8) is provable from the other axioms for an ordered integral domain.

9. If u and v are positive elements of an ordered integral domain such that $u \times v = 1$, prove that $u + v \geq 2$. Moreover, $u + v = 2$ only when $u = v = 1$.

Our aim now is to show how to make the integral domain of the integers $(Z, +_Z, \times_Z)$ into an ordered integral domain. In this case, as in others that will arise later, the best method is an indirect one, based upon the following theorem.

Theorem 4.3   Let $(R, +, \times)$ be an integral domain in which $0 \neq 1$. Assume that there is a subset $\mathscr{P}$ of R such that:

  i. $0 \notin \mathscr{P}$;
 ii. $x \in \mathscr{P}$ or $x = 0$ or $-x \in \mathscr{P}$ for any x in R;
iii. $[x \in \mathscr{P} \wedge y \in \mathscr{P}] \Rightarrow x + y \in \mathscr{P}$;
 iv. $[x \in \mathscr{P} \wedge y \in \mathscr{P}] \Rightarrow x \times y \in \mathscr{P}$.

If we define $x < y$ to mean $y - x \in \mathscr{P}$, then $(R, +, \times, <)$ is an ordered integral domain. Moreover, $\mathscr{P}$ coincides with the set of positive elements of $(R, +, \times, <)$.

Proof   We must verify Axioms O1–O5.

(O1)   Assume $x < x$. Then, by definition, $x - x \in \mathscr{P}$. But $x - x = 0$, contradicting (i). Hence, $x \not< x$.

[†] Remember that, by definition, $2 = 1 + 1$, $3 = 2 + 1$, etc.

(O2) Assume $x < y$ and $y < z$. Then $y - x \in \mathscr{P}$ and $z - y \in \mathscr{P}$. Hence, by (iii), $(y - x) + (z - y) \in \mathscr{P}$. But $(y - x) + (z - y) = z - x$. Thus, $z - x \in \mathscr{P}$, that is, $x < z$.

(O3) By (ii), $y - x \in \mathscr{P}$ or $y - x = 0$ or $-(y - x) \in \mathscr{P}$. If $y - x \in \mathscr{P}$, then $x < y$. If $y - x = 0$, then $x = y$. If $-(y - x) \in \mathscr{P}$, then, since $-(y - x) = x - y$, we have $x - y \in \mathscr{P}$, that is, $y < x$. Thus, $x < y$ or $x = y$ or $y < x$.

(O4) Assume $x < y$. Then $y - x \in \mathscr{P}$. But $(y + z) - (x + z) = y - x$. Hence, $x + z < y + z$.

(O5) Assume $x < y$ and $0 < z$. Then $y - x \in \mathscr{P}$ and $z - 0 \in \mathscr{P}$. Hence, $z \in \mathscr{P}$, since $z - 0 = z$. By (iv), $(y - x) \times z \in \mathscr{P}$. But $(y - x) \times z = (y \times z) - (x \times z)$ by Theorem 3.2d. Hence, $(y \times z) - (x \times z) \in \mathscr{P}$, that is, $x \times z < y \times z$.

Finally, observe that $0 < x \Leftrightarrow x \in \mathscr{P}$, so that $\mathscr{P}$ is the set of positive elements.   ∎

Theorem 4.3 enables us to define an order relation $<$ by first defining the set of "positive" elements. This is what we shall do now in the case of the integers.

Remember that the integers are equivalence classes with respect to the binary relation $\sim$ defined on ordered pairs of natural numbers. (Review pp. 86–89.) Again we appeal to the intuitive picture for an idea as to how to define a suitable order relation. The ordered pair (n, j) of natural numbers corresponds intuitively to $n - j$; if $n - j$ is to be "positive," then $j < n$. Thus, we should define an equivalence class α of ordered pairs (n, j) to be "positive" if all its ordered pairs have their first component n bigger than their second component j. This is what we shall do, but first we must confirm in the following lemma that such a definition is possible.

Lemma 4.4 Given $\alpha \in Z$, $(n, j) \in \alpha$, and $(k, i) \in \alpha$. Then

$$j < n \Leftrightarrow i < k.^\dagger$$

Proof Since $(n, j) \in \alpha$ and $(k, i) \in \alpha$, we have $(n, j) \sim (k, i)$, that is

$$(*)\quad n + i = k + j.$$

---

† The order relation $<$ used here is the relation $<$ defined on natural numbers (see Section 2.4).

Assume $j < n$. We must prove that $i < k$. Assume $i \not< k$. By the Trichotomy Law (Theorem 2.4.1c), $k \leq i$. Hence, since $j < n$ and $k \leq i$, we conclude, by Theorem 2.4.4f, that $k + j < n + i$. This contradicts (∗). We have shown that $j < n \Rightarrow i < k$. By a similar argument, $i < k \Rightarrow j < n$. ∎

Definition

$$\mathscr{P}_Z = \{\alpha : \alpha \in Z \wedge (\forall n)(\forall j)((n, j) \in \alpha \Rightarrow j < n)\}.$$

Thus, $\alpha \in \mathscr{P}_Z$ if and only if $j < n$ for all ordered pairs (n, j) in α. By Lemma 4.4, if this condition is met for one ordered pair in α, then it holds for all ordered pairs in α. In other words, $\alpha \in \mathscr{P}_Z$ if and only if $j < n$ for some ordered pair (n, j) in α.

Lemma 4.5 $\mathscr{P}_Z$ satisfies the conditions of Theorem 4.3, that is,

   i. $0_Z \notin \mathscr{P}_Z$;

  ii. $\alpha \in \mathscr{P}_Z$ or $\alpha = 0_Z$ or $-_Z\alpha \in \mathscr{P}_Z$ for all α in Z;

 iii. $[\alpha \in \mathscr{P}_Z \wedge \beta \in \mathscr{P}_Z] \Rightarrow \alpha +_Z \beta \in \mathscr{P}_Z$;

 iv. $[\alpha \in \mathscr{P}_Z \wedge \beta \in \mathscr{P}_Z] \Rightarrow \alpha \times_Z \beta \in \mathscr{P}_Z$.

Proof   (i) $(1, 1) \in 0_Z$ and $1 \not< 1$. Hence, $0_Z \notin \mathscr{P}_Z$.

   (ii)   Take any α in Z. Choose any ordered pair (n, j) in α. By the Trichotomy Law (Theorem 2.4.1c),

$$j < n \quad \text{or} \quad j = n \quad \text{or} \quad n < j.$$

If $j < n$, then $\alpha \in \mathscr{P}_Z$.

If $j = n$, then $\alpha = 0_Z$.

If $n < j$, then $-_Z\alpha \in \mathscr{P}_Z$, since $(j, n) \in -_Z\alpha$.

Hence, $\alpha \in \mathscr{P}_Z$ or $\alpha = 0_Z$ or $-_Z\alpha \in \mathscr{P}_Z$.

   (iii), (iv)   Assume $\alpha \in \mathscr{P}_Z$ and $\beta \in \mathscr{P}_Z$. Choose $(n, j) \in \alpha$ and $(k, i) \in \beta$. Then $j < n$ and $i < k$. Now, $\alpha +_Z \beta = [(n + k, j + i)]$. Since $j < n$ and $i < k$, it follows by Theorem 2.4.4e that $j + i < n + k$. Hence, $\alpha +_Z \beta \in \mathscr{P}_Z$. By definition, $\alpha \times_Z \beta = [(n \times k) + (j \times i), (n \times i) + (j \times k))]$. To prove that $\alpha \times_Z \beta \in \mathscr{P}_Z$, we must show that

$$(n \times i) + (j \times k) < (n \times k) + (j \times i).$$

From $i < k$, by definition, $k = i + u$ for some natural number u. From

$j < n$, we obtain $j \times u < n \times u$ by Theorem 2.5.6. Hence,

$$(n \times i) + (j \times k) = (n \times i) + (j \times (i + u)) = (n \times i) + (j \times i) + (j \times u)$$
$$< (n \times i) + (j \times i) + (n \times u) = (n \times i) + (n \times u) + (j \times i)$$
$$= (n \times (i + u)) + (j \times i) = (n \times k) + (j \times i). \quad \blacksquare$$

Definitions $\beta -_Z \alpha = \beta +_Z (-_Z\alpha)$. (This is a special case of the definition on p. 98.)

$$\alpha <_Z \beta \Leftrightarrow \beta -_Z \alpha \in \mathscr{P}_Z.$$

Corollary 4.6 $(Z, +_Z, \times_Z, <_Z)$ is an ordered integral domain, and $\mathscr{P}_Z$ is the set of positive elements.

Proof   By Theorem 4.3 and Lemma 4.5. $\blacksquare$

As a consequence of Corollary 4.6, all the results of Theorems 4.1 and 4.2 hold for the integers.

We shall now prove that the set of positive integers forms a Peano system. This will enable us to use mathematical induction for proving results about integers.

Theorem 4.7   Let $T(x) = x +_Z 1_Z$ for every x in $\mathscr{P}_Z$. Then $(\mathscr{P}_Z, T, 1_Z)$ is a Peano system.

Proof   Notice that $1_Z \in \mathscr{P}_Z$, since $(2, 1) \in 1_Z$ and $1 < 2$. Also, T is a singular operation in $\mathscr{P}_Z$ (by Lemma 4.5iii and the fact that $1_Z \in \mathscr{P}_Z$). Now let us check the axioms for Peano systems.

(P1) $1_Z \neq T(x) = x +_Z 1_Z$ for every x in $\mathscr{P}_Z$. For, if $1_Z = x +_Z 1_Z$, then $x = 0_Z$, contradicting Lemma 4.5i.

(P2) $T(x) = T(y) \Rightarrow x = y$. For, if $x +_Z 1_Z = y +_Z 1_Z$, then $x = y$, by the Cancellation Law for Addition (Theorem 3.1a).

(P3)   Assume $A \subseteq \mathscr{P}_Z$, $1_Z \in A$, and

$$(\forall x)(x \in A \Rightarrow x +_Z 1_Z \in A).$$

We must show that $A = \mathscr{P}_Z$. Remember that $(P, S, 1)$ is our original Peano system of natural numbers. Let $B = \{n : n \in P \wedge [(n + 1, 1)] \in A\}$. First, $1 \in B$, since $[(1 + 1, 1)] = [(2, 1)] = 1_Z \in A$. Next, assume $n \in B$, that is, $[(n + 1, 1)] \in A$. Hence, $[(n + 1, 1)] +_Z 1_Z \in A$. But,

$$[(n + 1, 1)] +_Z 1_Z = [(n + 1, 1)] +_Z [(2, 1)] = [(n + 1 + 2, 1 + 1)]$$
$$= [((n + 1) + 1, 1)].$$

Hence, $n + 1 \in B$. We have shown that $n \in B \Rightarrow n + 1 \in B$. Hence, by mathematical induction in the Peano system $(P, S, 1)$, $B = P$. Thus, for every n in P, $[(n + 1, 1)] \in A$. It suffices then to show that every element of $\mathscr{P}_Z$ is of the form $[(n + 1, 1)]$ for some n in P. To see this, take any α in $\mathscr{P}_Z$, and take any ordered pair (k, i) in α. Since α is in $\mathscr{P}_Z$, $k > i$. Hence, $k = i + j$ for some j in P. Thus, $(k, i) \sim (j + 1, 1)$, since $k + 1 = j + 1 + i$. Therefore, $\alpha = [(k, i)] = [(j + 1, 1)]$ for some j in P.   ∎

EXERCISES

1. Determine whether the following structures $(R, +, \times, <)$ are ordered integral domains.

    a. $R = Z$, $+$ is $+_Z$, $\times$ is $\times_Z$, and $\alpha < \beta \Leftrightarrow \beta <_Z \alpha$.

    b. R is the set of complex numbers, $+$ and $\times$ are ordinary addition and multiplication of complex numbers, and

    $$a + bi < c + di \Leftrightarrow (a <_r c) \vee (a = c \wedge b <_r d),$$

    where $<_r$ is the usual order relation on the real numbers.

    c. R is the set of all polynomials with integral coefficients $a_n x^n + \cdots + a_1 x + a_0$, $+$ and $\times$ are the usual addition and multiplication of polynomials (see Appendix C), and, if $f(x)$ and $g(x)$ are polynomials, then $f < g$ means that the leading coefficient of $g - f$ is positive. (For example, $2x - 4 < x^2 - 7$ because the leading coefficient 1 of $(x^2 - 7) - (2x - 4) = x^2 - 2x - 3$ is positive. Likewise, $2x < 5x$ because the leading coefficient 3 of $5x - 2x = 3x$ is positive.)

    d. R is the set of rational (respectively, real) numbers, $+$ and $\times$ are ordinary addition and multiplication of rational (respectively, real) numbers, and $<$ is the usual order relation.

2. Let R be the set of all polynomials $a_n x^n + \cdots + a_1 x + a_0$ with integral (respectively, rational or real) coefficients. (a) Define a subset $\mathscr{P}$ of R satisfying conditions (i)–(iv) of Theorem 4.3. (b) Describe the order relation $<$ which, according to Theorem 4.3, one can define in terms of $\mathscr{P}$.

3. Show that the order relation $<_Z$ is the only possible relation which will make the system of integers $(Z, +_Z, \times_Z)$ into an ordered integral domain, that is if $\ominus$ is a binary relation in Z such that $(Z, +_Z, \times_Z, \ominus)$ is an ordered integral domain, then $\ominus$ is identical with $<_Z$.

4. (Extension of Theorem 4.7) Show that the order relation $<_Z$, restricted to $\mathscr{P}_Z$, is the same as the order relation defined in the Peano system $(\mathscr{P}_Z, T, 1_Z)$. Hence, in particular, $x \in \mathscr{P}_Z \Leftrightarrow x \geq_Z 1_Z$.

There is another important mathematical idea which can be introduced within the context of the theory of ordered integral domains.

Definition   Let $(R, +, \times, <)$ be an arbitrary ordered integral domain. For any x in R, let

$$|x| = \begin{cases} x & \text{if } 0 \le x \\ -x & \text{if } x < 0. \end{cases}$$

$|x|$ is called the **absolute value** of x (or, sometimes, the **magnitude** of x).

Theorem 4.8   In any ordered integral domain:

a. $0 \le |x|$.

b. $|x| = 0 \Leftrightarrow x = 0$.

c. $|-x| = |x|$.

d. $|x - y| = |y - x|$.

e. $|x \times y| = |x| \times |y|$.

f. $-|x| \le x \le |x|$.

g. $|z| < u \Leftrightarrow -u < z < u$.

h. $|z| \le u \Leftrightarrow -u \le z \le u$.

i. $u \ge v \wedge u \ge -v \Rightarrow u \ge |v|$.

j. $|x + y| \le |x| + |y|$ (triangle inequality).

k. $|x - y| \ge ||x| - |y||$.

Proof   (a) If $0 \le x$, then $|x| = x \ge 0$. If $x < 0$, then $|x| = -x > 0$, by Theorem 4.2d.

(b) If $x = 0$, then $|x| = x = 0$. If $x \ne 0$, then $0 < x \vee x < 0$. If $0 < x$, then $|x| = x > 0$. If $x < 0$, then $|x| = -x > 0$.

(c) If $0 < x$, then $-x < 0$. Hence, $|x| = x$, and $|-x| = -(-x) = x$. If $0 = x$, then $-x = 0 = x$. If $x < 0$, then $-x > 0$. Hence, $|x| = -x$ and $|-x| = -x$.

(d) $x - y = -(y - x)$. Use (c).

(e) *Case i* $x = 0$ or $y = 0$. Then $x \times y = 0$. Therefore, $|x \times y| = 0$. Also, $|x| = 0$ or $|y| = 0$. Hence, $|x| \times |y| = 0$.

Case ii $0 < x$ and $0 < y$. Then, $0 < x \times y$, and $|x \times y| = x \times y$. Also, $|x| = x$ and $|y| = y$.

Case iii $x < 0$ and $y < 0$. Then $0 < x \times y$. Hence, $|x \times y| = (x \times y)$, $|x| = -x$, $|y| = -y$. But, $(-x) \times (-y) = x \times y$.

Case iv $x < 0$ and $y > 0$. Then $x \times y < 0$. Hence, $|x \times y| = -(x \times y)$, $|x| = -x$, $|y| = y$. But, $(-x) \times y = -(x \times y)$.

Case v $0 < x$ and $y < 0$. Similar to (iv).

(f)   *Case i* $0 < x$. Then $|x| = x$. Hence, $-|x| < 0$. Thus, $-|x| < x = |x|$.

Case ii $0 = x$. $-|x| = x = |x|$.

Case iii $x < 0$. $|x| = -x$. Hence, $-|x| = x$, and $0 < |x|$. Hence, $-|x| = x < |x|$.

(g)   Assume $|z| < u$. By (f), $z \leq |z|$. Hence, $z < u$. By (f), $-|z| \leq z$. But, $-|z| > -u$, since $|z| < u$. Hence, $-u < z$. Conversely, assume $-u < z < u$. If $|z| = z$, then $|z| < u$. If $|z| = -z$, then $-u < -|z|$. Hence, $|z| < u$.

(h)   Proof similar to that of (g).

(i)   Assume $u \geq v$ and $u \geq -v$. If $0 \leq v$, then $|v| = v$. If $v < 0$, then $|v| = -v$. In either case, $u \geq |v|$.

(j) $-|x| \leq x \leq |x|$, and $-|y| \leq y \leq |y|$. Adding,

$$(-|x|) + (-|y|) \leq x + y \leq |x| + |y|.$$

Hence, $-(|x| + |y|) \leq x + y \leq |x| + |y|$. By (h), taking $x + y$ to be z and $|x| + |y|$ to be u, we obtain $|x + y| \leq |x| + |y|$.

(k) $|x| = |y + (x - y)| \leq |y| + |x - y|$ by (j). Therefore $|x - y| \geq |x| - |y|$. Similarly, exchanging x and y, $|y - x| \geq |y| - |x|$. But, $|x - y| = |y - x|$ by (d). Hence,

$$|x - y| \geq |x| - |y| \qquad \text{and} \qquad |x - y| \geq |y| - |x|.$$

Notice that $|y| - |x| = -(|x| - |y|)$. Hence, taking u to be $|x - y|$ and v to be $|x| - |y|$ in (i), we obtain $|x - y| \geq ||x| - |y||$.

EXERCISES

1.   Prove: $|x|^2 = x^2$.

2.   Prove: $|x - y| \leq |x| + |y|$.

3.   Prove: $|x_1 \times x_2 \times \cdots \times x_k| = |x_1| \times |x_2| \times \cdots \times |x_k|$.

4.   Prove: $|x_1 + x_2 + \cdots + x_k| \leq |x_1| + |x_2| + \cdots + |x_k|$.

5.   Prove: $|x| > 2 \Leftrightarrow (x < -2 \vee x > 2)$, and $|x - 2| > 3 \Leftrightarrow (x < -1 \vee x > 5)$.

6.   For integers x, $|x - 4| > |x| \Leftrightarrow ?$; $|2x + 3| > 4 \Leftrightarrow ?$

7.   Prove: $|x| = |y| \Rightarrow (x = y \vee x = -y)$.

3.5   GREATEST COMMON DIVISOR. PRIMES

In this section, we shall develop a few of the simplest properties of the integers. To make the notation easier to read and to write, throughout this section (but only in this section) we shall employ the following conventions.

1.   Unless something is said to the contrary, we shall be discussing only integers. All variables will refer to integers.

2.   temporary notation   legal notation

$x + y$	will stand for	$x +_Z y.$
xy	will stand for	$x \times_Z y.$
$x < y$	will stand for	$x <_Z y.$
$x - y$	will stand for	$x -_Z y.$
0	will stand for	$0_Z.$
1	will stand for	$1_Z.$

The reader should remember that the positive integers form a Peano system (see Theorem 4.7). Hence, mathematical induction may be used to prove that certain properties hold for all positive integers; and, in addition, all the other results of Chapter 2 can be applied to positive integers.

Theorem 5.1   (*Division Theorem*)   Let α be any integer greater than 1. Then, for any integer β, there are unique integers q and r such that

i. $\beta = q\alpha + r,$

ii. $0 \leq r < \alpha.$

(q and r are called, respectively, the quotient and remainder upon division of β by α).

Proof   First we shall prove the existence of appropriate q and r for positive β. To this end, we use mathematical induction with respect to β.
   Let $B = \{\beta : \beta > 0 \wedge (\exists q)(\exists r)(\beta = q\alpha + r \wedge 0 \leq r < \alpha)\}$. When $\beta = 1$, choose $q = 0$ and $r = 1$. Thus, $1 \in B$. Now assume $\beta \in B$. Thus, there exist q and r such that $\beta = q\alpha + r \wedge 0 \leq r < \alpha$. From $r < \alpha$ it follows that $r + 1 \leq \alpha$ (Exercise 2.4.5, p. 67).
   Case 1 $r + 1 = \alpha$.   Then

$$\beta + 1 = q\alpha + (r + 1) = q\alpha + \alpha = (q + 1)\alpha.$$

Thus, the quotient for $\beta + 1$ is $q + 1$ and the remainder is 0.

Case 2 $r + 1 < \alpha$.   Since $\beta + 1 = q\alpha + (r + 1)$, the quotient for $\beta + 1$ is q and the remainder is $r + 1$.

Thus, in either case, $\beta + 1 \in B$. We have shown that $\beta \in B \Rightarrow \beta + 1 \in B$. Hence, by mathematical induction, $B = \mathscr{P}_Z = $ the set of all positive integers.

When $\beta = 0$, take $q = 0$ and $r = 0$. Now, assume $\beta < 0$. Then $-\beta > 0$. Hence, by what we have shown above, there exist q_1 and r_1 such that $-\beta = q_1\alpha + r_1$ and $0 \leq r_1 < \alpha$.

Case 1 $r_1 = 0$.   Then $-\beta = q_1\alpha$. Hence, $\beta = (-q_1)\alpha$. In this case, the quotient for β is $-q_1$ and the remainder is 0.

Case 2 $0 < r_1$.   Then $\beta = (-q_1)\alpha + (-r_1) = (-q_1)\alpha + (\alpha - r_1) - \alpha = (-q_1 - 1)\alpha + (\alpha - r_1)$. Then the quotient is $-q_1 - 1$ and the remainder is $\alpha - r_1$. (Observe that $\alpha - r_1 < \alpha$ since $0 < r_1$; and $0 < \alpha - r_1$ since $r_1 < \alpha$.)

Uniqueness   Assume $\beta = q\alpha + r = q^*\alpha + r^*$, $0 \leq r < \alpha$, and $0 \leq r^* < \alpha$. Then, $(q - q^*)\alpha = r^* - r$.

Case 1 $r^* = r$.   Then $r^* - r = 0$. Hence, $(q - q^*)\alpha = 0$. But $\alpha \neq 0$. Hence, $q - q^* = 0$. Therefore, $q = q^*$.

Case 2 $r \neq r^*$.   Hence, $q - q^* \neq 0$. Therefore, $|q - q^*| \geq 1$, and $|(q - q^*)\alpha| = |q - q^*| \, |\alpha| \geq |\alpha| = \alpha$. Thus, $|r^* - r| \geq \alpha$. But $|r^* - r| < \alpha$. (For, $r^* < \alpha$ and $-r \leq 0$. Hence, $r^* - r < \alpha$. Likewise, $r < \alpha$ and $-r^* \leq 0$. Hence, $r - r^* < \alpha$. But, $|r^* - r|$ is either $r - r^*$ or $r^* - r$.) This is a contradiction.

Thus, the only possibility is Case 1: $r^* = r$ and $q^* = q$.

EXAMPLES

1.   If $\beta = 7$ and $\alpha = 2$, then $q = 3$ and $r = 1$:

$$7 = 3 \cdot 2 + 1.$$

2.   If $\beta = 17$ and $\alpha = 3$, then $q = 5$ and $r = 2$:

$$17 = 3 \cdot 5 + 2.$$

3. If $\beta = 20$ and $\alpha = 4$, then $q = 5$ and $r = 0$:

$$20 = 5 \cdot 4.$$

4.   If $\beta = -11$ and $\alpha = 4$, then $q = -3$ and $r = 1$:

$$-11 = (-3) \cdot 4 + 1.$$

Definition $\alpha \mid \beta$ means $(\exists \gamma)(\beta = \alpha\gamma)$. ($\alpha \mid \beta$ is read: "α divides β" or "β is divisible by α".)

Notation $\alpha \nmid \beta$ stands for $\neg(\alpha \mid \beta)$.

Theorem 5.2

a. $\alpha \mid 0$.

b. $1 \mid \alpha$.

c. $\alpha \mid \alpha$.

d. $(\alpha \mid \beta \wedge \beta \mid \gamma) \Rightarrow \alpha \mid \gamma$.

e. $(\gamma \neq 0 \wedge \alpha \mid \gamma) \Rightarrow \mid \alpha \mid \, \leq \mid \gamma \mid$.

f. $0 \mid \gamma \Rightarrow \gamma = 0$.

g. $(\alpha \mid \beta \wedge \beta \mid \alpha) \Leftrightarrow (\alpha = \beta \vee \alpha = -\beta)$.

h. $\alpha \mid \alpha \times \beta$.

i. $(\alpha \mid \beta \wedge \alpha \mid \gamma) \Rightarrow \alpha \mid \beta + \gamma$.

j. $(\alpha \mid \beta \wedge \alpha \mid \gamma) \Rightarrow \alpha \mid \beta - \gamma$.

k. $\alpha \mid 1 \Rightarrow (\alpha = 1 \vee \alpha = -1)$.

Proof (a) $0 = \alpha \times 0$.

(b) $\alpha = 1 \times \alpha$.

(c) $\alpha = \alpha \times 1$.

(d) Assume $\beta = \alpha\mu$ and $\gamma = \beta\delta$. Then $\gamma = (\alpha\mu)\delta = \alpha(\mu\delta)$. Hence, $\alpha \mid \gamma$.

(e) Assume $\alpha \mid \gamma$ and $\gamma \neq 0$. Then $\gamma = \alpha\delta$. Hence, $\mid \gamma \mid \, = \mid \alpha \mid \mid \delta \mid$. Since $\gamma \neq 0$, $\delta \neq 0$. Therefore, $\mid \delta \mid \, \geq 1$. Hence, $\mid \gamma \mid \, = \mid \alpha \mid \mid \delta \mid \, \geq \mid \alpha \mid$.

(f) Assume $0 \mid \gamma$. Then $\gamma = 0 \cdot \delta = 0$.

(g) Assume $\alpha \mid \beta$ and $\beta \mid \alpha$. By (f), if $\alpha = 0$ or $\beta = 0$, then $\alpha = \beta = 0$. We may assume then that $\alpha \neq 0$ and $\beta \neq 0$. By (e), $\mid \alpha \mid \, \leq \mid \beta \mid$ and $\mid \beta \mid \, \leq \mid \alpha \mid$. Hence, $\mid \alpha \mid \, = \mid \beta \mid$. Therefore, $\alpha = \beta$ or $\alpha = -\beta$.

(h) $\alpha \times \beta = \alpha \times \gamma$ for some γ, namely, $\gamma = \beta$.

(i), (j) If $\beta = \delta\alpha$ and $\gamma = \tau\alpha$, then

$$\beta + \gamma = \delta\alpha + \tau\alpha = (\delta + \tau)\alpha \quad \text{and} \quad \beta - \gamma = \delta\alpha - \tau\alpha = (\delta - \tau)\alpha$$

(k) Assume $\alpha \mid 1$. By (e), $\mid \alpha \mid \, \leq 1$. Hence, $\mid \alpha \mid \, = 0$ or $\mid \alpha \mid \, = 1$. But $\mid \alpha \mid \, \neq 0$, since 0 does not divide 1. Hence, $\mid \alpha \mid \, = 1$. Thus, $\alpha = 1$ or $\alpha = -1$. ∎

Definition　We say that δ is a **greatest common divisor** (gcd) of α and β if and only if

  i.　$\delta \mid \alpha \wedge \delta \mid \beta$;

  ii.　$(\forall \gamma)(\gamma \mid \alpha \wedge \gamma \mid \beta \Rightarrow \gamma \mid \delta)$;

  iii.　$0 \leq \delta$.

Remarks　1.　There is at most one gcd of α and β. For, if δ and $\delta^{\#}$ were both gcd's of α and β, then, by (ii), $\delta \mid \delta^{\#}$ and $\delta^{\#} \mid \delta$. Hence, $\delta^{\#} = \pm \delta$ by Theorem 5.2g. If $\delta^{\#} = -\delta$, then, since $0 \leq \delta$, it would follow that $\delta^{\#} = -\delta \leq 0$. But, since $0 \leq \delta^{\#}$, $\delta^{\#} = 0$. Hence, in all cases, $\delta^{\#} = \delta$.

2.　0 is the gcd of 0 and 0, since 0 satisfies (i)–(iii).

3.　If $\alpha \neq 0$, the gcd of 0 and α is $\mid \alpha \mid$.

4.　The gcd of 1 and α is 1.

5.　If α and β are not both 0 and δ is the gcd of α and β, then $0 < \delta$.

Theorem 5.3　Given integers α and β which are not both 0, then α and β possess a gcd δ, which is positive and can be written in the form $x\alpha + y\beta$.

Proof　Let $A = \{x\alpha + y\beta : x \in Z \wedge y \in Z\}$. A contains at least one positive element. For, if $\alpha \neq 0$, take

$$x = \begin{cases} 1 & \text{if } 0 < \alpha \\ -1 & \text{if } \alpha < 0 \end{cases} \quad \text{and} \quad y = 0.$$

Then $x\alpha + y\beta = \mid \alpha \mid > 0$. If $\beta \neq 0$, take

$$x = 0 \quad \text{and} \quad y = \begin{cases} 1 & \text{if } 0 < \beta \\ -1 & \text{if } \beta < 0. \end{cases}$$

Then $x\alpha + y\beta = \mid \beta \mid > 0$. Let δ be the least positive element of A.[†] Then $\delta = x_0 \alpha + y_0 \beta$ for certain integers x_0 and y_0. Clearly, if $\gamma \mid \alpha$ and $\gamma \mid \beta$, then $\gamma \mid (x_0 \alpha + y_0 \beta)$. Thus, condition (ii) is satisfied. To prove condition (i), divide α by δ: $\alpha = q\delta + r$, where $0 \leq r < \delta$. Hence, $\alpha = q(x_0 \alpha + y_0 \beta) + r$. Then $r = (1 - qx_0)\alpha + (-y_0)\beta$. Thus, $r \in A$. Since $r < \delta$ and δ is the least positive element of A, r cannot be positive. Therefore, $r = 0$. Thus, $\alpha = q\delta$, which implies that $\delta \mid \alpha$. A similar proof shows that $\delta \mid \beta$. Thus, condition (i) is satisfied.　∎

---

  † Since the positive integers form a Peano system, we are entitled to use the Least Number Principle (Theorem 2.4.5).

EXAMPLE

4.   The gcd of 2 and 5 is 1. The gcd of 6 and 15 is 3.

Definition   We say that α and β are **relatively prime** if and only if the only positive common divisor of α and β is 1.

EXAMPLE

5.   6 and 35 are relatively prime. The only positive divisors of 6 are 1, 2, 3, and 6, while the positive divisors of 35 are 1, 5, 7, and 35.

Note α and β are relatively prime if and only if their gcd is 1. For, if α and β are relatively prime, their gcd could not be greater than 1; conversely, if their gcd is 1, then any positive common divisor would have to be a divisor of 1 and, therefore, would be equal to 1.

Lemma 5.4 α and β are relatively prime if and only if $1 = x\alpha + y\beta$ for some integers x and y.

Proof   If α and β are relatively prime, their gcd is 1. Hence $1 = x\alpha + y\beta$ for some x and y, by Theorem 5.3. Conversely, if $1 = x\alpha + y\beta$, then any common divisor of α and β, being also a divisor of $x\alpha + y\beta$, must be a divisor of 1. But the only positive divisor of 1 is 1 itself.   ∎

Definition α is **prime** if and only if $\alpha > 1$ and

$$(\forall \gamma)((\gamma \mid \alpha \wedge 0 < \gamma) \Rightarrow (\gamma = \alpha \vee \gamma = 1)).$$

Thus, the primes are the integers $\alpha > 1$ such that the only positive divisors of α are 1 and α itself.

Examples of primes: 2, 3, 5, 7, 11, 13, 17, 19, 23,. . . .

Terminology   The integers greater than 1 which are not prime are called **composite** integers.

Lemma 5.5   Every integer greater than 1 is divisible by a prime.

Proof   Let $A = \{\alpha : \alpha > 1 \wedge \alpha$ is not divisible by a prime$\}$. Assume $A \neq \varnothing$. Let β be the least element of A. Then, $1 < \beta$ and β is not divisible by a prime.

Case 1 β is a prime.   But, $\beta \mid \beta$, which contradicts the fact that β is not divisible by a prime.

Case 2 β is not a prime.   By definition, there is some positive γ such that $\gamma \mid \beta \wedge \gamma \neq \beta \wedge \gamma \neq 1$. Since $\gamma \mid \beta$, it follows by Theorem 5.2e that $\gamma < \beta$. Hence, $\gamma \notin A$, and so, γ is divisible by a prime, say, $\varrho \mid \gamma$, where ϱ is a prime. Then $\varrho \mid \beta$, contradicting the fact that $\beta \in A$.

In both cases we have obtained contradictions. Hence, A is empty.   ■

Theorem 5.6   (*Euclid's Theorem on the Infinitude of Primes*)   The set of primes is infinite.

Proof   Assume that there are only finitely many primes $\varrho_1, \varrho_2, \ldots, \varrho_k$. Consider the integer

$$\mu = (\varrho_1 \varrho_2 \cdots \varrho_k) + 1.$$

Clearly, $\mu > 1$. By Lemma 5.5, μ is divisible by a prime ϱ. Hence, $\varrho = \varrho_j$ for some j. Hence, $\varrho \mid \mu$ and $\varrho \mid (\varrho_1 \varrho_2 \cdots \varrho_k)$. Therefore, $\varrho \mid 1$, since $1 = \mu - (\varrho_1 \varrho_2 \cdots \varrho_k)$. Hence, $\varrho = 1$, contradicting the fact that ϱ is a prime. Therefore, there must be infinitely many primes.   ■

Theorem 5.7   If ϱ is a prime and $\varrho \mid \alpha\beta$, then $\varrho \mid \alpha$ or $\varrho \mid \beta$.

Proof   Assume ϱ is a prime and $\varrho \mid \alpha\beta$. Assume that ϱ does not divide α. We must show that $\varrho \mid \beta$. Since ϱ does not divide α, ϱ and α are relatively prime. (For, the only positive divisors of ϱ are 1 and ϱ, and ϱ is not a divisor of α.) Hence, $1 = x\varrho + y\alpha$ for some x, y, by Lemma 5.4. Therefore, $\beta = x\beta\varrho + y\alpha\beta$. Since $\varrho \mid \alpha\beta$, it follows that ϱ is a divisor of $x\beta\varrho + y\alpha\beta$, that is, $\varrho \mid \beta$.   ■

Note   An equivalent formulation of Theorem 5.7 reads: If a prime ϱ does not divide α and does not divide β, then ϱ does not divide $\alpha\beta$.

Corollary 5.8   If ϱ is a prime and $\varrho \mid \alpha_1\alpha_2 \cdots \alpha_k$, then $\varrho \mid \alpha_1$ or $\varrho \mid \alpha_2$ or $\cdots$ or $\varrho \mid \alpha_k$.

Proof   Exercise for the reader. (*Hint:* Induction on k.)   ■

Theorem 5.9   Every integer greater than 1 is a product of one or more primes, and the representation as a product is unique except for arbitrary permutations of the factors. (By a **product** of one prime, we simply mean that prime itself.)

Proof Assume that some integer greater than 1 is not a product of one or more primes, and let α be the least such integer. Clearly, α is itself not a prime. Hence, α is divisible by a positive integer β different from 1 and α. Thus, $\alpha = \beta\gamma$ for some γ. Obviously, γ is greater than 1. Since $\alpha = \beta\gamma$, it follows that $\beta < \alpha$ and $\gamma < \alpha$. By the minimality of α, β and γ must be products of one or more primes. But then $\alpha = \beta\gamma$ is a product of primes, contradicting our assumption.

Uniqueness Assume $\alpha = \varrho_1 \cdots \varrho_k = \sigma_1 \cdots \sigma_m$, where $\varrho_1, \ldots, \varrho_k$, σ_1, $\ldots, \sigma_m$ are primes. We must show that $k = m$ and $\varrho_1, \ldots, \varrho_k$ is a rearrangement of $\sigma_1, \ldots, \sigma_m$. Assume that this uniqueness of decomposition into prime factors does not hold for some α greater than 1, and let α be the least such integer for which it fails. Now, since $\varrho_1 \mid \alpha$, we have $\varrho_1 \mid \sigma_1 \cdots \sigma_m$. By Corollary 5.8, ϱ_1 divides at least one σ_i. By rearranging the σ's, we may assume $i = 1$. Then $\varrho_1 \mid \sigma_1$. But σ_1 is a prime and its only divisors are itself and 1. But ϱ_1 is a divisor of σ_1 and $\varrho_1 \neq 1$. (All primes are greater than 1.) Hence, $\varrho_1 = \sigma_1$. But, $\varrho_1\varrho_2 \cdots \varrho_k = \sigma_1\sigma_2 \cdots \sigma_m$. By the Cancellation Law for Multiplication, $\varrho_2 \cdots \varrho_k = \sigma_2 \cdots \sigma_m$. But, $\varrho_2 \cdots \varrho_k < \varrho_1\varrho_2 \cdots \varrho_k = \alpha$. Hence, by the minimality of α, the decompositions $\varrho_2 \cdots \varrho_k$ and $\sigma_2 \cdots \sigma_m$ are the same, that is, $k - 1 = m - 1$ and $\varrho_2 \cdots \varrho_k$ is a rearrangement of $\sigma_2 \cdots \sigma_m$. Hence, $k = m$ and $\varrho_1, \ldots, \varrho_k$ is a rearrangement of $\sigma_1, \ldots, \sigma_m$. ∎

EXAMPLES

6. $24 = 2 \cdot 2 \cdot 2 \cdot 3$. Of course, we also have

$$24 = 2 \cdot 2 \cdot 3 \cdot 2 = 2 \cdot 3 \cdot 2 \cdot 2 = 3 \cdot 2 \cdot 2 \cdot 2,$$

but these are just permutations of the given factorization.

7. $105 = 3 \cdot 5 \cdot 7$.

8. $100 = 2 \cdot 2 \cdot 5 \cdot 5$.

Given any integer α greater than 1, we can represent α as a product of primes. If we combine all the occurrences of each prime ϱ into a power of primes ϱ^j, then the resulting product is called the **prime power representation** of n:

$$n = \varrho_1^{j_1} \cdots \varrho_k^{j_k}.$$

EXAMPLES

9. $72 = 2^3 \cdot 3^2$.

10. $100 = 2^2 \cdot 5^2$.

11. $525 = 3 \cdot 5^2 \cdot 7 = 3^1 \cdot 5^2 \cdot 7^1$.

Sometimes it is convenient to allow zero powers of primes in a prime power representation, for example, $100 = 2^2 \cdot 3^0 \cdot 5^2$, $525 = 2^0 \cdot 3^1 \cdot 5^2 \cdot 7^1$, $98 = 2^1 \cdot 3^0 \cdot 5^0 \cdot 7^2$. If we have two prime power representations, $n = \varrho_1^{j_1} \cdots \varrho_k^{j_k}$ and $m = \sigma_1^{i_1} \cdots \sigma_r^{i_r}$, we may assume that the ϱ's and σ's are the same. For example, instead of $105 = 3 \cdot 5 \cdot 7$ and $132 = 2^2 \cdot 3 \cdot 11$, we may write $105 = 2^0 \cdot 3^1 \cdot 5^1 \cdot 7^1 \cdot 11^0$ and $132 = 2^2 \cdot 3^1 \cdot 5^0 \cdot 7^0 \cdot 11^1$.

EXERCISES

We let $\gcd(\alpha, \beta)$ designate the gcd of α and β.

1.   Prove: $\gcd(\mu\alpha, \mu\beta) = \mu \gcd(\alpha, \beta)$.

2.   If $\delta = \gcd(\alpha, \beta)$ and $\alpha = \delta\gamma$, $\beta = \delta\zeta$, show that $\gcd(\gamma, \zeta) = 1$.

3.   If $\alpha \mid \beta\gamma$ and α and β are relatively prime, prove that $\alpha \mid \gamma$.

4.   Euclidean algorithm for the greatest common divisor: Let $\beta > 0$. Consider any α.

Divide α by β: $\alpha = q_0\beta \;\; + r_0$, $0 \le r_0 < \beta$.

Divide β by r_0: $\beta = q_1 r_0 \;\; + r_1$, $0 \le r_0 < r_0$.

Divide r_0 by r_1: $r_0 = q_2 r_1 \;\; + r_2$, $0 \le r_2 < r_1$,   etc.

$$r_{k-2} = q_k r_{k-1} + r_k, \quad 0 \le r_k < r_{k-1}.$$
$$r_{k-1} = q_{k+1} r_k.$$

The division process continues until we finally obtain a remainder of zero.

Example: Let $\alpha = 316$ and $\beta = 70$.

$$316 = \;\; 4 \cdot 70 + 36.$$
$$70 = \;\; 1 \cdot 36 + 34.$$
$$36 = \;\; 1 \cdot 34 + 2.$$
$$34 = 17 \cdot 2.$$

a.   Show that the last nonzero remainder r_k is the gcd of α and β.

b.   Show how to represent r_k as a linear combination $x\alpha + y\beta$.

c.   Apply the Euclidean algorithm to find the gcd δ of 154 and 15, and write δ as a linear combination of 154 and 15.

d.   The same problem as (c) with 234 and 54 (instead of 154 and 15).

5.   If $n = \varrho_1^{j_1} \cdots \varrho_k^{j_k}$ and $m = \varrho_1^{i_1} \cdots \varrho_k^{i_k}$ (where the ϱ's are primes and the j's and i's are nonnegative), then

$$\gcd(m, n) = \varrho_1^{\min(j_1, i_1)} \cdots \varrho_k^{\min(j_k, i_k)},$$

where $\min(u, v)$ denotes the minimum of u and v.

3.6 INTEGERS MODULO n

Let n be a fixed integer greater than 1.

Definition $\alpha \equiv \beta \bmod n$ means $n \mid \alpha - \beta$. (This is read: α is congruent to β modulo n.)

Theorem 6.1 $\alpha \equiv \beta \bmod n$ if and only if α and β leave the same remainder upon division by n.

Proof   Let $\alpha = qn + r$ and $\beta = tn + s$, where $0 \le r < n$ and $0 \le s < n$. Then

$$(\#) \quad \alpha - \beta = (q - t)n + (r - s).$$

  i.   Assume $\alpha \equiv \beta \bmod n$. Then $n \mid \alpha - \beta$. Hence, by $(\#)$, since $n \mid (\alpha - \beta)$ and $n \mid (q - t)n$, it follows that $n \mid (r - s)$. Assume $r \ne s$. Then $r - s \ne 0$. Hence, $n \le |r - s|$ (by Theorem 5.2e). But $-n < r - s < n$. (For, $r < n$ and $-s < 0$. Hence, $r - s < n$. In addition, $s < n$. Hence, $-n < -s$. Since $0 \le r$, it follows that $-n < r - s$.) Therefore, $|r - s| < n$, which contradicts $n \le |r - s|$. Hence, $r = s$.
  ii.   Conversely, assume $r = s$. Then, by $(\#)$, $\alpha - \beta = (q - t)n$. Hence, $n \mid \alpha - \beta$, that is, $\alpha \equiv \beta \bmod n$.   ∎

EXAMPLES

$$2 \equiv 5 \bmod 3.$$
$$4 \equiv 10 \bmod 3.$$
$$7 \equiv 2 \bmod 5.$$
$$13 \equiv 3 \bmod 5.$$

Theorem 6.2   Congruence modulo n is an equivalence relation, that is,

a. $\alpha \equiv \alpha \bmod n$ (reflexivity);
b. $\alpha \equiv \beta \bmod n \Rightarrow \beta \equiv \alpha \bmod n$ (symmetry);
c. $(\alpha \equiv \beta \bmod n \wedge \beta \equiv \gamma \bmod n) \Rightarrow \alpha \equiv \gamma \bmod n$ (transitivity).

Proof (a) $\alpha - \alpha = 0$ and $n \mid 0$.

(b) By Theorem 6.1, if α and β leave the same remainder upon division by n, then so do β and α.

(c) Again we shall use Theorem 6.1. If α and β leave the same remainder upon division by n, and β and γ leave the same remainder upon division by n, then α and γ leave the same remainder upon division by n.   ■

EXERCISE

Prove Theorem 6.2b, c directly from the definition, without using Theorem 6.1.

Since congruence modulo n is an equivalence relation, we may form the equivalence classes with respect to this relation. These equivalence classes are called **residue classes modulo n**. We denote by Z_n the set of all residue classes modulo n. As usual, if α is in Z, we denote the equivalence class of α by $[\alpha]$.

EXAMPLES

1. $n = 2$. Any integer α leaves a remainder of 0 or 1 upon division by 2. Hence α is congruent modulo 2 to either 0 or 1. Thus, there are exactly two elements of Z_2, namely, [0] and [1].

2. $n = 3$. Any integer α leaves a remainder of 0, 1, or 2 upon division by 3. Hence, there are exactly three equivalence classes in Z_3, namely, [0], [1], [2].

Theorem 6.3 Z_n contains exactly n elements.

Proof Any integer leaves a remainder r upon division by n, where $0 \le r < n$. Hence, there are exactly n equivalence classes: [0], [1], ..., [$n - 1$].   ■

It is possible to "add" and "multiply" the residue classes of Z_n. The definition of these operations depends upon the following lemma.

Lemma 6.4 If $\alpha \equiv \beta \bmod n$ and $\gamma \equiv \delta \bmod n$, then

a. $\alpha + \gamma \equiv \beta + \delta \bmod n$;

b. $-\gamma \equiv -\delta \bmod n$;

c. $\alpha - \gamma \equiv \beta - \delta \bmod n$;

d. $\alpha\gamma \equiv \beta\delta \bmod n$.

Proof (a) $n \mid \beta - \alpha$ and $n \mid \delta - \gamma$. Hence, $n \mid (\beta - \alpha) + (\delta - \gamma)$, that is, $n \mid (\beta + \delta) - (\alpha + \gamma)$. Thus, $\alpha + \gamma \equiv \beta + \delta \bmod n$.

(b) Since $n \mid \gamma - \delta$, $n \mid -(\gamma - \delta)$. Hence, $n \mid (-\gamma) - (-\delta)$; thus, $-\gamma \equiv -\delta \bmod n$.

(c) This follows from (a) and (b).

(d) $\beta\delta - \alpha\gamma = \beta(\delta - \gamma) + \gamma(\beta - \alpha)$. Since $n \mid \delta - \gamma$ and $n \mid \beta - \alpha$, it follows that $n \mid \beta(\delta - \gamma) + \gamma(\beta - \alpha)$. Hence, $\alpha\gamma \equiv \beta\delta \bmod n$. ∎

Definition Given $X \in Z_n$ and $Y \in Z_n$. Choose $\alpha \in X$ and $\gamma \in Y$. Let

$$X +_n Y = [\alpha + \gamma];$$
$$-_n X = [-\alpha];$$
$$X \times_n Y = [\alpha\gamma].$$

That these definitions do not depend upon the choice of α in X and γ in Y follows from Lemma 6.4.

Theorem 6.5 $(Z_n, +_n, \times_n)$ is a commutative ring with unit element.

Proof Assume that X, Y, W are in Z_n. Choose $\alpha \in X$, $\gamma \in Y$, $\tau \in W$.

Axiom 1

$$
\begin{aligned}
(X +_n Y) +_n W &= ([\alpha] +_n [\gamma]) +_n [\tau] \\
&= [\alpha + \gamma] +_n [\tau] = [(\alpha + \gamma) + \tau] \\
&= [\alpha + (\gamma + \tau)] = [\alpha] +_n [\gamma + \tau] \\
&= [\alpha] +_n ([\gamma] +_n [\tau]) = X +_n (Y +_n W).
\end{aligned}
$$

Axiom 2

$$
\begin{aligned}
X +_n Y &= [\alpha] +_n [\gamma] = [\alpha + \gamma] = [\gamma + \alpha] \\
&= [\gamma] +_n [\alpha] = Y +_n X.
\end{aligned}
$$

Axiom 3 Let $0_n = [0]$. Then,

$$X +_n 0_n = [\alpha] +_n [0] = [\alpha + 0] = [\alpha] = X.$$

Moreover, $X +_n (-_n X) = [\alpha] +_n [-\alpha] = [\alpha + (-\alpha)] = [0] = 0_n$.

Axiom 4 $X \times_n (Y \times_n W) = (X \times_n Y) \times_n W.$ $\left.\right\}$ Exercises for

Axiom 5 $X \times_n (Y +_n W) = (X \times_n Y) +_n (X \times_n W)$ $\left.\right\}$ the reader.

Axiom 6   Let $1_n = [1]$. Then

$$X \times_n 1_n = [\alpha] \times_n [1]$$
$$= [\alpha \cdot 1] = [\alpha] = X.$$

Axiom 7 $X \times_n Y = Y \times_n X.$ Exercise for the reader.   ∎

EXAMPLES

1. $n = 2$. $Z_2 = \{[0], [1]\}$. Let 0_2, 1_2 stand for $[0]$, $[1]$, respectively.

$+_2$	0_2	1_2		$\times_2$	0_2	1_2
0_2	0_2	1_2		0_2	0_2	0_2
1_2	1_2	0_2		1_2	0_2	1_2

2. $n = 3$. $Z_3 = \{[0], [1], [2]\}$.

$+_3$	[0]	[1]	[2]		$\times_3$	[0]	[1]	[2]
[0]	[0]	[1]	[2]		[0]	[0]	[0]	[0]
[1]	[1]	[2]	[0]		[1]	[0]	[1]	[2]
[2]	[2]	[0]	[1]		[2]	[0]	[2]	[1]

3. $n = 4$. $Z_4 = \{[0], [1], [2], [3]\}$. Let 0_4, 1_4, 2_4, 3_4 stand for $[0]$, $[1]$, $[2]$, $[3]$, respectively.

$+_4$	0_4	1_4	2_4	3_4		$\times_4$	0_4	1_4	2_4	3_4
0_4						0_4				
1_4						1_4				
2_4						2_4				
3_4						3_4				

The reader should fill in the tables for $+_4$ and $\times_4$.

An important point to remember is that the symbols [0], [1], [2], etc. have different meanings depending upon the value of n. Thus, when $n = 2$, [0] stands for the set of all even integers, while, when $n = 3$, [0] stands for the set of all multiples of 3.

The sets Z_n are sometimes thought of as consisting of simply the integers $0, 1, \ldots, n - 1$. Addition and multiplication are performed by carrying out ordinary addition and multiplication and then taking the remainder upon division by n.

Theorem 6.6 $(Z_n, +_n, \times_n)$ is an integral domain if and only if n is a prime.

Proof   If n is not a prime, then $n = \alpha\beta$, where α and β are positive integers smaller than n. Hence, $[n] = [\alpha\beta] = [\alpha] \times_n [\beta]$. But $[n] = [0] = 0_n$, while $[\alpha] \neq 0_n$ and $[\beta] \neq 0_n$. Hence, Axiom 8 for an integral domain:

$$(x \neq 0 \wedge y \neq 0) \Rightarrow x \times y \neq 0$$

fails.

Conversely, assume n is a prime. Assume that X and Y are in Z_n, with $X \neq 0_n$ and $Y \neq 0_n$. Choose $\alpha \in X$ and $\beta \in Y$. Then $[\alpha] = X \neq 0_n$ and $[\beta] = Y \neq 0_n$. Hence, $\alpha \not\equiv 0 \bmod n$ and $\beta \not\equiv 0 \bmod n$, that is, $n \nmid \alpha$ and $n \nmid \beta$. Therefore, by Theorem 5.7, $n \nmid \alpha\beta$, that is, $\alpha\beta \not\equiv 0 \bmod n$. Hence, $[\alpha\beta] \neq 0_n$. But, $[\alpha\beta] = [\alpha] \times_n [\beta] = X \times_n Y$. Thus, Axiom 8 is satisfied and $(Z_n, +_n, \times_n)$ is an integral domain.   ∎

EXAMPLES

4. $(Z_2, +_2, \times_2)$ and $(Z_3, +_3, \times_3)$ are integral domains. $(Z_4, +_4, \times_4)$ is not an integral domain. (For, $[2] \neq 0_4$, while $[2] \times_4 [2] = [2 \times 2] = [4] = [0] = 0_4$.)

EXERCISE

Give examples of zero divisors in $(Z_6, +_6, \times_6)$.

3.7   CHARACTERISTIC OF AN INTEGRAL DOMAIN

In an integral domain, if we keep adding the unit element 1 to itself:

$$1, \quad 1 + 1, \quad 1 + 1 + 1, \quad 1 + 1 + 1 + 1, \quad \ldots,$$

we may or may not eventually obtain the value 0. For example, in the system of integers, we never reach a value 0, while, in the ring of integers modulo 2, $(Z_2, +_2, \times_2)$, the second term $1 + 1$ is 0. In order to make clear what the various possibilities are, we first make some of the ideas precise.

We shall work with an integral domain $\mathscr{D} = (D, +, \times)$, with zero element 0_D and unit element 1_D. Consider any x in D. Using the Peano system of positive integers, we apply the Iteration Theorem with $W = D$, $c = x$, and $g(u) = u + x$ for any u in D. This yields a unique function $F_x: P \to D$ such that $F_x(1) = x$ and $F_x(n + 1) = F_x(n) + x$ for every positive integer n. In particular,

$$F_x(2) = F_x(1) + x = x + x;$$
$$F_x(3) = F_x(2) + x = x + x + x;$$
$$F_x(4) = F_x(3) + x = x + x + x + x, \quad \text{etc.}$$

Intuitively,

$$F_x(n) = \underbrace{x + x + \cdots + x}_{n \text{ times}}.$$

In accord with the customary notation, we shall write nx instead of $F_x(n)$. Thus, $(n + 1)x = nx + x$. In particular,

$$1x = x,$$
$$2x = x + x,$$
$$3x = x + x + x, \quad \text{etc.}$$

Let us prove some plausible-looking facts about this new function nx.

Theorem 7.1   For any positive integers n and k and for any x and y in D:

a. $(n + k)x = (nx) + (kx)$.

b. $n(x + y) = (nx) + (ny)$.

c. $(n \times k)x = n(kx)$.

d. $n(x \times y) = (nx) \times y$.

e. $(nx) \times (ky) = (n \times k)(x \times y)$.

(Here, $n + k$ and $n \times k$ denote, respectively, the sum and product in $(P, S, 1)$. Strictly speaking, we should not use the same symbols for the addition and multiplication operations in P as for the addition and multiplication operations in D. We should use, for example, $+_D$ and $\times_D$ for the addition and multiplication operations in D. However, the text would then become cluttered with subscripts and would be difficult to read.)

Proof (a)   We shall use mathematical induction with respect to k. Assume that $n \in P$. Let $A = \{k : k \in P \land (n + k)x = nx + kx\}$. We must prove: $A = P$. First, $1 \in A$, since $(n + 1)x = nx + x = nx + 1x$. Now assume

$k \in A$. Then $(n + k)x = nx + kx$. Hence,

$$(n + (k + 1))x = ((n + k) + 1)x = (n + k)x + x = (nx + kx) + x$$
$$= nx + (kx + x) = nx + (k + 1)x.$$

Thus, $k + 1 \in A$. We have shown: $k \in A \Rightarrow k + 1 \in A$. By mathematical induction, $A = P$.

(b) Assume x and y are in D. Let

$$B = \{n : n \in P \wedge n(x + y) = nx + ny\}.$$

We must prove $B = P$. First, $1 \in B$, since $1(x + y) = x + y = 1x + 1y$. Now assume $n \in B$. Then $n(x + y) = nx + ny$. Therefore,

$$(n + 1)(x + y) = n(x + y) + (x + y)$$
$$= (nx + ny) + (x + y) \qquad \text{(by inductive hypothesis)}$$
$$= (nx + x) + (ny + y)$$
$$= (n + 1)x + (n + 1)y.$$

Thus, $n + 1 \in B$. We have shown: $n \in B \Rightarrow n + 1 \in B$. Hence, by mathematical induction, $B = P$.

(c) Assume $n \in P$. Let $C = \{k : k \in P \wedge (n \times k)x = n(kx)\}$. We must prove: $C = P$. First, $1 \in C$, since $(n \times 1)x = nx = n(1x)$. Now assume $k \in C$. Then $(n \times k)x = n(kx)$. Therefore,

$$(n \times (k + 1))x = ((n \times k) + n)x = (n \times k)x + nx \quad \text{(by (a))}$$
$$= n(kx) + nx \quad \text{(by inductive hypothesis)}$$
$$= n((kx) + x) \quad \text{(by (b))}$$
$$= n((k + 1)x).$$

Thus, $k + 1 \in C$. We have shown: $k \in C \Rightarrow k + 1 \in C$. By mathematical induction, $C = P$.

(d) Assume x and y are in D. Let $L = \{n : n \in P \wedge n(x \times y) = (nx) \times y\}$. We must show: $L = P$. First, $1 \in L$, since $1(x \times y) = x \times y = (1x) \times y$. Assume that $n \in L$. Then $n(x \times y) = (nx) \times y$. Hence,

$$(n + 1)(x \times y) = n(x \times y) + (x \times y)$$
$$= ((nx) \times y) + (x \times y) \qquad \text{(by inductive hypothesis)}$$
$$= ((nx) + x) \times y = ((n + 1)x) \times y.$$

Thus $n + 1 \in L$. We have shown: $n \in L \Rightarrow n + 1 \in L$. Hence, by mathematical induction, $L = P$.

(e) $(nx) \times (ky) = n(x \times (ky))$ (by (d))

$= n((ky) \times x)$

$= n(k(y \times x))$ (by (d))

$= (n \times k)(y \times x)$ (by (c))

$= (n \times k)(x \times y)$.   ∎

Notation   The values 1_D, $1_D + 1_D$, $1_D + 1_D + 1_D$, ... are of especial interest. For each positive integer n, we shall denote $n1_D$ by n_D. Thus,

$$1_D = 1(1_D) = 1_D,$$

$$2_D = 2(1_D) = 1_D + 1_D,$$

$$3_D = 3(1_D) = 1_D + 1_D + 1_D, \quad \text{etc.}$$

Definition   Let $\mathscr{D} = (D, +, \times)$ be an integral domain. The *characteristic* of $\mathscr{D}$ is

0   if, for all positive integers n, $n_D \neq 0_D$;

k if there is a positive integer n such that $n_D = 0_D$ and k is the least such n.

Thus, the characteristic is 0 if and only if 1_D, $1_D + 1_D$, $1_D + 1_D + 1_D$, ... are all different from 0_D. Otherwise, the characteristic is the smallest number of 1_D's which when added together yield 0_D. Notice that, in either case, the characteristic is a nonnegative integer.

EXAMPLES

1.   The characteristic of the system of integers $(Z, +_Z, \times_Z)$ is 0. For, by mathematical induction, it is easy to show that $n1_Z >_Z 0_Z$ for all positive integers n. (*Proof*   Let $A = \{n : n \in P \wedge n1_Z >_Z 0_Z\}$. Then $1 \in A$, since $1_Z >_Z 0_Z$. Assume now that $n \in A$. Then $n1_Z >_Z 0_Z$. Therefore, $(n + 1)1_Z = n1_Z + 1_Z >_Z 1_Z >_Z 0_Z$. Hence, by mathematical induction, $A = P$.) Observe that, in this example, $nx = n \times_Z x$ for any positive integer n and any integer x. (This is left as an exercise for the reader. Use mathematical induction with respect to n.)

2. The characteristic of $(Z_2, +_2, \times_2)$ is 2. For, $1_2 +_2 1_2 = [1] +_2 [1]$ $= [2] = 0_2$.

3. More generally, the characteristic of the system of residue classes modulo n, $(Z_n, +_n, \times_n)$, is n itself. To see this, first observe that, for any positive integer j and for any integer k, $j[k] = [j \times_Z k]$. (*Proof* Let $A = \{j : j \in P \wedge j[k] = [j \times_Z k]\}$. Then $1 \in A$, since $1[k] = [k] = [1 \times_Z k]$. Assume $j \in A$. Then, $j[k] = [j \times_Z k]$. Hence,

$$(j + 1)[k] = j[k] +_n [k] = [j \times_Z k] +_n [k]$$
$$= [(j \times_Z k) +_Z k] = [(j + 1) \times_Z k].$$

Thus, $j + 1 \in A$. We have shown: $j \in A \Rightarrow j + 1 \in A$. By mathematical induction, $A = P$.) Hence, $j[1_Z] = [j \times_Z 1_Z] = [j]$. Since $[n] = [0_Z]$ and $[j] \neq [0_Z]$ for all j such that $1 \leq j < n$, it follows that the least positive integer j such that $j[1_Z] = [0_Z]$ is n.

Theorem 7.2   The characteristic of any integral domain $\mathscr{D} = (D, +, \times)$ is either 0, 1, or a prime.

Proof   Assume that the characteristic n is not 0. Then n is a positive integer. If $n = 1$, then $1_D = 1(1_D) = 0_D$. Hence, the integral domain consists of just one element (by Theorem 3.4). Now assume that the integral domain contains more than one element. Then $n > 1$. Assume n is not a prime. Then $n = j \times k$, where $1 < j < n$ and $1 < k < n$. Now, $0_D = n1_D = (j \times k)1_D$ $= (j \times k)(1_D \times 1_D) = (j1_D) \times (k1_D)$ by Theorem 7.1e. Since $\mathscr{D}$ is an integral domain, either $j1_D = 0_D$ or $k1_D = 0_D$. But, $j < n$ and $k < n$, and this contradicts the fact that n is the least positive integer such that $n1_D = 0_D$. Therefore, n is a prime.   ■

Theorem 7.3   The characteristic of any ordered integral domain $\mathscr{D} = (D, +, \times, <)$ is 0.

Proof   Let $A = \{n : n \in P \wedge 0_D < n1_D\}$. A is the set of all positive integers n such that n_D is a positive element of D. First, $1 \in A$, since $1(1_D) = 1_D > 0_D$. Now assume $n \in A$. Thus, $0_D < n1_D$. Then, $(n + 1)1_D = n1_D + 1_D > 0_D$ by Theorem 4.2g. Thus, $n + 1 \in A$. We have shown: $n \in A \Rightarrow n + 1 \in A$. By mathematical induction, $A = P$. Therefore, $n1_D > 0_D$ for all positive integers n. Hence, $n1_D \neq 0_D$ for all positive integers n. This means that the characteristic of $\mathscr{D}$ is 0.   ■

EXERCISES

Let $\mathscr{D} = (D, +, \times)$ be an integral domain. Prove:

1. $nx = (n1_D) \times x$ for any positive integer n and any x in D.

2. $n0_D = 0_D$ for all positive integers n.

3. If $n1_D = 0_D$, then $nx = 0_D$ for all x in D.

4. If $nx = 0_D$ for some $x \neq 0_D$, then $n1_D = 0_D$.

5. If $x \neq 0_D$, then the characteristic of $\mathscr{D}$ is

$$\begin{array}{ll} 0 & \text{if } nx \neq 0_D \text{ for all positive integers } n; \\ p & \text{if } p \text{ is the least positive integer such that } px = 0_D. \end{array}$$

Let $\mathscr{D} = (D, +, \times)$ be any integral domain. For x in D, we now can extend the meaning of nx to the case where n is not just a positive integer but rather *any* integer.

Definition   Let $x \in D$ and $\alpha \in Z$. Then

$$\alpha x = \begin{cases} \alpha x & \text{if } 0_Z <_Z \alpha; \\ 0_D & \text{if } \alpha = 0_Z; \\ (-\alpha)(-x) & \text{if } \alpha <_Z 0_Z. \end{cases}$$

EXAMPLES

4.
$$0_Z x = 0_D.$$
$$(-1_Z)x = (-(-1_Z))(-x) = 1_Z(-x) = -x.$$
$$(-2_Z)x = (2_Z)(-x) = (-x) + (-x).$$
$$(-3_Z)x = (3_Z)(-x) = (-x) + (-x) + (-x).$$

Intuitively, if n is positive, then $(-n)x$ is the sum of n copies of $-x$.

5.   Consider the integral domain $(Z, +_Z, \times_Z)$. Then, for any integers α and β, $\alpha\beta = \alpha \times_Z \beta$. (We already have observed this fact when α is positive. For $\alpha = 0_Z$, $0_Z\beta = 0_Z = 0_Z \times_Z \beta$. When $\alpha <_Z 0_Z$, $\alpha\beta = (-\alpha)(-\beta) = (-\alpha) \times_Z (-\beta) = \alpha \times_Z \beta$.)

6.   Consider the integral domain $(Z_3, +_3, \times_3)$. Then

$$(-1)[1] = 1(-[1]) = -[1] = [2];$$
$$(-2)[1] = 2(-[1]) = 2[2] = [2] +_2 [2] = [1];$$
$$(-3)[1] = 3(-[1]) = 3[2] = [2] +_3 [2] +_3 [2] = [0].$$

7. $\alpha 0_D = 0_D$ (by Exercise (2) above and the definition).

Theorem 7.4 For any integers α, β and any x, y in D:

a. $(\alpha +_Z \beta)x = \alpha x + \beta x$.
b. $\alpha(x + y) = \alpha x + \alpha y$.
c. $(\alpha(-x)) = -(\alpha x)$.
d. $(\alpha \times_Z \beta)x = \alpha(\beta x)$.
e. $\alpha(x \times y) = (\alpha x) \times y$.
f. $(\alpha x) \times (\beta y) = (\alpha \times_Z \beta)(x \times y)$.
g. $(-\alpha)x = -(\alpha x)$.
h. $(\alpha - \beta)x = \alpha x - \beta x$.

Proof (a) *Case 1* $0_Z <_Z \alpha$ and $0_Z <_Z \beta$. This is just Theorem 7.1a.

Case 2 $\alpha = 0_Z$ or $\beta = 0_Z$. This is obvious.

Case 3 $\alpha <_Z 0_Z$ and $\beta <_Z 0_Z$. Then $\alpha +_Z \beta <_Z 0_Z$.

$$(\alpha +_Z \beta)x = (-(\alpha +_Z \beta))(-x) = ((-\alpha) +_Z (-\beta))(-x)$$
$$= (-\alpha)(-x) + (-\beta)(-x) \qquad \text{by Theorem 7.1a}$$
$$= \alpha x + \beta x.$$

Case 4 $\alpha <_Z 0_Z$ and $0_Z <_Z \beta$.

Case 4a $-\alpha = \beta$. Then $\alpha +_Z \beta = 0_Z$.

$$(\alpha +_Z \beta)x = 0_Z x = 0_D.$$
$$\alpha x + \beta x = (-\alpha)(-x) + \beta x = \beta(-x) + \beta x$$
$$= \beta((-x) + x) \qquad \text{by Theorem 7.1b}$$
$$= \beta 0_D = 0_D.$$

Case 4b $-\alpha <_Z \beta$. Then $0_Z <_Z \alpha + \beta$. Hence,

$$\alpha x + \beta x = \alpha x + ((\alpha +_Z \beta) +_Z (-\alpha))x$$
$$= \alpha x + (\alpha +_Z \beta)x + (-\alpha)x \qquad \text{by Theorem 7.1a}$$
$$= (\alpha +_Z \beta)x + (\alpha x) + (-\alpha)x$$
$$= (\alpha +_Z \beta)x + 0_D \qquad \text{by Case 4a}$$
$$= (\alpha +_Z \beta)x.$$

Case 4c $\beta <_Z -\alpha$. Then $\alpha +_Z \beta <_Z 0_Z$.

$$(\alpha +_Z \beta)x + (-\beta)x = ((\alpha +_Z \beta) +_Z (-\beta))x \qquad \text{by Case 3}$$
$$= \alpha x.$$

Therefore,

$$(\alpha +_z \beta)x = \alpha x - (-\beta)x = \alpha x - (\beta(-x))$$
$$= \alpha x + (-(\beta(-x))) = \alpha x + ((-1_D) \times \beta(-x)))$$
$$= \alpha x + (\beta((-1_D) \times (-x))) \qquad \text{by Theorem 7.1d}$$
$$= \alpha x + \beta(1_D \times (-(-x))) = \alpha x + \beta x.$$

Case 5 $\beta <_z 0_z$ and $0_z <_z \alpha$. This follows from Case 4 by switching α and β.

(b) *Case 1* $0_z <_z \alpha$. Use Theorem 7.1b.

Case 2 $0_z = \alpha$. $0_z(x + y) = 0_D$ and $0_z x + 0_z y = 0_D + 0_D = 0_D$.

Case 3 $\alpha <_z 0_z$.

$$\alpha(x + y) = (-\alpha)(-(x + y)) = (-\alpha)((-x) + (-y))$$
$$= (-\alpha)(-x) + (-\alpha)(-y) = \alpha x + \alpha y.$$

(c) $\alpha x + \alpha(-x) = \alpha(x + (-x)) \qquad \text{by (b)}$
$$= \alpha 0_D$$
$$= 0_D.$$

Hence, $\alpha x = -(\alpha(-x))$.

(d) *Case 1* $0_z <_z \alpha \wedge 0_z <_z \beta$. Use Theorem 7.1c.

Case 2 $0_z = \alpha$ or $0_z = \beta$. This is obvious.

Case 3 $\alpha <_z 0_z \wedge \beta <_z 0_z$. Then $\alpha \times_z \beta >_z 0_z$. Hence,

$$\alpha(\beta x) = \alpha((-\beta)(-x))$$
$$= (-\alpha)(-(-\beta)(-x))$$
$$= (-\alpha)((-\beta)x) \qquad \text{by (c)}$$
$$= ((-\alpha) \times_z (-\beta))x \qquad \text{by Theorem 7.1c}$$
$$= (\alpha \times_z \beta)x.$$

Case 4 $\alpha <_z 0_z \wedge 0_z <_z \beta$. Then $\alpha \times_z \beta <_z 0_z$. Hence,

$$(\alpha \times_z \beta)x = (-(\alpha \times_z \beta))(-x)$$
$$= ((-\alpha) \times_z \beta)(-x)$$
$$= (-\alpha)(\beta(-x)) \qquad \text{by Case 1}$$
$$= (-\alpha)(-(\beta x)) \qquad \text{by (c)}$$
$$= \alpha(\beta x) \qquad \text{by definition.}$$

Case 5 $\beta <_Z 0_Z \wedge 0_Z < \alpha$. Then $\alpha \times_Z \beta <_Z 0_Z$. Hence,

$$
\begin{aligned}
(\alpha \times_Z \beta)x &= (-(\alpha \times_Z \beta))(-x) \\
&= (\alpha \times_Z (-\beta))(-x) \\
&= \alpha((-\beta)(-x)) \qquad \text{by Case 1} \\
&= \alpha(\beta x) \qquad \text{by definition.}
\end{aligned}
$$

(e) *Case 1* $0_Z <_Z \alpha$. Use Theorem 7.1d.

Case 2 $0_Z = \alpha$. Obvious.

Case 3 $\alpha <_Z 0_Z$.

$$
\begin{aligned}
\alpha(x \times y) &= (-\alpha)(-(x \times y)) \\
&= (-\alpha)((-x) \times y) \\
&= ((-\alpha)(-x)) \times y \qquad \text{by Theorem 7.1d} \\
&= (\alpha x) \times y.
\end{aligned}
$$

(f)
$$
\begin{aligned}
\alpha x \times \beta y &= \alpha(x \times (\beta y)) \qquad \text{by (e)} \\
&= \alpha((\beta y) \times x) \\
&= \alpha(\beta(y \times x)) \qquad \text{by (e)} \\
&= (\alpha \times_Z \beta)(y \times x) \qquad \text{by (d)} \\
&= (\alpha \times_Z \beta)(x \times y).
\end{aligned}
$$

(g)
$$
\begin{aligned}
\alpha x + (-\alpha)x &= (\alpha + (-\alpha))x \qquad \text{by (a)} \\
&= 0_Z x \\
&= 0_D.
\end{aligned}
$$

Hence, $(-\alpha)x = -(\alpha x)$.

(h)
$$
\begin{aligned}
(\alpha - \beta)x &= (\alpha + (-\beta))x \\
&= \alpha x + (-\beta)x \qquad \text{by (a)} \\
&= \alpha x + (-(\beta x)) \qquad \text{by (g)} \\
&= \alpha x - \beta x. \qquad \blacksquare
\end{aligned}
$$

EXERCISES

1. Prove that $\alpha x = (\alpha 1_D) \times x$.

2.^{c} Exponentiation. Let $\mathscr{D} = (D, +, \times)$ be an integral domain. Assume $u \in D$. We define an operation which, for each natural number n, produces an element u^n in D. Let P be the set of positive integers. By the Iteration Theorem, with $W = D$,

$c = u$, and $g(v) = v \times u$ for any v in D, there is a unique mapping $\psi: P \to D$ such that

$$\psi(1) = u,$$

$$\psi(n + 1) = g(\psi(n)) = \psi(n) \times u \qquad \text{for all } n \text{ in } P.$$

We denote $\psi(n)$ by u^n. Thus, $u^1 = u$ and $u^{n+1} = u^n \times u$. In particular,

$$u^2 = u^{1+1} = u^1 \times u = u \times u,$$

$$u^3 = u^2 \times u = (u \times u) \times u, \text{ etc.}$$

Prove:

i. $u^{j+k} = u^j \times u^k$.

ii. $(u^j)^k = u^{j \times k}$.

iii. $(u \times v)^j = u^j \times v^j$.

iv. $(u + v)^2 = u^2 + 2(u \times v) + v^2$.

v. If $\mathscr{D} = (D, +, \times, <)$ is an ordered integral domain, and $u \in D$, $u > 0_D$, $n >_Z 1$, then $(1 + u)^n > 1 + nu$.

vi. $(-u)^n = \begin{cases} u^n & \text{if } n \text{ is even,} \\ -(u^n) & \text{if } n \text{ is odd.} \end{cases}$

3. *Radix Representation of Integers. Place Notation.* Let b be any integer greater than 1. Prove that, for any $a \geq 1$, there is a unique finite sequence $a_0, a_1, \ldots, a_n$ of integers such $0 \leq a_i < b$ and $a = a_0 + a_1 b + a_2 b^2 + \cdots + a_n b^n$. (*Hint*: To prove the existence of such a finite sequence, use mathematical induction and Theorem 5.1. Prove the uniqueness also by mathematical induction.) Our traditional decimal representation is obtained when we let the "radix" $b = 10$. We already have used decimal representations of numbers in previous examples. Another popular radix is $b = 2$; this **binary** representation is very convenient in work with computing machines.

Examples:

Decimal	1	2	3	4	5	6	7	8	9
Binary	1	10	11	100	101	110	111	1000	1001

Decimal	10	11	12	13	14	15	16
Binary	1010	1011	1100	1101	1110	1111	10000

3.8 NATURAL NUMBERS AND INTEGERS OF AN INTEGRAL DOMAIN

By the natural numbers of an integral domain $\mathscr{D} = (D, +, \times)$, we have in mind the elements 1_D, $1_D + 1_D$, $1_D + 1_D + 1_D$, etc. Since "etc." is not a legitimate part of a mathematical definition, we must find a different way to define the natural numbers of $\mathscr{D}$.

Definition   Let A be any subset of D. We say that A is **inductive** if and only if

i. $1_D \in A$,

ii. $(\forall x)(x \in A \Rightarrow x + 1_D \in A)$.

Thus, a set is inductive if and only if it contains 1_D and is closed under the operation of adding 1_D.

There is at least one inductive set, namely, D itself.

Definition $N_{\mathscr{D}}$ is the set of all elements of D which are contained in every inductive set.

Thus, $x \in N_{\mathscr{D}} \Leftrightarrow (\forall A)$ (A is inductive $\Rightarrow x \in A$). If $\mathscr{I}$ is the class of all inductive sets, then $N_{\mathscr{D}} = \bigcap_{A \in \mathscr{I}} A$. The elements of $N_{\mathscr{D}}$ are called **natural numbers of $\mathscr{D}$**.

Theorem 8.1

a. $(\forall A)$ (A is inductive $\Rightarrow N_{\mathscr{D}} \subseteq A$).

b. $N_{\mathscr{D}}$ is inductive.

Proof   (a)   This follows immediately from the definition of $N_{\mathscr{D}}$.

(b)   Clearly, 1_D belongs to every inductive set, and, therefore, also to $N_{\mathscr{D}}$. Assume $x \in N_{\mathscr{D}}$. Let A be any inductive set. Then $x \in A$. Since A is inductive, $x + 1_D \in A$. Thus, $x + 1_D$ is in every inductive set, that is, $x + 1_D \in N_{\mathscr{D}}$. We have shown that $x \in N_{\mathscr{D}} \Rightarrow x + 1_D \in N_{\mathscr{D}}$. Thus, $N_{\mathscr{D}}$ is inductive.   ■

Lemma 8.2

a. $\{n1_D : n \in P\}$ is inductive. (As before, P denotes the set of positive integers.)

b.   (A is inductive $\wedge n \in P) \Rightarrow n1_D \in A$.

Proof   (a)   Let $B = \{n1_D : n \in P\}$. Clearly, $1_D = 1(1_D) \in B$. Assume now that $x \in B$. Then $x = n1_D$ for some n in P. Then $x + 1_D = n1_D + 1_D = (n + 1)1_D \in B$. We have shown: $x \in B \Rightarrow x + 1_D \in B$. Hence, B is inductive.

(b)   Assume A inductive. Let $C = \{n : n \in P \wedge n1_D \in A\}$. We must show that $C = P$. First, $1 \in C$, since $1(1_D) = 1_D \in A$. Now assume that $n \in C$.

Thus, $n1_D \in A$. Then, $(n + 1)1_D = n1_D + 1_D \in A$, since A is closed under addition of 1_D. Thus, $n + 1 \in C$. We have shown: $n \in C \Rightarrow n + 1 \in C$. By mathematical induction, $C = P$.   ∎

Corollary 8.3 $N_{\mathscr{D}} = \{n1_D : n \in P\}$.

Proof By Lemma 8.2a, $N_{\mathscr{D}} \subseteq \{n1_D : n \in P\}$. By Lemma 8.2b,

$$\{n1_D : n \in P\} \subseteq N_{\mathscr{D}}.   ∎$$

EXAMPLES

1. In the integral domain $\mathscr{Z} = (Z, +_Z, \times_Z)$, $N_{\mathscr{Z}}$ is the set P of positive integers. To see this, observe that $n1_Z = n \times_Z 1_Z = n$ for any positive integer n (see Example 1, p. 128). Hence, by Corollary 8.3, $N_{\mathscr{Z}} = P$.

2. In the integral domain $\mathscr{Z}_2 = (Z_2, +_2, \times_2)$, $N_{\mathscr{Z}_2} = Z_2$. (Exercise for the reader.)

3. In the integral domain $\mathscr{Z}_p = (Z_p, +_p, \times_p)$, where p is a prime, $N_{\mathscr{Z}_p} = Z_p$.

Theorem 8.4 An integral domain $\mathscr{D} = (D, +, \times)$ has characteristic 0 if and only if $N_{\mathscr{D}}$ is infinite.

Proof (i) Assume $\mathscr{D}$ has characteristic 0. Then, $j1_D \neq 0_D$ for all positive integers j. Hence, $n1_D \neq k1_D$ for all distinct positive integers n and k. (For, assume $n < k$. Then $k = n + j$ for some positive integer j. Hence, $k1_D = (n + j)1_D = n1_D + j1_D$ by Theorem 7.1a. If $k1_D = n1_D$, then, by the Cancellation Law for Addition, $j1_D = 0$, contradicting the fact that $j1_D \neq 0$, for all positive integers j.) Therefore, the function G, such that $G(n) = n1_D$ for all positive integers n, is a one-one correspondence between P and $N_{\mathscr{D}}$. Since P is infinite (see Appendix D), $N_{\mathscr{D}}$ must also be infinite.

(ii) Assume $\mathscr{D}$ has characteristic $p \neq 0$. If H is the function such that $H(n) = n1_D$ for any positive integer $n \leq p$, let us show that H is a one-one correspondence between $\{n : n \in P \wedge n \leq p\}$ and $N_{\mathscr{D}}$. Assume, for the sake of contradiction, that H is not one-one. Then there exist n and k such that $n \leq p$, $k \leq p$, $n \neq k$, and $H(n) = H(k)$. Then, $n1_D = k1_D$. We may assume that $n < k$. Then $k = n + j$ for some positive integer j. Notice that $j < k \leq p$. Now, $k1_D = (n + j)1_D = n1_D + j1_D$. By the Cancellation Law for Addition, $j1_D = 0_D$. Since $j < p$, this contradicts the fact that p is the characteristic of $\mathscr{D}$. Hence, H is one-one. To prove that the range of H is $N_{\mathscr{D}}$, take any $n1_D$, where n is a positive integer. By the Division Theorem, $n = (q \times_Z p) +_Z r$, where q and r are integers and $0_Z \leq_Z r <_Z p$.

Case 1 $0_Z <_Z r$. Then

$$n1_D = ((q \times_Z p) +_Z r)1_D = (q \times_Z p)1_D +_Z r1_D$$
$$= q(p1_D) +_Z r1_D = q0_Z +_Z r1_D = r1_D \in \mathscr{R}(H).$$

Case 2 $r = 0_Z$. Then $n = q \times_Z p$. Hence,

$$n1_D = (q \times_Z p)1_D = q(p1_D) = q0_D = 0_D = p1_D \in \mathscr{R}(H).$$

We have proved that H is a one-one correspondence between $\{n : n \in P \wedge n \leq p\}$ and $N_{\mathscr{D}}$. Since $\{n : n \in P \wedge n \leq p\}$ is finite (see Appendix D), $N_{\mathscr{D}}$ must also be finite. ∎

Theorem 8.5

a. $x \in N_{\mathscr{D}} \wedge y \in N_{\mathscr{D}} \Rightarrow x + y \in N_{\mathscr{D}}$.
b. $x \in N_{\mathscr{D}} \wedge y \in N_{\mathscr{D}} \Rightarrow x \times y \in N_{\mathscr{D}}$.

(Thus, the sum and product of natural numbers of $\mathscr{D}$ are again natural numbers of $\mathscr{D}$.)

Proof Assume $x \in N_{\mathscr{D}}$ and $y \in N_{\mathscr{D}}$. Then, by Corollary 8.3, $x = n1_D$ and $y = k1_D$ for some positive integers n and k.

(a) $x + y = n1_D + k1_D = (n + k)1_D$ by Theorem 7.1a. But, $(n + k)1_D \in N_{\mathscr{D}}$ by Corollary 8.3.

(b) $x \times y = n1_D \times k1_D = (n \times k)1_D$ by Theorem 7.1e. But, $(n \times k1_D \in N_{\mathscr{D}}$ by Corollary 8.3. ∎

When $\mathscr{D}$ has characteristic 0 (in particular, when $\mathscr{D}$ is an ordered integral domain), the **natural numbers of** $\mathscr{D}$ possess the usual properties of the natural numbers. In fact, they form a Peano system. This is made explicit in the following theorem.

Theorem 8.6 Let $\mathscr{D} = (D, +, \times)$ be an integral domain having characteristic 0. Then

a. $0_D \notin N_{\mathscr{D}}$.

b. $(N_{\mathscr{G}}, T, 1_D)$ is a Peano system, where T is the function such that $T(x) = x + 1_D$ for all x in $N_{\mathscr{G}}$.

c. The sum and product definable (according to Theorems 2.3.1 and 2.5.1) within the Peano system $(N_{\mathscr{G}}, T, 1_D)$ are the same as the restrictions to $N_{\mathscr{G}}$ of the given operations $+$, $\times$ in $\mathscr{D}$.

Proof (a) By Corollary 8.3, $N_{\mathscr{G}} = \{n1_D : n \in P\}$. Since $\mathscr{D}$ has characteristic 0, $n1_D \neq 0_D$ for all n in P. Hence, $0_D \notin N_{\mathscr{G}}$.

(b) T is an operation in $N_{\mathscr{G}}$ by Theorem 8.5a.

Axiom P1 $1_D \neq T(x)$ for all x in $N_{\mathscr{G}}$. To see this, assume $1_D = T(x)$ $= x + 1_D$ for some x in $N_{\mathscr{G}}$. Hence, $x = 0_D$. But, $0_D \notin N_{\mathscr{G}}$ by (a). This contradicts the fact that $x \in N_{\mathscr{G}}$.

Axiom P2 $x \neq y \Rightarrow T(x) \neq T(y)$. This follows from the Cancellation Law for Addition, Theorem 3.1a.

Axiom P3   Assume $B \subseteq N_{\mathscr{G}} \wedge 1_D \in B \wedge (\forall x)(x \in B \Rightarrow T(x) \in B)$. Thus, B is inductive. Hence, $N_{\mathscr{G}} \subseteq B$ by Theorem 8.1a. Therefore, $N_{\mathscr{G}} = B$.

(c) According to Theorem 2.3.1, there is a *unique* binary operation $\oplus$ in the Peano system $(N_{\mathscr{G}}, T, 1_D)$ such that

$$x \oplus 1_D = T(x),$$

and

$$x \oplus T(y) = T(x \oplus y) \qquad \text{for all } x \text{ and } y \text{ in } N_{\mathscr{G}}.$$

By the definition of T,

$$x \oplus 1_D = x + 1_D,$$

and

$$x \oplus (y + 1_D) = (x \oplus y) + 1_D.$$

However, the restriction of the given addition operation $+$ of $\mathscr{D}$ satisfies these conditions:

$$x + 1_D = x + 1_D,$$

and

$$x + (y + 1_D) = (x + y) + 1_D.$$

Hence, $\oplus$ is identical with the restriction of the given addition $+$ inherited from $\mathscr{D}$.

A similar proof holds for multiplication and is left to the reader.   ■

EXERCISE

1. By Theorem 8.6, $(N_{\mathscr{D}}, T, 1_D)$ is a Peano system. Hence, by Theorem 2.7.1, there is a unique isomorphism between the Peano system $(P, S, 1)$ and $(N_{\mathscr{D}}, T, 1_D)$. Show that, in fact, this isomorphism Φ is given by the formula $\Phi(n) = n1_D$ for all positive integers n.

Since the characteristic of ordered integral domains must be 0 (by Theorem 7.3), the results of Theorem 8.6 apply to any ordered integral domain. But we also can obtain some additional information.

Theorem 8.7 Let $\mathscr{D} = (D, +, \times, <)$ be an ordered integral domain. Then

a. $x \in N_{\mathscr{D}} \Rightarrow 1_D \leq x$.

b. $x \in N_{\mathscr{D}} \land y \in N_{\mathscr{D}} \land x < y \Rightarrow y - x \in N_{\mathscr{D}}$.

c. The order relation $\lessgtr$, definable within the Peano system $(N_{\mathscr{D}}, T, 1_D)$ is the same as the restriction to $N_{\mathscr{D}}$ of the given order relation $<$ in $\mathscr{D}$.

Proof (a) $A = \{x : x \in D \land 1_D \leq x\}$ is clearly inductive. Hence, $N_{\mathscr{D}} \subseteq A$.

(b) Assume $x \in N_{\mathscr{D}} \land y \in N_{\mathscr{D}} \land x < y$. Then, $x = k1_D$ and $y = j1_D$ for some j and k in P. Since $x < y$, $k \neq j$. Therefore, $j < k$ or $k < j$. Assume $j < k$. Then $k = j + n$ for some positive integer n. Hence,

$$x = k1_D = (j + n)1_D = j1_D + n1_D \geq j1_D + 1_D \qquad \text{by Part (a)}$$
$$> j1_D = y.$$

This contradicts $x < y$. Hence, $k < j$, and so, $j - k \in P$. Therefore, by Theorem 7.4h,

$$y - x = j1_D - k1_D = (j - k)1_D \in N_{\mathscr{D}}.$$

(c) Let $\lessgtr$ be the order relation definable within the Peano system $(N_{\mathscr{D}}, T, 1_D)$: for any x, y in $N_{\mathscr{D}}$,

$$x \lessgtr y \Leftrightarrow (\exists z)(z \in N_{\mathscr{D}} \land x \oplus z = y).$$

By Theorem 8.6c, this becomes

$$x \lessgtr y \Leftrightarrow (\exists z)(z \in N_{\mathscr{D}} \land x + z = y).$$

Assume $x \lessgtr y$. Then $x + z = y$ for some z in $N_{\mathscr{D}}$. By (a), $0 < z$, and, therefore, $x < x + z = y$. Conversely, assume $x < y$. By (b), $y - x \in N_{\mathscr{D}}$. Hence, $x + z = y$ for some z in $N_{\mathscr{D}}$, namely, $z = y - x$. Therefore, $x \lessgtr y$. We have shown the equivalence of $x \lessgtr y$ and $x < y$. ∎

EXERCISE

2. Prove: $x \in N_{\mathscr{D}} \Rightarrow \neg(\exists y)(y \in N_{\mathscr{D}} \wedge x < y < x + 1_D)$.

Once the natural numbers of an integral domain have been defined, it is obvious how to define the notion of an integer of the domain.

Definition If $\mathscr{D} = (D, +, \times)$ is an integral domain, let the set $Z_{\mathscr{D}}$ consist of all elements x of D such that

$$x \in N_{\mathscr{D}} \quad \text{or} \quad x = 0_D \quad \text{or} \quad -x \in N_{\mathscr{D}}.$$

Thus,

$$x \in Z_{\mathscr{D}} \Leftrightarrow (x \in N_{\mathscr{D}} \vee x = 0_D \vee -x \in N_{\mathscr{D}}).$$

The elements of $Z_{\mathscr{D}}$ are called **integers of** $\mathscr{D}$. It is obvious from this definition that $N_{\mathscr{D}} \subseteq Z_{\mathscr{D}}$.

Theorem 8.8 $Z_{\mathscr{D}} = \{\alpha 1_D : \alpha \in Z\}$.

Proof (i) Assume $x \in Z_{\mathscr{D}}$. Then $x \in N_{\mathscr{D}}$ or $x = 0_D$ or $-x \in N_{\mathscr{D}}$. Hence, by Corollary 8.3, $x = n 1_D$ or $x = 0_D$ or $-x = n 1_D$ for some positive integer n. If $x = 0_D$, then $x = 0_Z 1_D$. If $-x = n 1_D$, then $x = (-n) 1_D$ by Theorem 7.4g. In all cases, $x = \alpha 1_D$ for some α in Z.

(ii) Assume $\alpha \in Z$. If $\alpha >_Z 0_Z$, then $\alpha 1_D \in N_{\mathscr{D}} \subseteq Z_{\mathscr{D}}$. If $\alpha = 0_Z$, then $\alpha 1_D = 0_D \in Z_{\mathscr{D}}$. If $\alpha <_Z 0_Z$, then $\alpha 1_D = (-\alpha)(-1_D) = -((-\alpha) 1_D)$ by Theorem 7.4c. Since $(-\alpha) 1_D \in N_{\mathscr{D}}$, $\alpha 1_D \in Z_{\mathscr{D}}$. In all cases, $\alpha 1_D \in Z_{\mathscr{D}}$. Hence, $\{\alpha 1_D : \alpha \in Z\} \subseteq Z_{\mathscr{D}}$.

From (i) and (ii), $Z_{\mathscr{D}} = \{\alpha 1_D : \alpha \in Z\}$. ∎

EXAMPLES

3. When $\mathscr{D}$ is $\mathscr{X} = (Z, |_Z, \times_Z)$, then $Z_{\mathscr{X}} = Z$. For, by Theorem 8.8, $Z_{\mathscr{X}} = \{\alpha 1_Z : \alpha \in Z\}$. By Example 5 on page 130, $\alpha x = \alpha \times_Z x$ for all α and x in Z. Hence, since $\alpha \times_Z 1_Z = \alpha$, $Z_{\mathscr{X}} = Z$.

4. When $\mathscr{D}$ is $\mathscr{X}_p = (Z_p, +_p, \times_p)$, then $Z_{\mathscr{X}_p} = Z_p$. (Exercise for the reader.)

Theorem 8.9

a. $x \in Z_{\mathscr{D}} \Leftrightarrow -x \in Z_{\mathscr{D}}$.

b. $x, y \in Z_{\mathscr{D}} \Rightarrow [x + y \in Z_{\mathscr{D}} \wedge x - y \in Z_{\mathscr{D}}]$.

c. $x, y \in Z_{\mathscr{D}} \Rightarrow x \times y \in Z_{\mathscr{D}}$.

Proof (a) Assume $x \in Z_{\mathscr{D}}$.

  Case 1 $x \in N_{\mathscr{D}}$. Then, $-(-x) \in N_{\mathscr{D}}$. Hence, $-x \in Z_{\mathscr{D}}$.

  Case 2 $x = 0_D$. Then, $-x = 0_D \in Z_{\mathscr{D}}$.

  Case 3 $-x \in N_{\mathscr{D}}$. Then $-x \in Z_{\mathscr{D}}$.

In all cases, $-x \in Z_{\mathscr{D}}$. Hence, $x \in Z_{\mathscr{D}} \Rightarrow -x \in Z_{\mathscr{D}}$. Therefore, substituting $-x$ for x, $-x \in Z_{\mathscr{D}} \Rightarrow -(-x) \in Z_{\mathscr{D}}$, that is, $-x \in Z_{\mathscr{D}} \Rightarrow x \in Z_{\mathscr{D}}$. Thus, $x \in Z_{\mathscr{D}} \Leftrightarrow -x \in Z_{\mathscr{D}}$.

  (b) Assume $x, y \in Z_{\mathscr{D}}$. By Theorem 8.8, $x = \alpha 1_D$ and $y = \beta 1_D$ for some α, β in Z. Then

$$x + y = \alpha 1_D + \beta 1_D = (\alpha + \beta) 1_D \qquad \text{by Theorem 7.4a.}$$

But, $(\alpha + \beta) 1_D \in Z_{\mathscr{D}}$, by Theorem 8.8. In addition, $x - y = x + (-y) \in Z_{\mathscr{D}}$, by what we have just proved and (a).

  (c) Assume $x, y \in Z_{\mathscr{D}}$. By Theorem 8.8, $x = \alpha 1_D$ and $y = \beta 1_D$ for some α, β in Z. Then

$$
\begin{aligned}
x \times y = (\alpha 1_D) \times (\beta 1_D) &= (\alpha \times_Z \beta)(1_D \times 1_D) \qquad \text{by Theorem 7.4f} \\
&= (\alpha \times_Z \beta) 1_D \in Z_{\mathscr{D}} \qquad \text{by Theorem 8.8.} \quad \blacksquare
\end{aligned}
$$

  We shall see in the next section that, when $\mathscr{D}$ is an integral domain of characteristic 0, then the integers of $\mathscr{D}$ are essentially the same as the standard system of integers $(Z, +_Z, \times_Z)$.

EXERCISE

3.   Prove that, if an integral domain $\mathscr{D} = (D, +, \times)$ is of characteristic 0, then, for any x in $Z_{\mathscr{D}}$, *exactly one* of the conditions $x \in N_{\mathscr{D}}$, $x = 0_D$, $-x \in N_{\mathscr{D}}$ holds.

4.   Prove that in an ordered integral domain $\mathscr{D} = (D, +, \times, <)$

$$x \in N_{\mathscr{D}} \Leftrightarrow (x \in Z_{\mathscr{D}} \wedge 0 < x)$$

    (*Hint:*   Theorem 8.7a.)

3.9 SUBDOMAINS. ISOMORPHISMS. CHARACTERIZATIONS OF THE INTEGERS

  When dealing with an integral domain $\mathscr{D}$, it is often important to consider parts of $\mathscr{D}$ which also form integral domains.

Definition   Let $\mathscr{E} = (E, \oplus, \otimes)$ and $\mathscr{D} = (D, +, \times)$ be integral domains, with unit elements 1_E and 1_D, respectively. We say that $\mathscr{E}$ is a **subdomain** of $\mathscr{D}$ if and only if

i. $E \subseteq D$,

ii. $1_D \in E$,

iii. $x \oplus y = x + y$ for all x, y in E,

iv. $x \otimes y = x \times y$ for all x, y in E.

We sometimes express (iii) by saying that $\oplus$ is the restriction of $+$ to E, and (iv) by saying that $\otimes$ is the restriction of $\times$ to E.

EXAMPLES

1.   The system of integers is a subdomain of the system of rational numbers (of real numbers, of complex numbers).

2.   The rational numbers form a subdomain of the system of real numbers (of complex numbers).

3.   Every integral domain is a subdomain of itself.

Theorem 9.1   Let $\mathscr{D} = (D, +, \times)$ be an integral domain and let $\mathscr{E} = (E, \oplus, \otimes)$ be a subdomain of $\mathscr{D}$. Then

a. E is closed under $+$ and $\times$;

b. $1_D = 1_E$;

c. $0_D = 0_E$;

d. $x -_E y = x -_D y$ for any x, y in E; hence, E is closed under the subtraction operation of $\mathscr{D}$.

e. $Z_{\mathscr{D}} \subseteq E$ (that is, any subdomain of $\mathscr{D}$ contains all the integers of $\mathscr{D}$).

Proof   (a)   For any x, y in E, $x \oplus y \in E$ and $x \otimes y \in E$, since $\oplus$ and $\otimes$ are binary operations in E. But, $x \oplus y = x + y$ and $x \otimes y = x \times y$ by (iii), (iv) of the definition of subdomain.

(b)   We know that $1_D \in E$. Then

$$1_D = 1_D \otimes 1_E \quad \text{since } 1_E \text{ is the unit element of } \mathscr{E}$$
$$= 1_D \times 1_E \quad \text{by (iv)}$$
$$= 1_E \quad \text{since } 1_D \text{ is the unit element of } \mathscr{D}.$$

(c) $0_E + 0_E = 0_E \oplus 0_E$ by (iii)

$\qquad\qquad = 0_E$ since 0_E is the zero element of $\mathscr{E}$.

Hence, if we subtract within $\mathscr{D}$ the quantity 0_E from both sides, we obtain $0_E = 0_D$.

(d) For any y in E, $y \oplus (-_E y) = 0_E$. Hence, $y + (-_E y) = 0_D$ by (iii) and (c). Therefore, $-_E y = -_D y$. It follows that $x -_E y = x \oplus (-_E y)$ $= x + (-_E y) = x + (-_D y) = x -_D y$ for any x, y in E.

(e) It is clear that E is inductive (by (ii), (iii) of the definition of subdomain). Hence, $N_{\mathscr{G}} \subseteq E$. If $y \in Z_{\mathscr{G}}$, then $y \in N_{\mathscr{G}}$ or $y = 0_D$ or $-y \in N_{\mathscr{G}}$. If $y \in N_{\mathscr{G}}$, then $y \in E$. If $y = 0_D$, then $y \in E$ (by (c)). If $-y \in N_{\mathscr{G}}$, then $y \in E$ (by (d) with $x = 0_D$). Thus, $Z_{\mathscr{G}} \subseteq E$. ■

There is an easy and extremely useful characterization of subdomains.

Theorem 9.2 Let $\mathscr{D} = (D, +, \times)$ be an integral domain. Let $E \subseteq D$ such that

1. $1_D \in E$,
2. E is closed under subtraction (that is, $x, y \in E \Rightarrow x -_D y \in E$),
3. E is closed under multiplication (that is, $x, y \in E \Rightarrow x \times y \in E$).

If we let $x \oplus y = x + y$ and $x \otimes y = x \times y$ for all x, y in E, then $\mathscr{E} = (E, \oplus, \otimes)$ is a subdomain of $\mathscr{D}$. ($\mathscr{E}$ is said to be the subdomain **determined** by E. Notice that, from the definitions and Theorem 9.1, every subdomain is determined by some subset of D.)

Proof Notice first that

$$(*)\quad 0_D \in E,$$

since $0_D = 1_D - 1_D \in E$, by (1), (2). Hence,

$$(**)\quad x \oplus 0_D = x + 0_D = x \qquad \text{for all } x \text{ in } E.$$

Observe also that $y \in E \Rightarrow -y \in E$. For, $-y = 0_D - y$ and we need only use (2). Hence, if $x, y \in E$, then $x + y = x - (-y) \in E$. Thus, E is closed under $+$ and $\times$, and, therefore, $\oplus$ and $\otimes$ are operations in E. We need only check now that the axioms for an integral domain are satisfied.

Axioms 1, 2, 4, 5, and 7 are obviously inherited from $\mathscr{D}$, since $\oplus$ and $\otimes$ are restrictions of $+$ and $\times$ to E. (For example, in the case of Axiom 2: for any x, y in E, $x \oplus y = x + y = y + x = y \oplus x$.) Axioms 3a,b follow from (∗) and (∗∗) above. (For, $x \oplus 0_D = x + 0_D = x$ and $x \oplus (-x)$ $= x + (-x) = 0_D$. Thus, $0_E = 0_D$.) Axiom 6 follows from assumptions (1) and (3), since $x \otimes 1_D = x \times 1_D = x$. Finally, Axiom 8 is inherited from $\mathscr{D}$, using the fact that $0_E = 0_D$.   ■

EXAMPLES

4.   Let $\mathscr{D}$ be the integral domain of rational numbers.

   a.   Let E be the set Z of integers. Then Z satisfies conditions (1)–(3) of Theorem 9.2, and, therefore, Z determines a subdomain of $\mathscr{D}$.

   b.   Let E be the set of all rational numbers $a/2^n$, where $a \in Z$ and n is a nonnegative integer. E determines a subdomain of $\mathscr{D}$.

5.   Let $\mathscr{D}$ be the integral domain of real numbers.

   a.   The set Z of integers determines a subdomain of $\mathscr{D}$.

   b.   The set Q of rational numbers determines a subdomain of $\mathscr{D}$.

   c.   The set of all numbers $a + b\sqrt{2}$, where a and b are integers, determines a subdomain of $\mathscr{D}$.

   d.   The set of all numbers $a + b\sqrt{2}$, where a and b are rational numbers, determines a subdomain of $\mathscr{D}$.

EXERCISES

1.   Let $\mathscr{D}$ be the integral domain $(Z_p, +_p, \times_p)$ of integers modulo p, where p is a prime. Find all the subdomains of $\mathscr{D}$.

2.   Let $\mathscr{D}$ be the system of integers $(Z, +_Z, \times_Z)$. Find all the subdomains of $\mathscr{D}$.

3.   Let $\mathscr{D}$ be the integral domain of rational numbers. Which of the following subsets E determine subdomains of $\mathscr{D}$?

   a. E is the set of all nonnegative rational numbers.

   b. E is the set of all rational numbers $a/2$, where $a \in Z$.

   c. E is the set of all even integers.

4.   Let $\mathscr{D}$ be the integral domain of real numbers. Which of the following subsets E determine subdomains of $\mathscr{D}$?

   a. E is the set of all $a + b\sqrt[3]{2}$, where a and b are rational numbers.

   b. E is the set of all $a + b\sqrt{5}$, where a and b are rational numbers.

   c.   Same as (b), except that a and b must be integers.

   d. E is the set of all $a + b\sqrt{6}$, where a and b are rational numbers.

5. Let $\mathscr{D}$ be the integral domain of complex numbers. Which of the following subsets determine subdomains of $\mathscr{D}$?

   a. E is the set of all $a + bi$, where a and b are integers.

   b. E is the set of all $a + bi$, where a is any real number and b is any integer.

   c. E is the set of all bi, where b is real.

6. In clause (2) of Theorem 9.2, show by an example that subtraction cannot be replaced by addition without destroying the validity of the result.

7. Show that, if $\mathscr{D}_1$ is a subdomain of $\mathscr{D}_2$, and $\mathscr{D}_2$ is a subdomain of $\mathscr{D}_3$, then $\mathscr{D}_1$ is a subdomain of $\mathscr{D}_3$.

8.^{C} Assume that $\mathscr{E}$ is a subdomain of $\mathscr{D}$.

   a. Prove that $N_{\mathscr{E}} = N_{\mathscr{D}}$ and $Z_{\mathscr{E}} = Z_{\mathscr{D}}$.

   b. Prove that $\mathscr{E}$ and $\mathscr{D}$ have the same characteristic.

When $\mathscr{D} = (D, +, \times, <)$ is an ordered integral domain, then any sub-domain $\mathscr{E} = (E, \oplus, \otimes)$ automatically becomes an ordered integral domain if we define the order relation $\ominus$ to be the restriction of $<$ to E. Thus, for any x, y in E,

$$x \ominus y \Leftrightarrow x < y.$$

Axioms O1–O5 automatically hold for $(E, \oplus, \otimes, \ominus)$, since they are in-herited from $\mathscr{D}$. For example, in the case of Axiom O1, $x \oplus x$. This follows from the fact that $x \not< x$ (since $\mathscr{D}$ satisfies Axiom O1) and the fact that $x \ominus x \Leftrightarrow x < x$ (by the definition of $\ominus$).

Definition Let $\mathscr{D} = (D, +, \times)$ and $\mathscr{D}^* = (D^*, +^*, \times^*)$ be integral domains. We say that $\mathscr{D}$ is **isomorphic** with $\mathscr{D}^*$ if and only if there is a function $\Phi \colon D \xrightarrow[\text{onto}]{1\text{-}1} D^*$ such that, for any x, y in D,

i. $\Phi(x + y) = \Phi(x) +^* \Phi(y)$,

ii. $\Phi(x \times y) = \Phi(x) \times^* \Phi(y)$.

Such a function Φ is called an **isomorphism** of $\mathscr{D}$ with $\mathscr{D}^*$.

Thus, $\mathscr{D}$ and $\mathscr{D}^*$ are isomorphic if and only if there is a one-one corre-spondence between D and D^* under which the addition and multiplication operations "correspond." Any property of integral domains which holds for $\mathscr{D}$ would also hold for $\mathscr{D}^*$ and vice versa.

Theorem 9.3 If Φ is an isomorphism of $\mathscr{D} = (D, +, \times)$ with $\mathscr{D}^* = (D^*, +^*, \times^*)$, then

a. $\Phi(0_D) = 0_{D*}$.

b. $\Phi(1_D) = 1_{D*}$.

c. $\Phi(-x) = -\Phi(x)$. (The first $-$ indicates the additive inverse operation of $\mathscr{D}$, and the second that of $\mathscr{D}*$.)

d. $\Phi(x - y) = \Phi(x) - \Phi(y)$. (The first $-$ indicates the subtraction operation of $\mathscr{D}$, and the second that of $\mathscr{D}*$.)

e. $\Phi(nx) = n\Phi(x)$ for all positive integers n and all x in D.

Proof (a) $\Phi(0_D) = \Phi(0_D + 0_D) = \Phi(0_D) +^* \Phi(0_D)$. Hence, $0_{D*} = \Phi(0_D)$.

(b) $\Phi(1_D) = \Phi(1_D \times 1_D) = \Phi(1_D) \times^* \Phi(1_D)$.

Case 1 $\Phi(1_D) \neq 0_{D*}$. Then, by the Cancellation Law for Multiplication, $1_{D*} = \Phi(1_D)$.

Case 2 $\Phi(1_D) = 0_{D*}$. But, $\Phi(0_D) = 0_{D*}$ by (a). Since Φ is one-one, $0_D = 1_D$. Hence, by Theorem 3.4, $D = \{0_D\}$. Since $\mathscr{R}(\Phi) = D^*$, $D^* = \{0_{D*}\}$. Hence, $0_{D*} = 1_{D*}$. Thus, $\Phi(1_D) = 0_{D*} = 1_{D*}$.

(c) $\Phi(x) +^* \Phi(-x) = \Phi(x + (-x)) = \Phi(0_D) = 0_{D*}$. Hence, $\Phi(-x) = -\Phi(x)$.

(d) $\Phi(x - y) = \Phi(x + (-y)) = \Phi(x) +^* \Phi(-y)$
$$= \Phi(x) +^* (-\Phi(y)) = \Phi(x) - \Phi(y).$$

(e) Assume $x \in D$. Let $A = \{n : n \in P \wedge \Phi(nx) = n\Phi(x)\}$. Clearly, $1 \in A$, since $\Phi(1x) = \Phi(x) = 1\Phi(x)$. Now, assume $n \in A$. Thus, $\Phi(nx) = n\Phi(x)$. Hence, $\Phi((n + 1)x) = \Phi(nx + x) = \Phi(nx) +^* \Phi(x) = n\Phi(x) +^* \Phi(x) = (n + 1)\Phi(x)$. Thus, $n + 1 \in A$. We have shown: $n \in A \Rightarrow n + 1 \in A$. By mathematical induction, $A = P$. ∎

EXAMPLES

6. If $\mathscr{D} = (D, +, \times)$ is any integral domain, then the identity function Φ, such that $\Phi(x) = x$ for all x in D, is an isomorphism of $\mathscr{D}$ with itself.

7. Let $D^* = \{\text{even, odd}\}$, with

$$\text{even} +^* \text{even} = \text{odd} +^* \text{odd} = \text{even},$$
$$\text{even} +^* \text{odd} = \text{odd} +^* \text{even} = \text{odd},$$
$$\text{even} \times^* \text{odd} = \text{odd} \times^* \text{even} = \text{even} \times^* \text{even} = \text{even},$$
$$\text{odd} \times^* \text{odd} = \text{odd}.$$

Then $(Z_2, +_2, \times_2)$ is isomorphic with $(D^*, +^*, \times^*)$. The required isomorphism Φ is: $\Phi([0]) =$ even, $\Phi([1]) =$ odd. Verification that Φ is an isomorphism is left to the reader.

8.   Let E be the set of all functions $f: Z \to Z$ such that $f(x +_z y) = f(x) +_z f(y)$ for all x, y in Z. For any functions f, g in E, let $(f \oplus g)(x) = f(x) + g(x)$ for all x in Z, and $(f \otimes g)(x) = f(g(x))$ for all x in Z. First, note that, if $f \in E$, there is a unique integer c such that $f(x) = c \times_z x$ for all x in Z. (To see this, let $c = f(1)$. Prove $f(x) = c \times_z x$ for all positive integers x, by mathematical induction, and then extend the result to negative integers.) Next, check that $(E, \oplus, \otimes)$ is an integral domain, and define a function Φ as follows: for any z in Z, let $\Phi(z)$ be the function f such that $f(x) = z \times_z x$ for all x in Z. Then Φ is an isomorphism of $(Z, +_z, \times_z)$ with $(E, \oplus, \otimes)$.

9.   Let $\mathscr{D}$ be the integral domain consisting of the set of all real numbers of the form $a + b \sqrt{2}$, where a and b are rational numbers, under ordinary addition and multiplication of real numbers. Let $\Phi(a + b \sqrt{2}) = a - b \sqrt{2}$ for all rational numbers a and b. One can check that Φ is an isomorphism of $\mathscr{D}$ with itself. In addition to Φ, there is the identity isomorphism of $\mathscr{D}$ with itself (see Example 6). Thus, there can be more than one isomorphism between an integral domain and itself.

Theorem 9.4   Let $\mathscr{D} = (D, +, \times)$ be any integral domain. Let $\oplus$ and $\otimes$ be the restrictions of $+$ and $\times$ to $Z_{\mathscr{D}}$. Then

a. $\mathscr{E} = (Z_{\mathscr{D}}, \oplus, \otimes)$ is a subdomain of $\mathscr{D}$.

b.   If $\mathscr{D}$ is of characteristic 0, then $(Z, +_z, \times_z)$ is isomorphic with $(Z_{\mathscr{D}}, \oplus, \otimes)$. (Thus, the integers of an integral domain of characteristic zero are essentially the same as the standard integers.)

Proof   (a) $\oplus$ and $\otimes$ are operations in $Z_{\mathscr{D}}$, by Theorem 8.9b,c. Moreover,

1. $1_D \in Z_{\mathscr{D}}$,
2. $Z_{\mathscr{D}}$ is closed under subtraction (Theorem 8.9b),
3. $Z_{\mathscr{D}}$ is closed under multiplication (Theorem 8.9c).

Hence, by Theorem 9.2, $(Z_{\mathscr{D}}, \oplus, \otimes)$ is a subdomain of $\mathscr{D}$.

   (b)   Let $\Phi(\alpha) = \alpha 1_D$ for all α in Z. Thus, by Theorem 8.8, $\Phi: Z \xrightarrow[\text{onto}]{} Z$. By Theorem 7.4a,

$$\Phi(\alpha +_z \beta) = (\alpha +_z \beta)1_D = \alpha 1_D \oplus \beta 1_D = \Phi(\alpha) \oplus \Phi(\beta),$$

and, by Theorem 7.4f,

$$\Phi(\alpha \times_Z \beta) = (\alpha \times_Z \beta) 1_D = \alpha 1_D \otimes \beta 1_D = \Phi(\alpha) \otimes \Phi(\beta).$$

It remains to show that Φ is one-one. Assume not. Then there exist $\alpha, \beta \in Z$ such that $\alpha \neq \beta$ and $\Phi(\alpha) = \Phi(\beta)$. Hence, $\alpha <_Z \beta$ or $\beta <_Z \alpha$. Say, $\alpha <_Z \beta$. Therefore, $\beta - \alpha$ is a positive integer. Then,

$$\Phi(\alpha) \oplus \Phi(\beta - \alpha) = \Phi(\alpha +_Z (\beta - \alpha)) = \Phi(\beta).$$

Since $\Phi(\alpha) = \Phi(\beta)$, $\Phi(\beta - \alpha) = 0_D$, that is, there is a positive integer n, namely, $n = \beta - \alpha$, such that $n 1_D = 0_D$. But this contradicts the fact that $\mathscr{D}$ is of characteristic 0. Hence, Φ is one-one, and Φ is an isomorphism between $(Z, +_Z, \times_Z)$ and $(Z_{\mathscr{D}}, \oplus, \otimes)$. ■ (Observe that, by Exercise 1 on p. 133, the operation αx defined on p. 130 is the same as multiplication within $\mathscr{D}$ by the corresponding integer of $\mathscr{D}$: $\alpha x = (\alpha 1_D) \times x$.)

EXERCISES

9. Prove that, for any integral domain $\mathscr{D}$ of characteristic 0, there is a unique isomorphism between $(Z, +_Z, \times_Z)$ and $(Z_{\mathscr{D}}, \oplus, \otimes)$. (*Hint*: Show that any isomorphism must be identical with that defined in the proof of Theorem 9.4b.)

10. Show that the relation *isomorphic with* is transitive, that is if $\mathscr{D}_1$ is isomorphic with $\mathscr{D}_2$, and $\mathscr{D}_2$ is isomorphic with $\mathscr{D}_3$, then $\mathscr{D}_1$ is isomorphic with $\mathscr{D}_3$.

11. Show that the relation *isomorphic with* is symmetric, that is, if $\mathscr{D}$ is isomorphic with $\mathscr{E}$, then $\mathscr{E}$ is isomorphic with $\mathscr{D}$.

12. Show that, if $\mathscr{D}$ and $\mathscr{E}$ are integral domains of characteristic 0, then the system of integers of $\mathscr{D}$ is isomorphic with the system of integers of $\mathscr{E}$. (*Hint*: Use Theorem 9.4 and Exercises 10, 11 above.)

13. Prove that, if $(Z, +_Z, \times_Z)$ is isomorphic with a subdomain of an integral domain $\mathscr{D}$, then $\mathscr{D}$ has characteristic 0.

14. Prove that, if $\mathscr{D}$ is isomorphic with $\mathscr{E}$, then $\mathscr{D}$ and $\mathscr{E}$ have the same characteristic.

15. Using the notation of Theorem 9.4, prove that, if $\mathscr{D}$ is an integral domain of characteristic p where p is a prime, then $(Z_p, +_p, \times_p)$ is isomorphic with $(Z_{\mathscr{D}}, \oplus, \otimes)$.

The notion of isomorphism can be extended to ordered integral domains.

Definition   Let $\mathscr{D} = (D, +, \times, <)$ and $\mathscr{D}^* = (D^*, +^*, \times^*, <^*)$ be ordered integral domains. We say that $\mathscr{D}$ is isomorphic with $\mathscr{D}^*$ if and only if there is a function $\Phi\colon D \xrightarrow[\text{onto}]{\text{1-1}} D^*$ such that, for any x, y in D,

i. $\Phi(x + y) = \Phi(x) +^* \Phi(y)$,

ii. $\Phi(x \times y) = \Phi(x) \times^* \Phi(y)$,

iii. $x < y \Leftrightarrow \Phi(x) <^* \Phi(y)$.

Thus, Φ is an isomorphism between the integral domains $(D, +, \times)$ and $(D^*, +^*, \times^*)$ which "preserves" the order, that is, satisfies (iii). Such a function Φ is called an **isomorphism** between $\mathscr{D}$ and $\mathscr{D}^*$.

By Theorem 9.3a, an isomorphism Φ takes 0_D into 0_{D^*}. Hence, taking $x = 0_D$ in (iii), we see that an isomorphism transforms positive elements into positive elements. Taking $y = 0_D$ in (iii), we see that an isomorphism transforms negative elements into negative elements. We also have the following extension of Theorem 9.4b.

Theorem 9.5 Let $\mathscr{D}$ be an integral domain $(D, +, \times)$ of characteristic 0. Let $\oplus$ and $\otimes$ be the restrictions of $+$ and $\times$ to $Z_{\mathscr{D}}$. Define a binary relation $\ominus$ in $Z_{\mathscr{D}}$ as follows: for any x, y in $Z_{\mathscr{D}}$,

$$x \ominus y \Leftrightarrow y - x \in N_{\mathscr{D}}.$$

Then

a. $(Z_{\mathscr{D}}, \oplus, \otimes, \ominus)$ is an ordered integral domain.

b. $(Z, +_Z, \times_Z, <_Z)$ is isomorphic with $(Z_{\mathscr{D}}, \oplus, \otimes, \ominus)$.

Proof (a) By Theorem 9.4a, we know that $(Z_{\mathscr{D}}, \oplus, \otimes)$ is an integral domain. Therefore, we need only prove the order axioms O1–O5.

(O1) $x \ominus\!\!\!\!/\ x$ for all x in $Z_{\mathscr{D}}$. (For, $x - x = 0_D \notin N_{\mathscr{D}}$, by Theorem 8.6a.)

(O2) Assume $x \ominus y$ and $y \ominus z$ for x, y, z in $Z_{\mathscr{D}}$. Then, $y - x \in N_{\mathscr{D}}$ and $z - y \in N_{\mathscr{D}}$. Therefore, by Theorem 8.5a,

$$z - x = (y - x) + (z - y) \in N_{\mathscr{D}}.$$

Hence, $x \ominus z$.

(O3) Assume x, y in $Z_{\mathscr{D}}$. By Theorem 8.9b, $y - x \in Z_{\mathscr{D}}$. Hence, by definition of $Z_{\mathscr{D}}$, either

$$y - x \in N_{\mathscr{D}} \quad \text{or} \quad y - x = 0_D \quad \text{or} \quad -(y - x) \in N_{\mathscr{D}}.$$

Hence, $x \ominus y$ or $x = y$ or $y \ominus x$.

(O4) Assume x, y, z in $Z_{\mathscr{D}}$ and $x \ominus y$. Thus, $y - x \in N_{\mathscr{D}}$. Hence, $(z + y) - (z + x) = y - x \in N_{\mathscr{D}}$, that is, $z + x \ominus z + y$.

(O5)   Assume x, y, z in $Z_{\mathscr{D}}$, with $x \otimes y$ and $0_{\mathscr{D}} \otimes z$. Thus, $y - x \in N_{\mathscr{D}}$ and $z = z - 0_D \in N_{\mathscr{D}}$. By Theorem 8.5b, $(y - x) \otimes z \in N_{\mathscr{D}}$. But,

$$(y - x) \otimes z = (y \otimes z) - (x \otimes z).$$

Hence, $x \otimes z \otimes y \otimes z$.

(b)   In the proof of Theorem 9.4b, we showed that the function $\Phi(\alpha) = \alpha 1_D$ is an isomorphism of $(Z, +_Z, \times_Z)$ with $(Z_{\mathscr{D}}, \oplus, \otimes)$. Therefore, it suffices to prove

$$\alpha <_Z \beta \Leftrightarrow \Phi(\alpha) \otimes \Phi(\beta) \qquad \text{for all} \quad \alpha, \beta \text{ in } Z,$$

that is,

$$\alpha <_Z \beta \Leftrightarrow \beta 1_D - \alpha 1_D \in N_{\mathscr{D}},$$

or, equivalently

$$0_Z <_Z \beta - \alpha \Leftrightarrow (\beta - \alpha)1_D \in N_{\mathscr{D}}, \qquad \text{by Theorem 7.4b.}$$

Thus, we must show: for any γ in Z,

$$0_Z <_Z \gamma \Leftrightarrow \gamma 1_D \in N_{\mathscr{D}}.$$

The implication $0_Z <_Z \gamma \Rightarrow \gamma 1_D \in N_{\mathscr{D}}$ follows from Corollary 8.3. For the converse, assume $0_Z \not<_Z \gamma$. Then, $\gamma \leq_Z 0_Z$.

Case 1 $\gamma = 0_Z$.   Then $\gamma 1_D = 0_D \notin N_{\mathscr{D}}$, by Theorem 8.6a.

Case 2 $\gamma <_Z 0_Z$.   Then, $0_Z <_Z (-\gamma)$. Hence, $(-\gamma)1_D \in N_{\mathscr{D}}$. By Theorem 7.4g, $-(\gamma 1_D) \in N_{\mathscr{D}}$. Hence, $\gamma 1_D \notin N_{\mathscr{D}}$. (Otherwise, by Theorem 8.5a, $0_D = (\gamma 1_D) \oplus (-(\gamma 1_D)) \in N_{\mathscr{D}}$, contradicting Theorem 8.6a.) Thus, we have shown: $0_Z \not<_Z \gamma \Rightarrow \gamma 1_D \notin N_{\mathscr{D}}$.   ∎

Remark   If $\mathscr{D}$ is an ordered integral domain $(D, +, \times, <)$, then, by Exercise 8.4, p. 141, the relation $\otimes$ defined in Theorem 9.5 is just the restriction of $<$ to $Z_{\mathscr{D}}$.

EXERCISE

16.   Let $\mathscr{D} = (D, +, \times, <)$ and $\mathscr{D}^* = (D^*, +^*, \times^*, <^*)$ be ordered integral domains. Let Ψ be an isomorphism between $(D, +, \times)$ and $(D^*, +^*, \times^*)$, that is, $\Psi: D \xrightarrow[\text{onto}]{1-1} D^*$ and Ψ preserves addition and multiplication. If Ψ takes positive elements of D into positive elements of D^*, prove that Ψ is order preserving and therefore, that Ψ is an isomorphism of $\mathscr{D}$ with $\mathscr{D}^*$.

Now let us turn to the problem of characterizing the system of integers (up to isomorphism[†]).

Theorem 9.5 Let $\mathcal{D} = (D, +, \times)$ be an integral domain. Then $\mathcal{D}$ is isomorphic with the system of integers $\mathcal{X} = (Z, +_Z, \times_Z)$ if and only if

a. $\mathcal{D}$ is of characteristic 0, and

b. $\mathcal{D}$ has no proper subdomains, that is, no subdomains different from $\mathcal{D}$ itself.

Proof (I) Assume $\mathcal{D}$ isomorphic with the system $\mathcal{X}$. Let Φ be an isomorphism between $\mathcal{D}$ and $\mathcal{X}$. Then, $\mathcal{D}$ is of characteristic 0. (For, by Theorem 9.3, if n is a positive integer, $\Phi(n1_D) = n\Phi(1_D) = n1_Z \neq 0_Z$. Hence, since $\Phi(0_D) = 0_Z$, it follows that $n1_D \neq 0_D$ for all positive integers n.) In addition, $\mathcal{D}$ cannot have a proper subdomain. For, if $\mathcal{E} = (E, \oplus, \otimes)$ were a proper subdomain of $\mathcal{D}$, then, since Φ is one-one, $\Phi[E] = \{\Phi(x) : x \in E\}$ would determine a proper subdomain of $\mathcal{X}$. (It is easy to check that conditions (1)–(3) of Theorem 9.2 hold for $\Phi[E]$.) But $\mathcal{X}$ cannot have a proper subdomain. For, by Theorem 9.1e, $Z_{\mathcal{X}}$ is included in any subdomain, and, by Example 3 on p. 140, $Z_{\mathcal{X}} = Z$. Hence, $\mathcal{D}$ cannot have a proper subdomain.

(II) Assume $\mathcal{D}$ satisfies (a) and (b). Then, by Theorem 9.4a, $\mathcal{E} = (Z_{\mathcal{D}}, \oplus, \otimes)$ is a subdomain of $\mathcal{D}$, which is isomorphic with $\mathcal{X}$. But, by (b), $\mathcal{E} = \mathcal{D}$. Hence, $\mathcal{D}$ is isomorphic with $\mathcal{X}$. ∎

We also can characterize the system of integers (up to isomorphism) in terms of ordered integral domains.

Theorem 9.6 Let $\mathcal{D} = (D, +, \times, <)$ be an ordered integral domain. Let D^+ denote the set of positive elements of $\mathcal{D}$. Then the following properties are equivalent.

a. $N_{\mathcal{D}} = D^+$.

b. $(D^+, S^+, 1_D)$ is a Peano system (where $S^+(x) = x + 1_D$ for all x in D^+).

c. D^+ satisfies the Least Number Principle, that is, every nonempty subset of D^+ contains a least element (with respect to the order relation $<$).

[†] To characterize the integers *up to isomorphism* we must find a simple property which is possessed by those and only those systems which are isomorphic to the system of integers.

Proof   (a) $\Rightarrow$ (c)   Assume $N_{\mathscr{D}} = D^+$. By Theorem 8.6b, $(N_{\mathscr{D}}, T, 1_D)$ is a Peano system, where $T(x) = x + 1_D$ for all x in $N_{\mathscr{D}}$. Hence, by Theorem 2.4.5, it satisfies the Least Number Principle. By Theorem 8.7c, the order relation in $(N_{\mathscr{D}}, T, 1_D)$ is the same as the order relation $<$ inherited from $\mathscr{D}$. Hence, the Least Number Principle holds for D^+.

(c) $\Rightarrow$ (b)   Assume every nonempty subset of D^+ contains a least element. By Theorem 8.7a, $N_{\mathscr{D}} \subseteq D^+$. Assume $N_{\mathscr{D}} \neq D^+$. Hence, $D^+ - N_{\mathscr{D}} \neq \varnothing$. Let z be the least element of $D^+ - N_{\mathscr{D}}$. Since $z \notin N_{\mathscr{D}}$, it follows that $z - 1_D \notin N_{\mathscr{D}}$. But, $z - 1_D < z$. Hence, by the minimality of z, $z - 1_D \notin D^+$. This means that $z - 1_D \leq 0_D$. Hence, $z \leq 1_D$. Since $z \notin N_{\mathscr{D}}$, $z \neq 1_D$. Hence, $0_D < z < 1_D$. Thus, $0_D < z^2$, and, multiplying both sides of $z < 1_D$ by the positive element z, we obtain $z^2 < z$. Therefore, $z^2 \in D^+ - N_{\mathscr{D}}$. (We know that $z^2 \notin N_{\mathscr{D}}$, because $z^2 < 1_D$.) This contradicts the assumption that z is the least element of $D^+ - N_{\mathscr{D}}$. Hence, $N_{\mathscr{D}} = D^+$. Therefore, $(D^+, S^+, 1_D)$ is $(N_{\mathscr{D}}, T, 1_D)$, and, thus, $(D^+, S^+, 1_D)$ is a Peano system by Theorem 8.6b.

(b) $\Rightarrow$ (a)   Assume $(D^+, S^+, 1_D)$ is a Peano system. But, $1_D \in N_{\mathscr{D}}$ and $(\forall x)(x \in N_{\mathscr{D}} \Rightarrow S^+(x) \in N_{\mathscr{D}})$. Hence, by mathematical induction for $(D^+, S^+, 1_D)$, $N_{\mathscr{D}} = D^+$.   ∎

Definition   By a **well-ordered integral domain** we mean an ordered integral domain $\mathscr{D} = (D, +, \times, <)$ satisfying any of the three equivalent conditions of Theorem 9.6. In particular, $\mathscr{D}$ is a well-ordered integral domain if and only if every nonempty subset of positive elements contains a least element.

Corollary 9.7   The system $(Z, +_Z, \times_Z, <_Z)$ is a well-ordered integral domain.

Proof   By Theorem 9.6 and Example 1, p. 136.   ∎

Theorem 9.8   Any well-ordered integral domain $\mathscr{D} = (D, +, \times, <)$ is isomorphic with $(Z, +_Z, \times_Z, <_Z)$. Hence, any two well-ordered integral domain are isomorphic.

Proof   By Theorem 9.7, $N_{\mathscr{D}} = D^+$. Hence, $Z_{\mathscr{D}} = D$. (For, $Z_{\mathscr{D}}$ consists of $N_{\mathscr{D}}$, 0_D, and all negatives of elements of $N_{\mathscr{D}}$, while D consists of D^+, 0_D, and all negatives of elements of D^+.) Then, by Theorem 9.5b, $\mathscr{D}$ is isomorphic with $(Z, +_Z, \times_Z, <_Z)$.   ∎

EXERCISES

17. An isomorphism between an ordered integral domain and itself is called an **automorphism**. Prove that the only automorphism of $(Z, +_z, \times_z, <_z)$ is the identity mapping.

18. Prove that there is a unique isomorphism between any two well-ordered integral domains.

SUPPLEMENTARY EXERCISES

1. *An alternative construction of the integers resembling the traditional development.* We start with a given Peano system $(P, S, 1)$. As before, the elements of P are called **natural numbers.** Take a set P^- which is disjoint from P and for which there is a one-one correspondence ψ between P and P^-. For each x in P, we let x^- denote $\psi(x)$. Let 0 be some object not in $P \cup P^-$. (The existence of P^- and 0 follows easily from the axioms of set theory.)

    Let $J = P \cup P^- \cup \{0\}$. J is called the set of **integers**. This is a more natural and, perhaps, historically accurate definition of the integers.

    Definition
    $$-x = \begin{cases} x^- & \text{if } x \in P, \\ 0 & \text{if } x = 0, \\ u & \text{if } x = u^-. \end{cases}$$

    Thus, $-x$ is defined for each integer x, and its value is an integer.

    I. Prove: (i) $-(-x) = x$.   (ii) $x = -y \Leftrightarrow -x = y$.
    (iii) $-x = 0 \Leftrightarrow x = 0$.   (iv) $x \in P \Leftrightarrow -x \in P^-$.

    Definition
    $$|x| = \begin{cases} x & \text{if } x \in P \text{ or } x = 0, \\ -x & \text{if } x \in P^-. \end{cases}$$

    II. Prove: (i) $|-x| = |x|$.   (ii) $|x| = 0 \Leftrightarrow x = 0$.

    Definition $x <_J y$ for $(x \in P \wedge y \in P \wedge x < y)$, or
    $(y \in P \wedge (x = 0 \text{ or } x \in P^-))$, or
    $(x \in P^- \wedge y = 0)$, or
    $(x \in P^- \wedge y \in P^- \wedge -y < -x)$.
    $x \leq_J y$ for $(x <_J y \text{ or } x = y)$.

    III. Prove:

    (i) $x \not<_J x$.
    (ii) $x <_J y \wedge y <_J z \Rightarrow x <_J z$.
    (iii) Exactly one of $x <_J y$, $x = y$, $y <_J x$ holds.
    (iv) $x \in P \Leftrightarrow 0 <_J x$.
    (v) $x \in P^- \Leftrightarrow x <_J 0$.
    (vi) $x <_J 0 \Leftrightarrow 0 <_J -x$.

(vii) $0 <_J x \Leftrightarrow -x <_J 0$.

(viii) $x \leq_J x$.

(ix) $x \leq_J y \wedge y <_J z \Rightarrow x <_J z$.

(x) $x <_J y \wedge y \leq_J z \Rightarrow x <_J z$.

(xi) $x \leq_J y \wedge y \leq_J z \Rightarrow x \leq_J z$.

(xii) $x \leq_J y$ or $y \leq_J x$.

Notation If $x, y \in P$ and $x < y$, there is a unique v in P such that $x + v = y$. Denote this v by $y - x$.

Definition

$$
x +_J y = \begin{cases}
x + y & \text{if } x \in P, y \in P, \\
x & \text{if } y = 0, \\
y & \text{if } x = 0, \\
-(|x| + |y|) & \text{if } x \in P^-, y \in P^-, \\
\left.\begin{array}{l} |y| - |x| \\ 0 \\ -(|x| - |y|) \end{array}\right\} & \text{if } x <_J 0 <_J y \text{ and } \left\{\begin{array}{l} |x| < |y| \\ |x| = |y| \\ |x| > |y| \end{array}\right. \\
y +_J x & \text{if } y <_J 0 <_J x.
\end{cases}
$$

This complicated definition corresponds to the intuitive notion of addition of integers. The analysis into cases seems unavoidable.

IV. Prove:

   (i) $x +_J 0 = 0 +_J x = x$.

   (ii) $x +_J y = y +_J x$.

   (iii) $x +_J (-x) = 0$.

   (iv) $x +_J y = 0 \Leftrightarrow x = -y$.

   (v) $-(x +_J y) = (-x) +_J (-y)$. (*Hint*: Divide into cases as follows.

Case 1: $0 <_J y$. *Case 2*: $x = 0$. *Case 3*: $y = 0 \wedge 0 <_J x$.
Case 4: $0 <_J x \wedge y <_J 0$. *Case 4A*: $x > |y|$. *Case 4B*: $x = |y|$.
Case 4C: $x < |y|$. *Case 5*: $x <_J 0$.)

Definition $x -_J y = x +_J (-y)$.

(*Note*: When $x, y \in P$ and $y < x$, then

$$x -_J y = x +_J (-y) = x - |y| = x - y.)$$

   (vi) $x -_J x = 0$.

   (vii) $-(x -_J y) = y -_J x$.

   (viii) $x <_J y \Leftrightarrow x -_J y <_J 0$.

   (ix) $x = y \Leftrightarrow x -_J y = 0$.

   (x) $x >_J y \Leftrightarrow x -_J y >_J 0$. (Of course, $x >_J y$ stands for $y <_J x$.)

   (xi) $x <_J y \Leftrightarrow -y <_J -x$.

   (xii) $x = y \Leftrightarrow -y = -x$.

(xiii) $(\forall x)(x \in J \Rightarrow (\exists u)(\exists v)(u \in P \wedge v \in P \wedge x = u -_J v))$.

(xiv) If $x = a -_J b$ and $y = c -_J d$, where $a, b, c, d \in P$, then $x +_J y = (a + c)$
$-_J (b + d)$. (*Hint:* Analysis into cases.)

(xv) $(x +_J y) +_J z = x +_J (y +_J z)$.

(xvi) $x +_J y = z \Leftrightarrow y = z -_J x$.

Definition

$$x \times_J y = \begin{cases} x \times y & \text{if } x, y \in P, \\ 0 & \text{if } x = 0 \text{ or } y = 0, \\ |x| \times |y| & \text{if } x \in P^- \text{ and } y \in P^-, \\ -(|x| \times |y|) & \text{if one of } x \text{ and } y \text{ is in } P \text{ and the other is in } P^-. \end{cases}$$

V. Prove:

(i) $x \times_J y = 0 \Leftrightarrow x = 0$ or $y = 0$.

(ii) $x \times_J y = y \times_J x$.

(iii) $|x \times_J y| = |x| \times_J |y|$.

(iv) $x \times_J 1 = x$.

(v) $(x \neq 0 \wedge y \neq 0) \Rightarrow x \times_J y = \begin{cases} |x| \times |y| & \text{when } x \text{ and } y \text{ are both in } P \\ & \text{or both in } P^-, \\ -(|x| \times |y|) & \text{when one of } x \text{ and } y \text{ is in } P \\ & \text{and the other in } P^-. \end{cases}$

(vi) $(-x) \times_J y = x \times_J (-y) = -(x \times_J y)$.

(vii) $(-x) \times_J (-y) = x \times_J y$.

(viii) $(x \times_J y) \times_J z = x \times_J (y \times_J z)$.

(ix) $u, v, w \in P \Rightarrow u \times_J (v -_J w) = (u \times_J v) -_J (u \times_J w)$.
(*Hint:* Cases. $v < w, v = w, w < v$.)

(x) $x \times_J (y +_J z) = (x \times_J y) +_J (x \times_J z)$.

(xi) $x \times_J (y -_J z) = (x \times_J y) -_J (x \times_J z)$.

(xii) $0 <_J z \Rightarrow [x <_J y \Leftrightarrow x \times_J z <_J y \times_J z]$.

(xiii) $z <_J 0 \Rightarrow [x <_J y \Leftrightarrow x \times_J z >_J y \times_J z]$.

VI. Prove that $(J, +_J, \times_J, <_J)$ is a well-ordered integral domain.

VII. Prove that $(J, +_J, \times_J, <_J)$ is isomorphic with $(Z, +_Z, \times_Z, <_Z)$ (This shows that this alternative approach gives essentially the same result as that used in the text.)

2. We constructed the system of integers $(Z, +_Z, \times_Z)$ by starting with a Peano system $(P, S, 1)$. Show that, if we start with a different Peano system $(P^*, S^*, 1^*)$, then the resulting system of integers is isomorphic with $(Z, +_Z, \times_Z)$.

CHAPTER 4

RATIONAL NUMBERS AND ORDERED FIELDS

4.1 RATIONAL NUMBERS

By a rational number one ordinarily understands a quotient α/β of two integers α and β, with $\beta \neq 0$. Of course, we all are quite capable of using rational numbers in practical affairs, but our elementary and secondary-school education did not give us any precise definition of the notion of rational number. The reader should convince himself of this by attempting to formulate a *noncircular* definition. By a noncircular definition we mean one that does not use disguised synonyms for the idea being defined and does not use other undefined notions. For example, the word "ratio" is a synonym for "quotient" and should not be used to explain the latter. Likewise, we cannot say that α/β is the *number* which when *multiplied* by β yields α. What do we mean here by "number" or by "multiplied"?

The situation is quite analogous to the problem we encountered in defining the integers in terms of the natural numbers. Our solution of that problem amounted to thinking of an ordered pair (n, j) of natural numbers as a *representative* of the integer $n - j$, then defining a relation between two ordered pairs (n, j) and (k, i) corresponding to the intuitive idea that the "integers" $n - j$ and $k - i$ were equal, and, finally, taking the integers to be equivalence classes with respect to that relation. We shall construct the rational numbers in a similar manner.

We start with the system of integers $(Z, +_Z, \times_Z, <_Z)$. In this section, we shall use the same conventions agreed upon in Section 3.5, namely, we shall write

$$\alpha + \beta \quad \text{instead of} \quad \alpha +_Z \beta, \qquad 0 \quad \text{instead of} \quad 0_Z,$$
$$\alpha\beta \text{ (or, sometimes, } \alpha \cdot \beta) \quad \text{instead of} \quad \alpha \times_Z \beta, \qquad 1 \quad \text{instead of} \quad 1_Z.$$
$$\alpha < \beta \quad \text{instead of} \quad \alpha <_Z \beta,$$

Lowercase Greek letters α, β, γ, δ, ... will stand for arbitrary integers.

Intuitively, let us think of an ordered pair (α, β) of integers, with $\beta \neq 0$, as *representing* the rational number α/β. For example, $(1, 2)$ represents $\frac{1}{2}$, $(4, 12)$ represents $\frac{1}{3}$, etc. Of course, many different ordered pairs represent the same rational number; for example, $\frac{1}{2}$ is represented by $(1, 2)$, $(-1, -2)$, $(2, 4)$, $(-2, -4)$, $(3, 6)$, $(-3, -6)$, etc. This suggests that we should define a relation which would hold between two ordered pairs of integers when and only when they represent the same rational number according to the usual intuitive picture.

Notation Let $W = Z \times (Z - \{0\})$. W is the set of all ordered pairs of integers (α, β) such that $\beta \neq 0$. You may think of the pairs (α, β) in W as **fractions** and call α the **numerator** and β the **denominator**.

Definition

$$(\alpha, \beta) \approx (\gamma, \delta) \quad \text{means} \quad (\alpha, \beta) \in W \wedge (\gamma, \delta) \in W \wedge \alpha\delta = \gamma\beta.$$

If we look again at our intuitive picture, we see that this definition corresponds to the equivalence

$$\frac{\alpha}{\beta} = \frac{\gamma}{\delta} \Leftrightarrow \alpha\delta = \gamma\beta,$$

that is, to check equality of fractions, one "cross-multiplies."

EXAMPLES

1. $(1, 2) \approx (2, 4)$ since $1 \cdot 4 = 2 \cdot 2$.
2. $(-3, -2) \approx (6, 4)$ since $(-3) \cdot 4 = 6 \cdot (-2)$.
3. $(0, 1) \approx (0, 2)$ since $0 \cdot 2 = 0 \cdot 1$.
4. $(2, 2) \approx (3, 3)$ since $2 \cdot 3 = 3 \cdot 2$.
5. $(3, -2) \approx (-6, 4)$ since $3 \cdot 4 = (-6) \cdot (-2)$.

Theorem 1.1 $\approx$ is an equivalence relation in W, that is, for any $(\alpha, \beta) \in W$, $(\gamma, \delta) \in W$, $(\mu, \tau) \in W$:

a. $(\alpha, \beta) \approx (\alpha, \beta)$ (reflexivity).
b. $(\alpha, \beta) \approx (\gamma, \delta) \Rightarrow (\gamma, \delta) \approx (\alpha, \beta)$ (symmetry).
c. $(\alpha, \beta) \approx (\gamma, \delta) \wedge (\gamma, \delta) \approx (\mu, \tau) \Rightarrow (\alpha, \beta) \approx (\mu, \tau)$ (transitivity).

Proof (a) $\alpha\beta = \alpha\beta$.

(b) $\alpha\delta = \gamma\beta \Rightarrow \gamma\beta = \alpha\delta$.

(c)   Assume

$$(*)\quad \alpha\delta = \gamma\beta \qquad \text{and} \qquad (**)\quad \gamma\tau = \mu\delta.$$

We must show that $\alpha\tau = \mu\beta$. By $(*)$, $(\alpha\delta)\tau = (\gamma\beta)\tau$, and so, $(\alpha\tau)\delta = \beta(\gamma\tau)$. By $(**)$, this implies that $(\alpha\tau)\delta = \beta(\mu\delta)$. Since $(\gamma, \delta) \in W$, $\delta \neq 0$. Therefore, by the Cancellation Law, $\alpha\tau = \beta\mu$.   ∎

Definition Q is the set of all equivalence classes with respect to the equivalence relation $\approx$. The elements of Q are called **rational numbers**.

We shall use lowercase italic letters $r, s, t, \ldots, r_1, s_1, t_1, \ldots$ as variables for rational numbers. As usual, we shall denote the equivalence class of an ordered pair (α, β) by $[(\alpha, \beta)]$.

EXAMPLES

6. $\{(1, 2), (-1, -2), (2, 4), (-2, -4), \ldots\}$ $= [(1, 2)] \in Q$.

7. $\{(2, 3), (-2, -3), (4, 6), (-4, -6), (6, 9), (-6, -9), \ldots\}$
$= [(2, 3)] \in Q$.

The rational numbers for zero and one can be represented intuitively by $\frac{0}{1}$ and $\frac{1}{1}$, respectively. We are thus led to the following definitions.

Definitions $0_Q = [(0, 1)]$.
$1_Q = [(1, 1)]$.

Lemma 1.2   For any ordered pair (α, β) in W:

a. $(\alpha, \beta) \in 0_Q \Leftrightarrow \alpha = 0$.

b. $(\alpha, \beta) \in 1_Q \Leftrightarrow \alpha = \beta$.

(Thus, a fraction represents the zero rational if and only if its numerator is 0, and a fraction represents the rational unit element if and only if its numerator and denominator are equal.)

Proof   (a) $(\alpha, \beta) \in 0_Q \Leftrightarrow (\alpha, \beta) \approx (0, 1)$
$\Leftrightarrow \alpha \cdot 1 = 0 \cdot \beta$
$\Leftrightarrow \alpha = 0$.

(b) $(\alpha, \beta) \in 1_Q \Leftrightarrow (\alpha, \beta) \approx (1, 1)$
$$\Leftrightarrow \alpha \cdot 1 = 1 \cdot \beta$$
$$\Leftrightarrow \alpha = \beta. \quad \blacksquare$$

We wish to define appropriate addition and multiplication operations in Q. In this, we shall be guided by our experience with fractions.

Ordinarily we add two fractions as follows:

$$\frac{\alpha}{\beta} + \frac{\gamma}{\delta} = \frac{\alpha\delta + \beta\gamma}{\beta\delta}.$$

Using the ordered-pair notation:

$$(\alpha, \beta) + (\gamma, \delta) = (\alpha\delta + \beta\gamma, \beta\delta).$$

Since rational numbers are not particular ordered pairs but rather equivalence classes of ordered pairs, we have to check that the "sum" we have just obtained is really independent of the particular ordered pairs (α, β) and (γ, δ) chosen from their respective equivalence classes.

Lemma 1.3

$$(\alpha, \beta) \approx (\alpha_1, \beta_1) \wedge (\gamma, \delta) \approx (\gamma_1, \delta_1) \Rightarrow (\alpha\delta + \beta\gamma, \beta\delta) \approx (\alpha_1\delta_1 + \beta_1\gamma_1, \beta_1\delta_1)$$

Proof   Assume

$$(\#)\quad \alpha\beta_1 = \alpha_1\beta \quad\text{and}\quad (\#\#)\quad \gamma\delta_1 = \gamma_1\delta.$$

Now,

$$(\alpha\delta + \beta\gamma)\beta_1\delta_1 = \alpha\delta\beta_1\delta_1 + \beta\gamma\beta_1\delta_1$$
$$= \alpha_1\beta\delta\delta_1 + \gamma_1\delta\beta\beta_1 \quad \text{(by ($\#$) and ($\#\#$))}$$
$$= \alpha_1\delta_1(\beta\delta) + \beta_1\gamma_1(\beta\delta)$$
$$= (\alpha_1\delta_1 + \beta_1\gamma_1)\beta\delta.$$

Thus, $(\alpha\delta + \beta\gamma, \beta\delta) \approx (\alpha_1\delta_1 + \beta_1\gamma_1, \beta_1\delta_1)$. $\quad \blacksquare$

Definition   Assume r and s are in Q. Choose (α, β) in r and (γ, δ) in s. Then

$$r +_Q s = [(\alpha\delta + \beta\gamma, \beta\delta)].$$

Lemma 1.3 tells us that the sum $r +_Q s$ does not depend upon the arbitrary choices of (α, β) in r and of (γ, δ) in s. The rule for addition can be written as

$$[(\alpha, \beta)] +_Q [(\gamma, \delta)] = [(\alpha\delta + \beta\gamma, \beta\delta)].$$

Let us turn now to multiplication. The ordinary rule for multiplying fraction is

$$\frac{\alpha}{\beta} \times \frac{\gamma}{\delta} = \frac{\alpha\gamma}{\beta\delta}.$$

In terms of ordered pairs, this would read

$$(\alpha, \beta) \times (\gamma, \delta) = (\alpha\gamma, \beta\delta).$$

To obtain the corresponding definition of multiplication in Q, we must first prove the following.

Lemma 1.4

$$(\alpha, \beta) \approx (\alpha_1, \beta_1) \wedge (\gamma, \delta) \approx (\gamma_1, \delta_1) \Rightarrow (\alpha\gamma, \beta\delta) \approx (\alpha_1\gamma_1, \beta_1\delta_1).$$

Proof Assume $\alpha\beta_1 = \alpha_1\beta$ and $\gamma\delta_1 = \gamma_1\delta$. Multiplying these equations, we obtain $\alpha\beta_1\gamma\delta_1 = \alpha_1\beta\gamma_1\delta$, and, therefore, $(\alpha\gamma)(\beta_1\delta_1) = (\alpha_1\gamma_1)(\beta\delta)$, that is, $(\alpha\gamma, \beta\delta) \approx (\alpha_1\gamma_1, \beta_1\delta_1)$. ∎

Definition Assume r and s are in Q. Choose (α, β) in r and (γ, δ) in s. Then

$$r \times_Q s = [(\alpha\gamma, \beta\delta)].$$

Lemma 1.4 is the basis for this definition, since it tells us that the product $r \times_Q s$ does not depend upon the arbitrary choices of (α, β) in r and of (γ, δ) in s. The rule for multiplication can be written as

$$[(\alpha, \beta)] \times_Q [(\gamma, \delta)] = [(\alpha\gamma, \beta\delta)].$$

Theorem 1.5 $\mathscr{Q} = (Q, +_Q, \times_Q)$ is an integral domain. 0_Q is the zero element, 1_Q is the unit element, and $0_Q \neq 1_Q$. Moreover, $\mathscr{Q}$ satisfies the following important property:

9.† $$(\forall x)(x \neq 0 \Rightarrow (\exists y)(x \times y = 1)).$$

---

† This continues the numbering of Chapter 3. Axioms 1–8 (pp. 95–96, 100) were the axioms for an integral domain.

Proof Assume r, s, t are in Q. Choose $(\alpha, \beta) \in r$, $(\gamma, \delta) \in s$, $(\mu, \tau) \in t$.

Axiom 1 $(r +_Q s) +_Q t = [(\alpha\delta + \beta\gamma, \beta\delta)] +_Q [(\mu, \tau)]$
$$= [((\alpha\delta + \beta\gamma)\tau + \beta\delta\mu, \beta\delta\tau)].$$

On the other hand,

$$r +_Q (s +_Q t) = [(\alpha, \beta)] +_Q [(\gamma\tau + \delta\mu, \delta\tau)]$$
$$= [(\alpha\delta\tau + \beta(\gamma\tau + \delta\mu), \beta\delta\tau)].$$

But $(\alpha\delta + \beta\gamma)\tau + \beta\delta\mu = \alpha\delta\tau + \beta(\gamma\tau + \delta\mu)$. Hence,

$$(r +_Q s) +_Q t = r +_Q (s +_Q t).$$

Axiom 2 $r +_Q s = [(\alpha\delta + \beta\gamma, \beta\delta)].$
$s +_Q r = [(\gamma\beta + \delta\alpha, \delta\beta)].$

Hence, $r +_Q s = s +_Q r$.

Axiom 3 (3a) $r +_Q 0_Q = [(\alpha, \beta)] +_Q [(0, 1)]$
$$= [(\alpha \cdot 1 + \beta \cdot 0, \beta \cdot 1)] = [(\alpha, \beta)] = r.$$

(3b) $r +_Q [(-\alpha, \beta)] = [(\alpha, \beta)] +_Q [(-\alpha, \beta)]$
$$= [(\alpha\beta + \beta(-\alpha), \beta^2)]$$
$$= [(\alpha\beta - \alpha\beta, \beta^2)] = [(0, \beta^2)] = 0_Q.$$

(Thus, $-_Q[(\alpha, \beta)] = [(-\alpha, \beta)]$.)

Axiom 4 $r \times_Q (s \times_Q t) = [(\alpha, \beta)] \times_Q [(\gamma\mu, \delta\tau)]$
$$= [(\alpha(\gamma\mu), \beta(\delta\tau))].$$

$(r \times_Q s) \times_Q t = [(\alpha\gamma, \beta\delta)] \times_Q [(\mu, \tau)] = [((\alpha\gamma)\mu, (\beta\delta)\tau)].$

Hence, $r \times_Q (s \times_Q t) = (r \times_Q s) \times_Q t$.

Skip Axiom 5 for a moment.

Axiom 6 $r \times_Q 1_Q = [(\alpha, \beta)] \times_Q [(1, 1)] = [(\alpha \cdot 1, \beta \cdot 1)]$
$$= [(\alpha, \beta)] = r.$$

Axiom 7 $r \times_Q s = [(\alpha\gamma, \beta\delta)].$
$s \times_Q r = [(\gamma\alpha, \delta\beta)] = [(\alpha\gamma, \beta\delta)].$

Axiom 5   We need only prove 5a by virtue of the commutativity of multiplication (Axiom 7).

$$r \times_Q (s +_Q t) = [(\alpha, \beta) \times_Q [(\gamma\tau + \delta\mu, \delta\tau)]$$
$$= [(\alpha(\gamma\tau + \delta\mu), \beta(\delta\tau))]$$
$$= [(\alpha\gamma\tau + \alpha\delta\mu, \beta\delta\tau)];$$

$$(r \times_Q s) +_Q (r \times_Q t) = [(\alpha\gamma, \beta\delta)] +_Q [(\alpha\mu, \beta\tau)]$$
$$= [(\alpha\gamma\beta\tau + \alpha\mu\beta\delta, \beta^2\delta\tau)]$$
$$= [(\beta(\alpha\gamma\tau + \alpha\delta\mu), \beta(\beta\delta\tau))].$$

But[†]

$$(\alpha\gamma\tau + \alpha\delta\mu, \beta(\delta\tau)) \approx (\beta(\alpha\gamma\tau + \alpha\delta\mu), \beta^2\delta\tau)$$

and, therefore,

$$[(\alpha\gamma\tau + \alpha\delta\mu, \beta(\delta\tau))] = [(\beta(\alpha\gamma\tau + \alpha\delta\mu), \beta^2\delta\tau)].$$

Axiom 8   Assume $r \neq 0_Q$ and $s \neq 0_Q$. We must show that $r \times_Q s \neq 0_Q$. But,

$$r \times_Q s = [(\alpha, \beta)] \times_Q [(\gamma, \delta)] = [(\alpha\gamma, \beta\delta)].$$

Since $r \neq 0_Q$ and $(\alpha, \beta) \in r$, it follows, by Theorem 1.2a, that $\alpha \neq 0$. Likewise, since $s \neq 0_Q$ and $(\gamma, \delta) \in s$, $\gamma \neq 0$. But, from $\alpha \neq 0$ and $\gamma \neq 0$ we conclude that $\alpha\gamma \neq 0$, since $(Z, +_Z, \times_Z)$ is an integral domain. Hence, $(\alpha\gamma, \beta\delta) \notin 0_Q$, that is, $r \times_Q s = [(\alpha\gamma, \beta\delta)] \neq 0_Q$.

$0_Q \neq 1_Q$ For, $0_Q = [(0, 1)]$ and $1_Q = [(1, 1)]$. Hence,

$$0_Q = 1_Q \Leftrightarrow (0, 1) \approx (1, 1)$$
$$\Leftrightarrow 0 \cdot 1 = 1 \cdot 1$$
$$\Leftrightarrow 0 = 1.$$

But $0 \neq 1$ (see p. 89).

Axiom 9   Assume $r \neq 0_Q$. Hence, $\alpha \neq 0$ by Theorem 1.2a. Therefore, $(\beta, \alpha) \in W$, and $[(\beta, \alpha)] \in Q$. Then

$$r \times_Q [(\beta, \alpha)] = [(\alpha, \beta)] \times_Q [(\beta, \alpha)] = [(\alpha\beta, \beta\alpha)] = 1_Q$$

by Theorem 1.2b. (This proof corresponds to the familiar result that $\alpha/\beta \cdot \beta/\alpha = 1$.)   ■

[†] In general, $(\varrho, \sigma) \approx (\beta\varrho, \beta\sigma)$ for any integers ϱ, σ, β, with $\sigma \neq 0$ and $\beta \neq 0$. This is clear, since $\varrho(\beta\sigma) = (\beta\varrho)\sigma$. This fact corresponds to the familiar result that one can cancel a common factor from both numerator and denominator of a fraction.

4.2 FIELDS

Instead of deriving the familiar properties of the rational numbers directly, we shall study a class of algebraic systems, fields, of which the system of rational numbers is a special case. The reason for doing things this way is that many other important mathematical structures, such as the real number system and the complex number system, are fields.

Definition A **field** is an integral domain $(F, +, \times)$ satisfying the additional properties:

9. $(\forall x)(x \neq 0 \Rightarrow (\exists y)(x \times y = 1))$.
10. $0 \neq 1$.

Since a field is an integral domain, all theorems and concepts concerning integral domains automatically apply to fields (especially Theorems 3.3.1–3.3.3).

Corollary 2.1 The rational number system $\mathcal{Q} = (Q, +_Q, \times_Q)$ is a field.

Proof This is the content of Theorem 1.5. ∎

Theorem 2.2 In a field:

$$(\forall x)(x \neq 0 \Rightarrow (\exists! y)(x \times y = 1)).$$

(For every nonzero x, there is a *unique* y such that $x \times y = 1$.)

Proof By Axiom 9 we know that there is at least one y such that $x \times y = 1$. Assume now that $x \times z = 1$. Then

$$y = 1 \times y = (x \times z) \times y = (z \times x) \times y$$
$$= z \times (x \times y) = z \times 1 = z. \quad ∎$$

Notation Let $(F, +, \times)$ be a field. For any $x \neq 0$ in F, the unique y such that $x \times y = 1$ is denoted x^{-1} and is called the **multiplicative inverse** of x.[†] Thus, if $x \neq 0$,

$$x \times y = 1 \Leftrightarrow y = x^{-1}.$$

[†] According to this definition, x^{-1} is not defined when $x = 0$. If one wishes to have a meaning for x^{-1} in all cases, then one can choose an arbitrary value for 0^{-1}, say 0. Then, of course, $x \times x^{-1} = 1$ would not hold for $x = 0$.

Notation   Let $(F, +, \times)$ be a field. For any x in F and for any $y \neq 0$ in F,

$$\text{let } x/y \quad \text{stand for} \quad x \times y^{-1}.$$

The element x/y is called the **quotient** of x and y.

Theorem 2.3   (*Properties of fields*)

a. $1^{-1} = 1$.

b. $x \neq 0 \Rightarrow x^{-1} \neq 0$.

c. $x \neq 0 \Rightarrow (x^{-1})^{-1} = x$.

d. $x \neq 0 \Rightarrow x^{-1} = 1/x$.

e. $x \neq 0 \wedge y \neq 0 \Rightarrow (x \times y)^{-1} = x^{-1} \times y^{-1}$
       (or, by virtue of (d), $1/x \times y = (1/x) \times (1/y)$).

f. $x/1 = x$.

g. $x \neq 0 \Rightarrow 0/x = 0$.

h. $x \neq 0 \Rightarrow x/x = 1$.

i. $y \neq 0 \rightarrow (z \times y = x \Leftrightarrow z = x/y)$.

j. $y \neq 0 \wedge v \neq 0 \Rightarrow (x/y) \times (u/v) = (x \times u)/(y \times v)$.
       (rule for multiplying quotients).

k. $y \neq 0 \wedge v \neq 0 \Rightarrow x/y = (x \times v)/(y \times v)$
       (cancellation of common factor in numerator and denominator).

l. $y \neq 0 \wedge v \neq 0 \Rightarrow x/y + u/v = ((x \times v) + (y \times u))/(y \times v)$
       (rule for adding quotients).

m. $y \neq 0 \wedge v \neq 0 \Rightarrow (x/y = u/v \Leftrightarrow x \times v = y \times u)$
       (cross-multiplication test for equality).

n. $x \neq 0 \wedge y \neq 0 \Rightarrow 1/(x/y) = y/x$ (that is, $(x/y)^{-1} = y/x$).

o. $y \neq 0 \wedge u \neq 0 \wedge v \neq 0 \Rightarrow (x/y)/(u/v) = (x \times v)/(y \times u)$.

Proof   (a) $1 \times 1 = 1$. Hence, $1 = 1^{-1}$.

   (b)   Assume $x \neq 0$. Then, $x \times x^{-1} = 1$. If $x^{-1} = 0$, then $1 = x \times x^{-1}$
$= x \times 0 = 0$, contradicting Axiom 10.

   (c)   Assume $x \neq 0$. Then $x^{-1} \times x = 1$. Hence, x is an object which when
multiplied by x^{-1} yields 1. But any such object is equal to $(x^{-1})^{-1}$.

   (d)   Assume $x \neq 0$. Then $1/x = 1 \times x^{-1} = x^{-1}$.

   (e)   Assume $x \neq 0$ and $y \neq 0$. Then $(x \times y) \times (x^{-1} \times y^{-1}) = (x \times x^{-1})$
$\times (y \times y^{-1}) = 1 \times 1 = 1$. Hence, $x^{-1} \times y^{-1} = (x \times y)^{-1}$.

   (f) $x/1 = x \times 1^{-1} = x \times 1 = x$.

(g)   Assume $x \neq 0$. Then $0/x = 0 \times x^{-1} = 0$.

(h)   Assume $x \neq 0$. Then $x/x = x \times x^{-1} = 1$.

(i)   Assume $y \neq 0$. Then $(x/y) \times y = (x \times y^{-1}) \times y = x \times (y \times y^{-1}) = x \times 1 = x$. Moreover, if $z \times y = x$, then, multiplying by y^{-1}, we obtain $z = x \times y^{-1} = x/y$.

(j)   Assume $y \neq 0$ and $v \neq 0$.

$$x/y \times u/v = (x \times y^{-1}) \times (u \times v^{-1}) = (x \times u) \times (y^{-1} \times v^{-1})$$
$$= (x \times u) \times (y \times v)^{-1} \quad \text{by (e)}$$
$$= (x \times u)/(y \times v).$$

(k)   Assume $y \neq 0$ and $v \neq 0$.

$$(x \times v)/(y \times v) = x/y \times v/v \quad \text{by (j)}$$
$$= x/y \times 1 \quad \text{by (h)}$$
$$= x/y.$$

(l)   Assume $y \neq 0$ and $v \neq 0$. Then

$$((x \times v) + (y \times u))/(y \times v) = ((x \times v) + (y \times u))(y \times v)^{-1}$$
$$= ((x \times v) + (y \times u))(y^{-1} \times v^{-1})$$
$$= (x \times v \times y^{-1} \times v^{-1}) + (y \times u \times y^{-1} \times v^{-1})$$
$$= (x \times y^{-1} \times v \times v^{-1}) + (u \times v^{-1} \times y \times y^{-1})$$
$$= (x \times y^{-1} \times 1) + (u \times v^{-1} \times 1)$$
$$= (x \times y^{-1}) + (u \times v^{-1}) = x/y + u/v.$$

(m)   Assume $y \neq 0$ and $v \neq 0$. Now assume $x/y = u/v$. Then $x \times y^- = u \times v^{-1}$. Hence,

$$x \times y^{-1} \times y \times v = u \times v^{-1} \times y \times v$$
$$x \times 1 \times v = u \times v^{-1} \times v \times y$$
$$x \times v = u \times 1 \times y = y \times u.$$

All these steps are reversible. Hence, conversely, if $x \times v = y \times u$, then $x/y = u/v$.

(n)   Assume $x \neq 0$ and $y \neq 0$. Notice that

$$(x/y) \times (y/x) = (x \times y)/(y \times x) \quad \text{by (j)}$$
$$= 1 \quad \text{by (h)}.$$

Hence, $y/x = (x/y)^{-1}$.

(o)   Assume $y \neq 0$, $u \neq 0$, and $v \neq 0$. Then

$$(x/y)/(u/v) = (x/y) \times (u/v)^{-1} = (x/y) \times (v/u) \qquad \text{by (n)}$$
$$= (x \times v)/(y \times u). \qquad \blacksquare$$

EXERCISES

Prove:

1.   If $y \neq 0$, then $-(x/y) = (-x)/y = x/(-y)$.

2.   If $y \neq 0$ and $v \neq 0$, then $x/y - u/v = ((x \times v) - (u \times y))/(y \times v)$.

Theorem 2.4   Among the axioms for a field,

$$\text{Axiom 8} \quad x \neq 0 \wedge y \neq 0 \Rightarrow x \times y \neq 0$$

is superfluous; that is, it can be proved from the other axioms.

Proof   Assume $x \times y = 0$ and $x \neq 0$. We must show that $y = 0$. Since $x \neq 0$, we have $x \times x^{-1} = 1$. Hence, multiplying $x \times y = 0$ on both sides by x^{-1}, we obtain

$$x^{-1} \times (x \times y) = x^{-1} \times 0$$
$$(x^{-1} \times x) \times y = 0$$
$$(x \times x^{-1}) \times y = 0$$
$$1 \times y = 0$$
$$y = 0. \qquad \blacksquare$$

For easy reference, we shall list below the axioms for a field, omitting the superfluous axiom (8).

Axiom 1 $x + (y + z) = (x + y) + z$.

Axiom 2 $x + y = y + x$.

Axiom 3   There is an element 0 in R such that

3a. $(\forall x)(x + 0 = x)$;

3b. $(\forall x)(\exists y)(x + y = 0)$.

Axiom 4 $x \times (y \times z) = (x \times y) \times z$.

Axiom 5 $x \times (y + z) = (x \times y) + (x \times z)$.[†]

Axiom 6 There is an element 1 such that

$$x = x \times 1$$

for all x.[‡]

Axiom 7 $x \times y = y \times x$.

Axiom 9 $(\forall x)(x \neq 0 \Rightarrow (\exists y)(x \times y = 1))$.

Axiom 10 $0 \neq 1$.

EXAMPLES OF FIELDS

1. The field of rational numbers $(Q, +_Q, \times_Q)$ (Corollary 2.1).

2. The real numbers under ordinary addition and multiplication. (Here again we rely upon the reader's previous mathematical experience. However, we shall give a rigorous proof later.)

3. The complex numbers under ordinary addition and multiplication. (The same remarks apply here as in Example 2.)

4. The set $\{0, 1\}$ under the following operations:

$$\begin{cases} 0 + 0 = 1 + 1 = 0 \\ 0 + 1 = 1 + 0 = 1 \end{cases}$$

+	0	1
0	0	1
1	1	0

$$\begin{cases} 0 \times 0 = 1 \times 0 = 0 \times 1 = 0 \\ 1 \times 1 = 1 \end{cases}$$

×	0	1
0	0	0
1	0	1

Verification of the field axioms is left to the reader.

[†] This is our old Axiom 5a. Axiom 5b is superfluous since it follows from Axiom 5a and Axiom 7 (commutativity of multiplication).

[‡] In our original formulation of Axiom 6, we demanded that $x = x \times 1 = 1 \times x$. However, by virtue of Axiom 7 (commutativity of multiplication), we know that $x \times 1 = 1 \times x$.

5.   The set $\{0, 1, 2\}$ under the following operations:

+	0	1	2
0	0	1	2
1	1	2	0
2	2	0	1

×	0	1	2
0	0	0	0
1	0	1	2
2	0	2	1

Verification of the field axioms is again left to the reader.

Examples 4 and 5 resemble the systems of integers modulo 2 and 3, respectively. In fact, they are easily seen to be isomorphic to $(Z_2, +_2, \times_2)$ and $(Z_3, +_3, \times_3)$, respectively (see Section 3.6). This suggests the question: Which of the rings $(Z_n, +_n, \times_n)$ are fields?

Theorem 2.5   Every finite integral domain with at least two elements is a field.

Proof   Let $\mathscr{D} = (D, +, \times)$ be a finite integral domain, where $D = \{d_1, d_2, \ldots, d_n\}, n \geq 2$. Since D has at least two elements, $0 \neq 1$ by Theorem 3.3.4. This is Axiom 10. To prove Axiom 9, we assume $x \neq 0$ and we must show that there is some y such that $x \times y = 1$. Now, consider the elements $x \times d_1, x \times d_2, \ldots, x \times d_n$. All of them are distinct, for, if $x \times d_i = x \times d_j$, then $d_i = d_j$ by the Cancellation Law for Multiplication. (Remember that $x \neq 0$.) Since $x \times d_1, x \times d_2, \ldots, x \times d_n$ are n distinct elements and there are precisely n elements in D, the set $\{x \times d_1, x \times d_2, \ldots, x \times d_n\}$ is all of D. Hence, one element of that set, say $x \times d_k$, is the unit element 1. Thus, $x \times d_k = 1$. This establishes Axiom 9.   ∎

Corollary 2.6   Assume that n is an integer greater than one. Then, $(Z_n, +_n, \times_n)$ is a field if and only if n is a prime.

Proof   By Theorem 2.5, if $(Z_n, +_n, \times_n)$ is an integral domain, then $(Z_n, +_n, \times_n)$ is a field. The converse, that if $(Z_n, +_n, \times_n)$ is a field then it is an integral domain, is a consequence of the definition of a field. Hence, $(Z_n, +_n, \times_n)$ is a field if and only if it is an integral domain. But, by Theorem 3.6.6, $(Z_n, +_n, \times_n)$ is an integral domain if and only if n is a prime.   ∎

The fields $(Z_n, +_n, \times_n)$, where n is a prime, are not the only finite fields.

EXAMPLE

6. A field with four elements, $F = \{0, 1, t, s\}$.

+	0	1	t	s
0	0	1	t	s
1	1	0	s	t
t	t	s	0	1
s	s	t	1	0

×	0	1	t	s
0	0	0	0	0
1	0	1	t	s
t	0	t	s	1
s	0	s	1	t

The reader should make a careful check of all the field axioms. He should also show that $(F, +, \times)$ is not isomorphic with $(Z_4, +_4, \times_4)$.

EXERCISES

Determine whether the following structures $(F, +, \times)$ are fields.

3. $(F, +, \times) = (Z, +_Z, \times_Z)$.

4. $(F, +, \times) = (Z_6, +_6, \times_6)$.

5. F is the set of all polynomials with integral coefficients, and $+$ and $\times$ are ordinary addition and multiplication of polynomials.

6. F is the set of all polynomials with rational coefficients, and $+$ and $\times$ are ordinary addition and multiplication of polynomials.

7. $(F, +, \times) = (Z_5, +_5, \times_5)$.

8. F is the set of all "quotients" of polynomials with rational coefficients

$$\frac{a_n x^n + \cdots + a_1 x + a_0}{b_k x^k + \cdots + b_1 x + b_0} \quad (b_k \neq 0)$$

with $+$ and $\times$ as "ordinary" addition and multiplication (that is, the usual rules for adding and multiplying fractions).

9. F is the set of all complex numbers $a + bi$, where a and b are rational numbers, and $+$ and $\times$ are the usual addition and multiplication of complex numbers.

10. F is the set of all real numbers of the form $a + b\sqrt{2}$, where a and b are rational numbers.

11. F is the set of all real numbers of the form $a + b\sqrt[3]{2}$, where a and b are rational numbers.

4.3 QUOTIENT FIELD OF AN INTEGRAL DOMAIN

The construction, carried out in Section 4.1, of the rational numbers from the integers, is a special case of a general algebraic process, which we shall describe now.

Let $\mathscr{D} = (D, +, \times)$ be an integral domain in which $0 \neq 1$. Let $W = D \times (D - \{0\})$. W is the set of all ordered pairs (x, y) of elements of D, where $y \neq 0$. Define a relation $\approx$ in W as follows:

For any (x, y) and (u, v) in W:

$$(x, y) \approx (u, v) \text{ means that } x \times v = u \times y.$$

Theorem 3.1 $\approx$ is an equivalence relation in W.

Proof Same as the proof of Theorem 1.1. ∎

Let F be the set of equivalence classes with respect to $\approx$. We denote the equivalence class of an ordered pair (x, y) of W by $[(x, y)]$.

Definitions $0_F = [(0, 1)]$.
$1_F = [(1, 1)]$.

Here, 0 and 1 denote the zero and unit elements, respectively, of $\mathscr{D}$.

Lemma 3.2 For any ordered pair (x, y) in W,

a. $(x, y) \in 0_F \Leftrightarrow x = 0$.
b. $(x, y) \in 1_F \Leftrightarrow x = y$.

Proof Same as for Lemma 1.2. ∎

Lemma 3.3

$$(x, y) \approx (x_1, y_1) \wedge (u, v) \approx (u_1, v_1) \Rightarrow ((x \times v) + (y \times u), y \times v)$$
$$\approx ((x_1 \times v_1) + (y_1 \times u_1), y_1 \times v_1).$$

Proof Same as for Lemma 1.3. ∎

Definition Assume r and s are in F. Choose any (x, y) in r and any (u, v) in s. Let

$$r +_F s = [((x \times v) + (y \times u), y \times v)].$$

Lemma 3.4

$(x, y) \approx (x_1, y_1) \wedge (u, v) \approx (u_1, v_1) \Rightarrow (x \times u, y \times v) \approx (x_1 \times u_1, y_1 \times v_1).$

Proof   Same as for Lemma 1.4.   ∎

Definition   Assume r and s are in F. Choose any (x, y) in r and any (u, v) in s. Let

$$r \times_F s = [(x \times u, y \times v)].$$

Theorem 3.5 $\mathfrak{F} = (F, +_F, \times_F)$ is a field. 0_F and 1_F are the zero and unit elements, respectively.

Proof   The reader should check carefully that the proof of Theorem 1.5 can serve as a proof of Theorem 3.5, with $(Z, +_Z, \times_Z)$ being replaced throughout by $(D, +, \times)$.   ∎

The field $\mathfrak{F}$ of Theorem 3.5 is called the **quotient field** of $\mathscr{D}$. Clearly, the rational number system $\mathscr{Q} = (Q, +_Q, \times_Q)$ is the quotient field of the system of integers $(Z, +_Z, \times_Z)$.

Theorem 3.6   Let H be the set of all elements of F of the form $[(x, 1)]$, where $x \in D$. Then H is closed under the operations $+_F$ and $\times_F$. If we denote by $+_H$ and $\times_H$ the restrictions of $+_F$ and $\times_F$ to H, $(H, +_H, \times_H)$ is then an integral domain, and the function $\Psi(x) = [(x, 1)]$ is an isomorphism of $\mathscr{D}$ with $(H, +_H, \times_H)$. Moreover, every element r of F is a quotient s/t of elements s and t of H. (This explains the name **quotient field**. The elements of F are quotients of elements of H, and H is an isomorphic copy of the original integral domain $\mathscr{D}$.)

Proof   (i)   Consider any $[(x, 1)]$ and $[(y, 1)]$, with x and y in D. Then,

$[(x, 1)] +_F [(y, 1)] = [((x \times 1) + (y \times 1), 1 \times 1)] = [(x + y, 1)] \in H;$
$[(x, 1)] \times_F [(y, 1)] = [(x \times y, 1 \times 1)] = [(x \times y, 1)] \in H.$

Thus, H is closed under $+_F$ and $\times_F$.

(ii)   To verify that $(H, +_H, \times_H)$ is an integral domain, we need only check that $1_F \in H$ and that H is closed under subtraction (by Theorem 3.9.2). Clearly, $1_F = [(1, 1)] \in H$. In addition, if $[(x, 1)] \in H$ and $[(y, 1)] \in H$ then

$[(x, 1) -_F [(y, 1)] = [(x, 1)] +_F (-_F[(y, 1))]$
$= [(x, 1)] +_F [(-y, 1)] = [(x - y, 1)] \in H.$

(iii) Clearly, $\Psi: D \xrightarrow[\text{onto}]{} H$. To show that Ψ is one-one, assume $\Psi(x) = \Psi(y)$. Then, $[(x, 1)] = [(y, 1)]$. Hence, $(x, 1) \approx (y, 1)$, that is, $x = x \times 1 = y \times 1 = y$. Thus, Ψ is one-one. Moreover,

$$\Psi(x) +_F \Psi(y) = [(x, 1)] +_F [(y, 1)] = [(x + y, 1)] = \Psi(x + y),$$

and

$$\Psi(x) \times_F \Psi(y) = [(x, 1)] \times_F [(y, 1)] = [(x \times y, 1)] = \Psi(x \times y).$$

Thus, Ψ is an isomorphism.

(iv) Assume r in F. Take any (x, y) in r. Now, $[(x, 1)] \in H$ and $[(y, 1)] \in H$, and

$$[(x, 1)]/[(y, 1)] = [(x, 1)] \times_F [(y, 1)]^{-1} = [(x, 1)] \times_F [(1, y)]$$
$$= [(x \times 1, 1 \times y)] = [(x, y)] = r.$$

Thus, r is the quotient of $[(x, 1)]$ and $[(y, 1)]$.   ∎

Remark In the special case of the rational numbers and the integers, the elements of H are of the form $[(\alpha, 1)]$, where α is an integer. In our intuitive picture, $[(\alpha, 1)]$ is represented by the fraction $\alpha/1$, and ordinarily we draw no distinction between α and $\alpha/1$. It is customary to think of the integers as forming part of the set of rational numbers. We can agree then to call the elements of H **integers** (or **rational integers**). There is no danger of confusion since $(H, +_H, \times_H)$ is isomorphic with $(Z, +_Z, \times_Z)$, and two isomorphic structures are essentially the same. As a matter of fact, the elements $[(\alpha, 1)]$, where $\alpha \in Z$, are the integers of the field $\mathcal{Q}$, in the sense of Section 3.8. This follows by Theorem 3.8.8 (that is, $Z_{\mathcal{Q}} = \{\alpha 1_D : \alpha \in Z\}$), since $[(\alpha, 1)] = \alpha[(1, 1)] = \alpha 1_D$. (One proves $[(\alpha, 1)] = \alpha[(1, 1)]$ as follows: We first consider positive integers α and use mathematical induction. When $\alpha = 1$, $1[(1, 1)] = [(1, 1)]$. Assume, as inductive hypothesis, that $[(\alpha, 1)] = \alpha[(1, 1)]$ for a positive integer α. Then,

$$(\alpha +_Z 1)[(1, 1)] = \alpha[(1, 1)] +_Q [(1, 1)] = [(\alpha, 1)] +_Q [(1, 1)]$$
$$= [((\alpha \times 1) + (1 \times 1), 1 \times 1)] = [(\alpha + 1, 1)].$$

Thus, the equation holds for all positive integers α. When $\alpha = 0$, $0[(1, 1)] = 0_Q = [(0, 1)]$. Finally, if α is negative,

$$\alpha[(1, 1)] = (-\alpha)(-[(1, 1)]) = -((-\alpha)[(1, 1)]) = -[(-\alpha, 1)] = [(\alpha, 1)].)$$

In what follows, we shall sometimes write an arbitrary rational number in the form α/β, where α and β are integers, instead of the more precise form $\alpha 1_Q/\beta 1_Q$. For example, we can write $[(1, 2)]$ as $1/2$, $[(3, 4)]$ as $3/4$, etc. This brings us back to the familiar notation for rational numbers.

EXERCISES

1. In the quotient field $\mathfrak{F}$, prove
    $$[(x, y)] -_F [(u, v)] = [((x \times v) - (u \times y), y \times v)].$$

2. If we take the integral domain $\mathscr{D}$ consisting of all polynomials $a_n x^n + \cdots + a_1 x + a_0$ with integral coefficients, under ordinary addition and multiplication of polynomials, describe the quotient field of $\mathscr{D}$.

3. If $\mathscr{D}$ is a field, describe the quotient field $\mathfrak{F}$ of $\mathscr{D}$, and show that $\mathfrak{F}$ is isomorphic with $\mathscr{D}$.

4.^{D} Can there be an infinite field in which $2 = 0$?

4.4 ORDERED FIELDS

We say that a structure $(F, +, \times, <)$ is an **ordered field** if and only if $(F, +, \times, <)$ is an ordered integral domain and $(F, +, \times)$ is a field. This means that, in addition to the field axioms (see pp. 166–167), the following axioms must hold.

O1. $x \not< x$.
O2. $x < y \wedge y < z \Rightarrow x < z$.
O3. $x < y$ or $x = y$ or $y < x$.
O4. $x < y \Rightarrow x + z < y + z$.
O5. $x < y \wedge 0 < z \Rightarrow x \times z < y \times z$.

By Theorem 3.4.3, to make a field $(F, +, \times)$ into an ordered field it suffices to define a subset $\mathscr{P}$ of F such that

i. $0 \notin \mathscr{P}$,
ii. $x \in \mathscr{P}$ or $x = 0$ or $-x \in \mathscr{P}$ for any x in F,
iii. $x \in \mathscr{P} \wedge y \in \mathscr{P} \Rightarrow x + y \in \mathscr{P}$,
iv. $x \in \mathscr{P} \wedge y \in \mathscr{P} \Rightarrow x \times y \in \mathscr{P}$.

If we define $x < y$ to mean that $y - x \in \mathscr{P}$, then $(F, +, \times, <)$ must be an ordered field. We shall now show how this method is applied in the case of the quotient field of an ordered integral domain.

Theorem 4.1 Let $\mathscr{D} = (D, +, \times, <)$ be an ordered integral domain, and let $\mathscr{F} = (F, +_F, \times_F)$ be the quotient field of $\mathscr{D}$.

a.   If $r \in F$, $(x, y) \in r$, and $(u, v) \in r$, then $x \times y > 0$ if and only if $u \times v > 0$.

b.   Define a subset $\mathscr{P}$ of F as follows: For any r in F, $r \in \mathscr{P}$ if and only if, for any ordered pair (x, y) in r, $x \times y > 0$. (Notice that, by Theorem 3.4.2g–i, $x \times y > 0$ if and only if x and y have the same sign.) Then $\mathscr{P}$ satisfies conditions (i)–(iv) of Theorem 3.4.3. Hence, if $r <_F s$ is defined as $s -_F r \in \mathscr{P}$, then $(F, +_F, \times_F, <_F)$ is an ordered field. Moreover, the isomorphism Ψ of Theorem 3.6 is order-preserving, that is, $x < y \Leftrightarrow \Psi(x) <_F \Psi(y)$.

Proof (a)   Since $(x, y) \in r$ and $(u, v) \in r$, $(x, y) \approx (u, v)$, that is, $x \times v = u \times y$. Assume $x \times y > 0$. Then, by Theorem 3.4.2i, either $x > 0 \wedge y > 0$ or $x < 0 \wedge y < 0$.

Case i $x > 0 \wedge y > 0$.   We know that $v \neq 0$. Hence, $u \neq 0$. (Otherwise, $0 \neq x \times v = u \times y = 0 \times y = 0$.) If u and v have opposite signs (that is, one positive and the other negative), one side of $x \times v = u \times y$ would be positive and the other negative (by Theorem 3.4.2g,i).

Case ii $x < 0 \wedge y < 0$.   As in case (i), $v \neq 0$ and $u \neq 0$. If u and v were of opposite sign, then one side of $x \times v = u \times y$ would be positive and the other negative. Hence, $u \times v > 0$.

The converse, $u \times v > 0 \Rightarrow x \times y > 0$, follows by symmetry.

(b)   i. $0_F = [(0, 1)]$. Since $0 \times 1 = 0 \not> 0$, $0_F \notin \mathscr{P}$.

ii.   Assume $r \in F$. Take any (x, y) in r.

Case 1 $x = 0$. Then, $r = 0_F$.

Case 2 x and y have the same sign. Then $x \times y > 0$. Hence, $r \in \mathscr{P}$.

Case 3 x and y have opposite signs. Hence, $-x$ and y have the same sign. Then $(-x) \times y > 0$. Thus, $[(-x, y)] \in \mathscr{P}$. But, $[(-x, y)] = -r$.

(iii)–(iv)   Assume r and s in $\mathscr{P}$. Choose (x, y) in r and (u, v) in s. Then $x \times y > 0$ and $u \times v > 0$. Now,

$$r +_F s = [((x \times v) + (y \times u), y \times v)].$$

But, $((x \times v) + (y \times u)) \times (y \times v) = ((x \times y) \times v^2) + (y^2 \times (u \times v))$. Since $v \neq 0$ and $y \neq 0$, it follows that $v^2 > 0$ and $y^2 > 0$. Hence, $(x \times y) \times v^2 > 0$ and $y^2 \times (u \times v) > 0$, and, therefore, $((x \times y) \times v^2) + (y^2 \times (u \times v)) > 0$. Thus, $r +_F s \in \mathscr{P}$.

$r \times_F s = [(x \times u, y \times v)]$. Now, $(x \times u) \times (y \times v) = (x \times y) \times (u \times v)$. Since $x \times y > 0$ and $u \times v > 0$, we obtain $(x \times y) \times (u \times v) > 0$. Hence, $r \times_F s \in \mathscr{P}$.

To prove that Ψ preserves order, note that

$$\Psi(y) -_F \Psi(x) = \Psi(y) +_F (-_F\Psi(x)) = [(y, 1)] +_F (-_F[(x, 1)])$$
$$= [(y, 1)] +_F [(-x, 1)] = [((y\times 1) + ((-x)\times 1), 1\times 1)]$$
$$= [(y - x, 1)].$$

Hence,

$$\Psi(x) <_F \Psi(y) \Leftrightarrow \Psi(y) -_F \Psi(x) \in \mathscr{P}$$
$$\Leftrightarrow [(y - x, 1)] \in \mathscr{P}$$
$$\Leftrightarrow (y - x)\times 1 > 0$$
$$\Leftrightarrow y - x > 0$$
$$\Leftrightarrow x < y. \qquad \blacksquare$$

Corollary 4.2 There is a binary relation $<_Q$ in Q such that $(Q, +_Q, \times_Q, <_Q)$ is an ordered field. (This order relation is an extension of the order relation on the integers of Q.)

EXAMPLES

1. $[(3, 5)] <_Q [(2, 3)]$, since

$$[(2, 3)] -_Q [(3, 5)] = [(2, 3)] +_Q [(-3, 5)] = [((2\times 5) + ((-3)\times 3), 3\times 5)]$$
$$= [(10 - 9, 15)] = [(1, 15)],$$

and 1 and 15 have the same sign. (This corresponds to the familiar inequality: $3/5 < 2/3$.)

2. $[(2, -3)] <_Q [(1, 2)]$, since

$$[(1, 2)] -_Q [(2, -3)] = [(1, 2)] +_Q [(-2, -3)]$$
$$= [((1\times(-3)) + ((-2)\times 2), 2\times(-3))]$$
$$= [(3 - 4, -6)] = [(-1, -6)],$$

and -1 and -6 have the same sign.

Let r be in the quotient field. Then there is some (x, y) in r with $y > 0$. For, take any (x, y) in r. We know that $y \neq 0$. If $y > 0$, we are through. If $y < 0$, then $(-x, -y) \in r$ (since $(x, y) \approx (-x, -y)$), and $-y > 0$.

Now assume that r and s are in the quotient field. We may choose $(x, y) \in r$ and $(u, v) \in s$, with $y > 0$ and $v > 0$. Then,

$$r <_F s \Leftrightarrow x \times v < u \times y \qquad \text{(cross-multiplication rule)}.$$

For,

$$
\begin{aligned}
r <_F s &\Leftrightarrow s -_F r \in \mathscr{P} \\
&\Leftrightarrow [(u, v)] -_F [(x, y)] \in \mathscr{P} \\
&\Leftrightarrow [((u \times y) - (v \times x)), v \times y)] \in \mathscr{P} \\
&\Leftrightarrow ((u \times y) - (v \times x)) \times (v \times y) > 0 \\
&\Leftrightarrow (u \times y) - (v \times x) > 0 \qquad \text{(since } v \times y > 0) \\
&\Leftrightarrow u \times y > v \times x.
\end{aligned}
$$

EXERCISE

1.   Prove that $<_Q$ is the only possible relation which would make the rational number system into an ordered field.

EXAMPLE

3.   Let $\mathscr{D}$ be the ordered integral domain of polynomials with integral coefficients (see Exercise 1c, p. 110). Remember that, for polynomials f and g,

$$f < g \Leftrightarrow \text{ the leading coefficient of } g - f \text{ is positive.}$$

Let $\mathfrak{F} = (F, +_F, \times_F, <_F)$ be the quotient field of $\mathscr{D}$. The elements of F are equivalence classes $[(f, g)]$, where f and g are polynomials and $g \neq 0$. Thus, $[(f, g)]$ is a positive element of F if and only if $f \times g$ is a positive polynomial, that is, if and only if the leading coefficient of $f \times g$ is positive. This, in turn, is equivalent to the condition that the leading coefficients of f and g have the same sign. Hence, $[(f, g)] <_F [(f_1, g_1)]$ if and only if $[(f_1, g_1)] - [(f, g)] = [(f_1 g - f g_1, g g_1)]$ is positive, that is, if and only if the leading coefficients of $f_1 g - f g_1$ and $g g_1$ have the same sign. In particular, using familiar notation we have

$$0 < \cdots < \frac{1}{x^2} < \frac{1}{x} < 1 < x < x^2 < \cdots.$$

(More precisely, $[(0, 1)] <_F \cdots <_F [(1, x^2)] <_F [(1, x)] <_F [(1, 1)] <_F [(x, 1)] <_F [(x^2, 1)] <_F \cdots.$)

The reader should arrange the following elements of F in their proper order: $[(x^2 + 2, x^3 - 3)]$, $[(x, 1)]$, $[(x^4, 2)]$, $[(-x^3, x^2)]$.

In addition to the standard properties of ordered integral domains, ordered fields have certain special properties.

Theorem 4.3   Let $\mathcal{F} = (F, +, \times, <)$ be any ordered field. Then

a. $x \neq 0 \Rightarrow (x > 0 \Leftrightarrow x^{-1} > 0)$.

b. $x \neq 0 \Rightarrow (x < 0 \Leftrightarrow x^{-1} < 0)$.

c. $y > 0 \wedge v > 0 \Rightarrow (x/y < u/v \Leftrightarrow x \times v < y \times u)$.

d. $y \neq 0 \Rightarrow |x/y| = |x|/|y|$.

e. $y \neq 0 \Rightarrow |y^{-1}| = 1/|y|$.

f.   (Density) $x < y \Rightarrow x < (x + y)/2 < y$. (Remember that $2 = 1 + 1$.)

Proof   (a), (b)   Assume $x \neq 0$. By Theorem 2.3b, $x^{-1} \neq 0$. But $x \times x^{-1} = 1$ and $1 > 0$. Hence, x and x^{-1} have the same sign, by Theorem 3.4.2g–i.

(c)   Assume $y > 0$ and $v > 0$. Then $y \times v > 0$. Hence,

$$x/y < u/v \Leftrightarrow x/y \times (y \times v) < u/v \times (y \times v)$$
$$\Leftrightarrow x \times v < u \times y.$$

(d)   Assume $y \neq 0$. Then $|x/y| \times |y| = |(x/y) \times y| = |x|$. Hence, $|x/y| = |x|/|y|$.

(e)   Take $x = 1$ in (d).

(f)   Since $x < y$, $x + x < x + y < y + y$. Hence, $2x < x + y < 2y$. If we multiply these inequalities by the positive element $\frac{1}{2}$, we obtain

$$x < (x + y)/2 < y. \quad \blacksquare$$

EXERCISES

In any ordered field, prove:

2. $y \neq 0 \Rightarrow (x/y > 0 \Leftrightarrow x \times y > 0)$.

3. $x > 0 \wedge y > 1 \Rightarrow x/y < x$.

4. $x > 0 \wedge 0 < y < 1 \Rightarrow x/y > x$.

5.   There are infinitely many elements between any two elements.

6.^{D} $(a_1 b_1 + \cdots + a_n b_n)^2 \leq (a_1^2 + \cdots + a_n^2)(b_1^2 + \cdots + b_n^2)$ (Cauchy–Schwarz Inequality). (*Hint:* For any u,

$$(a_1 + u b_1)^2 + \cdots + (a_n + u b_n)^2 \geq 0.$$

Expand the left side and choose a suitable value for u.)

4.5 SUBFIELDS. RATIONAL NUMBERS OF A FIELD

Let $\mathfrak{F} = (F, +, \times)$ and $\mathscr{E} = (E, \oplus, \otimes)$ be fields. We say that $\mathscr{E}$ is a *subfield* of $\mathfrak{F}$ if and only if

1. $E \subseteq F$.
2. $x \oplus y = x + y$ for all x and y in E (that is, $\oplus$ is the restriction of $+$ to E).
3. $x \otimes y = x \times y$ for all x and y in E (that is, $\otimes$ is the restriction of $\times$ to E).

EXAMPLES

1. The field of rational numbers is a subfield of the field of real numbers.

2. The field of real numbers is a subfield of the field of complex numbers.

EXERCISES

1. Show that every field is a subfield of itself.

2. If $\mathfrak{F}_1$ is a subfield of $\mathfrak{F}_2$ and $\mathfrak{F}_2$ is a subfield of $\mathfrak{F}_3$, prove that $\mathfrak{F}_1$ is a subfield of $\mathfrak{F}_3$.

Theorem 5.1 Let $\mathscr{E} = (E, \oplus, \otimes)$ be a subfield of $\mathfrak{F} = (F, +, \times)$. Then

a. E is closed under $+$ and $\times$.
b. $0_E = 0_F$.
c. $1_E = 1_F$.
d. $\mathscr{E}$ is a subdomain of $\mathfrak{F}$.
e. $x -_E y = x -_F y$ for all x and y in E.
f. $x/^E y = x/^F y$ for all x and y in E with $y \neq 0$ (where $/^E$ is the quotient in E and $/^F$ is the quotient in F).

Proof (a) By (2), (3) of the definition of subfield.

(b) $0_E + 0_E = 0_E \oplus 0_E = 0_E$. Hence, by the Cancellation Law for $+$ in $\mathfrak{F}$, $0_E = 0_F$.

(c) $1_E \times 1_E = 1_E \otimes 1_E = 1_E = 1_E \times 1_F$. Now, since $\mathscr{E}$ is a field, $1_E \neq 0_E$. But, $0_E = 0_F$ by (b). Thus, $1_E \neq 0_F$. Hence, by the Cancellation Law for $\times$ in $\mathfrak{F}$, $1_E = 1_F$.

(d)   From the definition. (Note that $1_F \in E$ by (c).)

(e)   By Theorem 3.9.1d and (d) above.

(f)   Assume $y \in E$ and $y \neq 0_E$. Hence,

$$y \times y^{(-1)_E} = y \otimes y^{(-1)_E} = 1_E = 1_F,$$

where $y^{(-1)_E}$ is the multiplicative inverse of y in $\mathscr{E}$. Therefore, $y^{(-1)_E} = y^{(-1)_F}$ (the multiplicative inverse of y in $\mathscr{F}$). Hence,

$$x/^E y = x \otimes y^{(-1)_E} = x \times y^{(-1)_F} = x/^F y. \quad \blacksquare$$

There is a very useful characterization of subfields.

Theorem 5.2   Let $\mathscr{F} = (F, +, \times)$ be a field, and assume that E is a subset of F satisfying the following conditions:

a. E contains at least two elements.

b. E is closed under subtraction (that is, $x - y \in E$ for all x, y in E).

c. $x \in E \wedge y \in E \wedge y \neq 0_F \Rightarrow x/y \in E$.

If we define $\oplus$ and $\otimes$ as the restriction of $+$ and $\times$ to E, then $(E, \oplus, \otimes)$ is a subfield of $\mathscr{F}$ (called the **subfield determined** by E).

Proof   Since E has at least two elements, there is an element z in E such that $z \neq 0_F$. Hence, by (c), $1_F = z/z \in E$. E is closed under taking of multiplicative inverses of nonzero elements: $x \in E \wedge x \neq 0_F \Rightarrow x^{-1} \in E$. (For, $x^{-1} = 1_F/x \in E$, by (c).) Therefore, E is closed under multiplication. (For, assume $x, y \in E$. If $y = 0_F$, then $x \times y = x \times 0_F = 0_F = y \in E$. If $y \neq 0_F$, then $x \times y = x \times (y^{-1})^{-1} = x/y^{-1} \in E$ by (c)). Now, by Theorem 3.9.2, $\mathscr{E} = (E, \oplus, \otimes)$ is a subdomain of $\mathscr{F}$. It only remains to verify that $\mathscr{E}$ is a field. Since $\mathscr{E}$ is an integral domain with at least two elements, it suffices to verify Axiom 9:

$$y \in E \wedge y \neq 0 \Rightarrow (\exists z)(z \in E \wedge y \times z = 1_F).$$

But this follows from the fact that E is closed under taking of multiplicative inverses of nonzero elements. $\blacksquare$

EXAMPLES

3.   Let $\mathscr{F} = (F, +, \times)$ be the field of real numbers. The following subsets E determine subfields of $\mathscr{F}$.

i. E is the set of rational numbers.

ii. E is the set of real numbers of the form $a + b\sqrt{2}$, where a and b are rational.

iii. E is the set of real numbers of the form $a + b\sqrt{3}$, where a and b are rational.

4.   Let $\mathfrak{F} = (F, +, \times)$ be the field of complex numbers. The following subsets E determine subfields of $\mathfrak{F}$.

i. F is the set of all real numbers.

ii. F is the set of all $a + bi$, where a and b are rational.

iii. F is the set of all $a + b\sqrt{2}$, where a and b are rational.

5.   Let $\mathfrak{F} = (F, +, \times)$ be the four-element field given in the example on p. 169. Then the subset $\{0, 1\}$ determines a subfield of $\mathfrak{F}$ (in fact, the field given in Example 4 on p. 167).

EXERCISES

3.   Let $\mathfrak{F}$ be the field of real numbers. Which of the following sets determine subfields of $\mathfrak{F}$?

   a.   The set of integers.

   b.   The set of all real numbers $a + b\sqrt[3]{2}$, where a and b are rational.

   c.   The set of all real numbers $a + b\sqrt{6}$, where a and b are rational.

   d.   The set of all real numbers $a + b\sqrt{2}$, where a and b are integers.

   e.   The set of all $a/2^n$, where a is an integer and n is a positive integer.

   f.   The set of all irrational numbers.

4.   Let $\mathfrak{F}$ be the field of complex numbers. Which of the following sets determine subfields of $\mathfrak{F}$?

   a.   The set of all bi, where b is real.

   b.   The set of all $a + bi$, where a and b are integers.

   c.   The set of all $a + bi$, where a and b are nonnegative real numbers.

   d.   The set of all $a + bu$, where a, b are rational numbers, and $u = (-1 + \sqrt{-3})/2$. (Notice that $u^2 + u + 1 = 0$, and, therefore, $u^3 = 1$.)

5.   a.   Prove that, if K_1 and K_2 determine subfields of a field $\mathfrak{F}$, then the intersection $K_1 \cap K_2$ also determines a subfield of $\mathfrak{F}$. More generally, show that the interaction of any collection of sets which determine subfields of $\mathfrak{F}$ again determines a subfield of $\mathfrak{F}$.

   b.   If K_1 (respectively, K_2) consists of all real numbers of the form $a + b\sqrt{2}$ (respectively, $a + b\sqrt{3}$), where a and b are rational, what is the intersection $K_1 \cap K_2$?

6. If $\mathfrak{F}$ is a field of characteristic p, where p is a prime, show that $N_{\mathfrak{F}}$, the set of natural numbers of $\mathfrak{F}$, contains p elements and determines a subfield of $\mathfrak{F}$ which is contained in every other subfield of $\mathfrak{F}$.

7. By a **prime field** we mean a field which contains no proper subfield (that is, no subfield other than itself). Find all prime fields. (*Hint*: Use the characteristic.)

In an integral domain $\mathscr{D}$, we defined the set $N_{\mathscr{D}}$ of natural numbers of $\mathscr{D}$ and the set $Z_{\mathscr{D}}$ of integers of $\mathscr{D}$, which were analogues of the standard natural numbers and integers. In a field $\mathfrak{F}$ we also have a quotient operation, which we can use to obtain an analogue of the system of rational numbers.

Definition Let $\mathfrak{F} = (F, +, \times)$ be a field. The set $Q_{\mathfrak{F}}$ consists of all elements of F of the form x/y, where $x \in Z_{\mathfrak{F}}$, $y \in Z_{\mathfrak{F}}$, and $y \neq 0_F$.

The elements of $Q_{\mathfrak{F}}$ are called **rational numbers** of $\mathfrak{F}$. Thus, the rational numbers of $\mathfrak{F}$ are the quotients of integers of $\mathfrak{F}$. Observe that $Z_{\mathfrak{F}} \subseteq Q_{\mathfrak{F}}$. (For, if $x \in Z_{\mathfrak{F}}$, then $x = x/1_F \in Q_{\mathfrak{F}}$.)

Theorem 5.3

a. $Q_{\mathfrak{F}}$ determines a subfield $\mathscr{Q}_{\mathfrak{F}} = (Q_{\mathfrak{F}}, \oplus, \otimes)$ of $\mathfrak{F}$.

b. Every subfield of $\mathfrak{F}$ contains $\mathscr{Q}_{\mathfrak{F}}$ as a subfield.

c. If $\mathfrak{F}$ has characteristic 0, then $\mathscr{Q} = (Q, +_Q, \times_Q)$ is isomorphic with $\mathscr{Q}_{\mathfrak{F}} = (Q_{\mathfrak{F}}, \oplus, \otimes)$.

Proof (a) $0_F \in Z_{\mathfrak{F}} \subseteq Q_{\mathfrak{F}}$ and $1_F \in Z_{\mathfrak{F}} \subseteq Q_{\mathfrak{F}}$. Since $0_F \neq 1_F$, $Q_{\mathfrak{F}}$ has at least two elements.

In addition, $Q_{\mathfrak{F}}$ is closed under subtraction. (For, assume $x, y, u, v \in Z_{\mathfrak{F}}$ with $y \neq 0_F$ and $v \neq 0_F$. Then, $x/y - u/v = ((x \times v) - (y \times u))/(y \times v)$. But, $(x \times v) - (y \times u) \in Z_{\mathfrak{F}}$ and $y \times v \in Z_{\mathfrak{F}}$.

Finally, if $r, s \in Q_{\mathfrak{F}}$ and $s \neq 0_F$, then $r/s \in Q_{\mathfrak{F}}$. (For, let $r = x/y$ and $s = u/v$, where $x, y, u, v \in Z_{\mathfrak{F}}$, and $y \neq 0_F$, $v \neq 0_F$. Then, $r/s = (x/y)/(u/v) = (x \times v)/(y \times u) \in Q_{\mathfrak{F}}$, since $x \times v \in Z_{\mathfrak{F}}$ and $y \times u \in Z_{\mathfrak{F}}$.)

Hence, by Theorem 5.2, $Q_{\mathfrak{F}}$ determines a subfield of $\mathfrak{F}$.

(b) Let $\mathscr{E} = (E, \oplus, \otimes)$ be a subfield of $\mathfrak{F}$. Since E contains 1_F and is closed under $+$, E is inductive, and, therefore, $N_{\mathfrak{F}} \subseteq E$. Since E contains 0_F and is closed under subtraction, it follows that $Z_{\mathfrak{F}} \subseteq E$. In turn, since E is closed under taking of quotients of nonzero elements, $Q_{\mathfrak{F}} \subseteq E$. By Theorems 5.1 and 5.2, $Q_{\mathfrak{F}}$ determines a subfield of $\mathscr{E}$.

(c)   Assume $\mathfrak{F}$ has characteristic 0. We already know that the function $\Phi(\alpha) = \alpha 1_F$, for any α in Z, is an isomorphism of $(Z, +_Z, \times_Z)$ with the system of integers of $\mathfrak{F}$ (Theorem 3.9.4b). Observe that, if α, β, γ, δ are in Z, $\beta \neq 0_Z$, $\delta \neq 0_Z$, and $(\alpha, \beta) \approx (\gamma, \delta)$ (that is, $\alpha \times_Z \delta = \gamma \times_Z \beta$), then

$$\frac{\Phi(\alpha)}{\Phi(\beta)} = \frac{\Phi(\gamma)}{\Phi(\delta)}.$$

(For, from $\alpha \times_Z \delta = \gamma \times_Z \beta$, $\Phi(\alpha \times_Z \delta) = \Phi(\gamma \times_Z \beta)$. Then

$$\Phi(\alpha) \times_Z \Phi(\delta) = \Phi(\gamma) \times_Z \Phi(\beta),$$

and, therefore,

$$\frac{\Phi(\alpha)}{\Phi(\beta)} = \frac{\Phi(\gamma)}{\Phi(\delta)}.$$

Thus, for any r in Q, we choose some pair (α, β) in r and define $\Psi(r) = \Phi(\alpha)/\Phi(\beta)$. Hence, $\Psi: Q \to Q_{\mathfrak{F}}$. Let us check that Ψ is an isomorphism of $\mathcal{Q}$ with $\mathcal{Q}_{\mathfrak{F}}$.

(i) Ψ is one-one.   (Assume $\Psi(r) = \Psi(s)$. Choose (α, β) in r and (γ, δ) in s. Then $\Phi(\alpha)/\Phi(\beta) = \Phi(\gamma)/\Phi(\delta)$.) Hence, $\Phi(\alpha) \times \Phi(\delta) = \Phi(\gamma) \times \Phi(\beta)$. Since Φ preserves multiplication, $\Phi(\alpha \times_Z \delta) = \Phi(\gamma \times_Z \beta)$, and, since Φ is one-one, $\alpha \times_Z \delta = \gamma \times_Z \beta$. Thus, $(\alpha, \beta) \approx (\gamma, \delta)$, and, therefore, $r = s$.)

(ii)   The range $\mathcal{R}(\Psi)$ is $Q_{\mathfrak{F}}$.   (Assume $z \in Q_{\mathfrak{F}}$. Then $z = x/y$, with x, y in $Z_{\mathfrak{F}}$ and $y \neq 0_F$. Since $\mathcal{R}(\Phi) = Z_{\mathfrak{F}}$, there are α, β in Z with $\beta \neq 0_Z$ such that $\Phi(\alpha) = x$ and $\Phi(\beta) = y$. Let r be the rational number $[(\alpha, \beta)]$. Then $\Psi(r) = \Phi(\alpha)/\Phi(\beta) = x/y = z$.)

(iii) $\Psi(r +_Q s) = \Psi(r) + \Psi(s)$.   (Let $(\alpha, \beta) \in r$ and $(\gamma, \delta) \in s$. Then $(\alpha\delta + \beta\gamma, \beta\delta) \in r +_Q s$. Hence,

$$\Psi(r +_Q s) = \frac{\Phi(\alpha\delta + \beta\gamma)}{\Phi(\beta\delta)} = \frac{\Phi(\alpha\delta) + \Phi(\beta\gamma)}{\Phi(\beta)\Phi(\delta)} = \frac{\Phi(\alpha)\Phi(\delta) + \Phi(\beta)\Phi(\gamma)}{\Phi(\beta)\Phi(\delta)}$$

$$= \frac{\Phi(\alpha)}{\Phi(\beta)} + \frac{\Phi(\gamma)}{\Phi(\delta)}$$

$$= \Psi(r) + \Psi(s).)$$

(iv) $\Psi(r \times_Q s) = \Psi(r) \times \Psi(s)$.   (Let $(\alpha, \beta) \in r$ and $(\gamma, \delta) \in s$. Then, $(\alpha\gamma, \beta\delta) \in r \times_Q s$.

$$\Psi(r \times_Q s) = \frac{\Phi(\alpha\gamma)}{\Phi(\beta\delta)} = \frac{\Phi(\alpha)\Phi(\gamma)}{\Phi(\beta)\Phi(\delta)} = \frac{\Phi(\alpha)}{\Phi(\beta)} \times \frac{\Phi(\gamma)}{\Phi(\delta)}$$

$$= \Psi(r) \times \Psi(s).) \quad \blacksquare$$

Corollary 5.4

a.   Let $\mathfrak{F} = (F, +, \times, <)$ be an ordered field, and let $\lessgtr$ denote the restriction of $<$ to $Q_{\mathfrak{F}}$. Then the ordered field of rational numbers $(Q, +_Q, \times_Q, <_Q)$ is isomorphic with $(Q_{\mathfrak{F}}, \oplus, \otimes, \lessgtr)$.

b.   For any two ordered fields $\mathcal{K}$ and $\mathfrak{F}$, the ordered rational subfields $\mathcal{Q}_{\mathcal{K}}$ and $\mathcal{Q}_{\mathfrak{F}}$ are isomorphic.

Proof   (a)   It suffices to show that the function Ψ defined in Theorem 5.3 is order preserving. We know, by the Remark to Theorem 3.9.5b on p. 150, that the isomorphism Φ is order preserving. Now, let r be a positive rational. We may take $(\alpha, \beta) \in r$, where α and β are positive integers. Hence, $\Phi(\alpha)$ and $\Phi(\beta)$ are positive. Therefore, $\Psi(r) = \Phi(\alpha)/\Phi(\beta)$ is positive. Thus, Ψ takes positive elements into positive elements. Hence,

$$r < s \Rightarrow s - r > 0 \Rightarrow \Psi(s - r) > 0 \Rightarrow \Psi(s) - \Psi(r) > 0 \Rightarrow \Psi(s) < \Psi(s).$$

Conversely, $r \not< s \Rightarrow s \leq r \Rightarrow \Psi(s) \leq \Psi(r) \Rightarrow \Psi(r) \not< \Psi(s)$. Thus, Ψ is order preserving.

(b)   The rational number systems of $\mathcal{K}$ and $\mathfrak{F}$ are each order isomorphic with $(Q, +_Q, \times_Q, <_Q)$, by virtue of Part (a). Therefore, they are isomorphic with each other (by the transitivity and symmetry of the isomorphism relation).

EXERCISES

8.   Let Θ be an isomorphism between fields $\mathfrak{F} = (F, +, \times)$ and $F^* = (F^*, +^*, \times^*)$. Since fields are integral domains, the properties of isomorphisms given in Theorem 3.9.3 hold for Θ. In addition, prove that, if $y \neq 0_F$, then $\Theta(y^{-1}) = (\Theta(y))^{-1}$ and $\Theta(x/y) = \Theta(x)/\Theta(y)$.

9.   Let $\mathfrak{F} = (F, +, \times)$ be a field. Show that $\mathfrak{F}$ is isomorphic with $\mathcal{Q} = (Q, +_Q, \times_Q)$ if and only if $\mathfrak{F}$ has characteristic 0 and $\mathfrak{F}$ has no proper subfield.

10.   a.   Let an integral domain $\mathcal{D} = (D, +, \times)$ be a subdomain of a field $\mathcal{K} = (K, +_K, \times_K)$. Show that the set $F = \{x/y : x \in D \wedge y \in D \wedge y \neq 0\}$ determines a subfield of $\mathcal{K}$ which is isomorphic with the quotient field of $\mathcal{D}$.

   b.   Show that the set $\{a/b : a \text{ and } b \text{ are integers} \wedge b \text{ is odd}\}$ determines a subdomain $\mathcal{D}$ of the field of rational numbers. Describe the quotient field of $\mathcal{D}$. (It suffices to find a field isomorphic to the quotient field of $\mathcal{D}$.)

   c.   Show that the set $\{a + b\sqrt{2} : a \text{ and } b \text{ are integers}\}$ determines a subdomain $\mathcal{D}$ of the field of real numbers. Describe the quotient field of $\mathcal{D}$.

11. Let $\mathfrak{F}_1$ (respectively, $\mathfrak{F}_2$) be the subfield of the field of real numbers determined by the set of all elements $a + b\sqrt{2}$ (respectively, $a + b\sqrt{3}$), where a and b are rational numbers. Is $\mathfrak{F}_1$ isomorphic with $\mathfrak{F}_2$?

12. a. Prove that any two two-element fields are isomorphic.

    b. Prove that any two three-element fields are isomorphic.

13. a. Find all subfields of $\mathcal{Q} = (Q, +_Q, \times_Q)$.

    b. Find all subfields of the field of all real numbers of the form $a + b\sqrt{2}$, where a and b are rational numbers.

    c. Find all subfields of the four-element field on p. 169.

    d. If p is a prime, find all subfields of the field $(Z_p, +_p, \times_p)$.

CHAPTER 5

THE REAL NUMBER SYSTEM

5.1 INADEQUACY OF THE RATIONALS

Although the system of rational numbers is suitable for ordinary every-day mathematical applications, such as bookkeeping or carpentry, it is inadequate for most scientific purposes. To see this, let us look at the familiar geometric interpretation of numbers as points on a line (Figure 5.1). We choose a point on the line and label it 0. This point is called the **origin**.

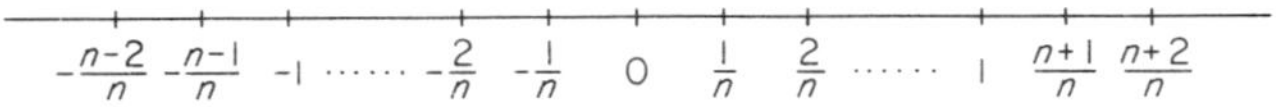

Figure 5.1

We also establish a fixed distance as the **unit length**. Take some positive integer n. Let us divide the unit length into n parts. (There is a simple ruler-and-compass construction enabling us to do this.) Taking one of these n parts, we lay it off k times to the right of 0, obtaining a point at a distance k/n from 0. If we lay off the same part k times to the left of 0, the point obtained is associated with the negative rational number $-k/n$.

Do *all* points on the line correspond to rational numbers? That this is not so was known by the mathematicians of ancient Greece. Merely construct a right triangle with both legs of unit length, as in Figure 5.2. By the Pythagorean Theorem,

$$c^2 = 1^2 + 1^2 = 1 + 1 = 2.$$

If we lay off the segment AB to the right of the origin on our line in Figure 5.1, the right-hand end point is at a distance c from the origin. But c cannot be a rational number, by virtue of the following result.

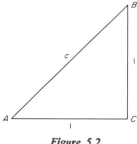

Figure 5.2

Theorem 1.1   There is no rational number c such that $c^2 = 2$.

Proof   Assume the contrary. Let c be a rational number such that $c^2 = 2$. We may assume that c is positive. (For, if c is negative, we consider $-c$ instead.) Now, c can be represented as a quotient of positive integers k/n. Of course, such a representation is not unique. (For example, $2k/2n$ also equals c.) However, we take the least positive integer n for which there is some k such that $c = k/n$. Now,

$$2 = c^2 = \left(\frac{k}{n}\right)^2 = \frac{k^2}{n^2}.$$

Therefore, $2n^2 = k^2$. Thus, k^2 is even, and, therefore, k must be even also. Hence, $k = 2j$ for some positive integer j. Thus,

$$2n^2 = (2j)^2 = 4j^2.$$

Dividing by 2, we obtain

$$n^2 = 2j^2.$$

Thus, n^2 is even, and, therefore, n is also even. Hence, $n = 2m$ for some positive integer m. Then,

$$c = \frac{k}{n} = \frac{2j}{2m} = \frac{j}{m}.$$

But, $m = n/2 < n$. Thus, c is representable as j/m, where m is a positive integer less than n. This contradicts the definition of n. Hence, the assumption that there is a rational number c such that $c^2 = 2$ has led to a contradiction.   ■

Sometimes one expresses the result of Theorem 1.1 by saying that the number $\sqrt{2}$ is irrational. We must realize, however, that we have not assigned any meaning yet to the expression $\sqrt{2}$. We are inclined to do so by the geometric picture given above: There are points on the line whose distance from the origin 0 are not rational numbers. We shall extend the rational number system to a larger system, the system of real numbers. The distances of all points on the line from the origin 0 will correspond to the real numbers.

The number $\sqrt{2}$ is not the only "irrational" number. For example, the number $\sqrt{3}$ is not rational, that is, there is no rational number c such that $c^2 = 3$. The proof of this fact is quite analogous to that of Theorem 1.1. In fact, assume that there is a rational number c such that $c^2 = 3$. Let us show that this assumption leads to a contradiction. Take the least positive integer n such that $c = k/n$ for some positive integer k. Then,

$$3 = c^2 = (k/n)^2 = k^2/n^2.$$

Hence, $3n^2 = k^2$. Thus, k^2 is divisible by 3. Since 3 is a prime, k must be divisible by 3. Let $k = 3j$, where j is a positive integer. Then,

$$3n^2 = k^2 = (3j)^2 = 9j^2.$$

Dividing by 3, we obtain $n^2 = 3j^2$. Hence, n^2 is divisible by 3, and, therefore, so is n. Let $n = 3m$, where m is a positive integer. Thus,

$$c = k/n = 3j/3m = j/m.$$

But, $m = n/3 < n$, contradicting the definition of n.

EXERCISES

1. Prove the assertion made above that, if n^2 is even, then n is even.

2. Show that $\sqrt{5}$ is irrational, that is, there is no rational number c such that $c^2 = 5$.

3. Show that $\sqrt[3]{2}$ is irrational, that is, there is no rational number c such that $c^3 = 2$.

4. Prove that $\sqrt{6}$ is irrational.

5. Prove that the following real numbers are irrational:

   a. $\sqrt{2} + \sqrt{3}$.
   b. $\sqrt{2} + \sqrt{3} + \sqrt{5}$.
   c. $\sqrt{2} + \sqrt[3]{2}$.

   (In this exercise we are assuming that the reader is intuitively familiar with the real numbers.)

6. Prove: If n and k are positive integers and $\sqrt[k]{n}$ is not an integer, then $\sqrt[k]{n}$ is irrational, that is, if there is no integer c such that $c^k = n$, then there is no rational number c such that $c^k = n$. Note: Theorem 1.1 and Exercises 2–4 are special cases of this result.

7. Prove: Any rational solution c of

$$x^k + a_{k-1}x^{k-1} + \cdots + a_1x + a_0 = 0$$

(where all the a_i's are integers) must be an integer. (Hence, if there is no integral solution, there is no rational solution. The result of Exercise 6 is the special case where the polynomial is $x^k - n$.)

Real numbers will have to be defined in such a way that, not only are the ordinary arithmetic operations of addition, subtraction, multiplication, and division (by a nonzero number) performable, but also such that every distance of a point on the line from the origin corresponds to some real number, and, vice versa, every real number corresponds to some distance. (Positive real numbers are to correspond to distances from the origin of points to the right of the origin and negative real numbers to distances from the origin of points to the left of the origin.) To formulate these conditions more precisely, we must introduce a few new ideas.

Definition   Let $\mathcal{F} = (F, +, \times, <)$ be an ordered field. By a cut[†] in $\mathcal{F}$ we mean an ordered pair (A, B) of subsets A and B of F such that

1. A and B are nonempty;
2. $A \cup B = F$;
3. $(x \in A \wedge y \in B) \Rightarrow x < y$.

Thus, a cut is a division of the whole set F into two nonempty parts such that every member of the first is smaller than every member of the second. Condition (3) implies that A and B are disjoint, that is, $A \cap B = \varnothing$. (For, if $u \in A \cap B$, then $u < u$, contradicting Axiom O1 for ordered fields.)

By a **gap** in $\mathcal{F}$ we mean a cut (A, B) in $\mathcal{F}$ such that A contains no maximum and B contains no minimum element. (We say that z is a **maximum** element of a set C if and only if $z \in C$ and no member of C is greater than z (equivalently, $(\forall u)(u \in C \Rightarrow u \leq z)$). Likewise, z is a **minimum** element of C if and only if $z \in C$ and $(\forall u)(u \in C \Rightarrow z \leq u)$.)

[†] Sometimes called a **Dedekind cut** in honor of Richard Dedekind (1831–1916), who first introduced this idea into mathematics.

EXAMPLES

1. Let $\mathfrak{F} = (F, +, \times, <)$ be any ordered field, and let $c \in F$. Define $A_1 = \{u : u \in F \wedge u \leq c\}$ and $B_1 = \{u : u \in F \wedge u > c\}$. Then (A_1, B_1) is a cut in $\mathfrak{F}$, but it is not a gap in $\mathfrak{F}$, since A_1 has a maximum element (namely, c itself).

Define

$$A_2 = \{u : u \in F \wedge u < c\} \quad \text{and} \quad B_2 = \{u : u \in F \wedge u \geq c\}.$$

Then (A_2, B_2) is a cut in $\mathfrak{F}$ but not a gap in $\mathfrak{F}$, since B_2 has a minimum element (namely, c itself).

Notice that, if we define

$$A_3 = \{u : u \in F \wedge u < c\} \quad \text{and} \quad B_3 = \{u : u \in F \wedge u > c\},$$

then (A_3, B_3) is not even a cut in $\mathfrak{F}$. (Why?)

The cuts (A_1, B_1) and (A_2, B_2) in $\mathfrak{F}$ are said to be **determined by the element c** of F. In the next example, we shall find a cut (A, B) which is not determined by any element of F; in other words, A does not have a maximum element and B does not have a minimum element, that is, (A, B) is a gap.

2. In this example, we shall be working with the ordered field, $\mathcal{Q}$, of rational numbers. Let $A = \{u : u \in Q \wedge (u < 0 \vee u^2 < 2)\}$ and $B = \{u : u \in Q \wedge 0 < u \wedge u^2 > 2\}$. Now, it is easy to see as follows that (A, B) is a cut in $\mathcal{Q}$.

   i. $A \neq \varnothing$ (since $1 \in A$); $B \neq \varnothing$ (since $2 \in B$).

   ii. Every negative rational and 0 are in A. For every positive rational u, either $u^2 < 2$ or $u^2 > 2$. (We already have proved that there is no rational number u such that $u^2 = 2$.) Thus, $A \cup B = Q$.

   iii. Assume $x \in A \wedge y \in B$. Since $y \in B$, $0 < y$. If $x < 0$, then $x < y$. If, on the other hand, $0 \leq x$, then $x^2 < 2$. However, $2 < y^2$, and, therefore, $x^2 < y^2$. It follows that $x < y$. (For, if $0 < y \leq x$, then $y^2 \leq x^2$.)

Thus, (A, B) is a cut in $\mathcal{Q}$. In addition, as we shall now show, (A, B) is a gap in $\mathcal{Q}$.

First, let us show that A does not contain a maximum element. To do this, we assume that $x \in A$ and we must find another element w in A such that $x < w$. If x is negative, we can take $w = 0$. So, we may confine our attention to the case where $0 \leq x$. We *guess* that w can be taken in the form $x + 1/n$, where n is a natural number of $\mathcal{Q}$, and we then try to find a suitable value of n. Now, in order for $x + 1/n$ to belong to A, we must satisfy the inequality $(x + 1/n)^2 < 2$. This is equivalent to

$$x^2 + 2x/n + 1/n^2 < 2,$$

or

$$2x/n + 1/n^2 < 2 - x^2.$$

(Notice that $2 - x^2 > 0$, since $x \in A$ and $0 \leq x$.) We know that $x < 2$ (for, if $x \geq 2$, then $x^2 > 2$, and, therefore, $x \notin A$). In addition, $1/n^2 \leq 1/n$ (since every natural number n is ≥ 1). Thus,

$$2x/n + 1/n^2 < 4/n + 1/n = 5/n.$$

Hence, it suffices to find n such that $5/n < 2 - x^2$, or, equivalently, $n > 5/(2 - x^2)$. But, given any rational number r (in particular, $r = 5/(2 - x^2)$), it is always possible to find a natural number n greater than r. (For, if $r \leq 0$, simply take $n = 1$. If $r > 0$, r can be written in the form j/k, where j and k are natural numbers. Then, we let $n = j + 1$.) Thus, A contains no maximum element.

Let us show that B contains no minimum element. Assume $y \in B$, that is, $y > 0$ and $y^2 > 2$. We must find v in B such that $v < y$. We *guess* that such a v can be found in the form $y - 1/n$, where n is a natural number. The condition that $y - 1/n \in B$ is equivalent to the inequality

$$(y - 1/n)^2 > 2,$$

that is,

$$y^2 - 2y/n + 1/n^2 > 2,$$

or

$$y^2 - 2 > 2y/n - 1/n^2.$$

(Notice that $y^2 - 2 > 0$, since $y \in B$.) Now, $2y/n > 2y/n - 1/n^2$. Hence, it suffices to satisfy:

$$y^2 - 2 > 2y/n,$$

or

$$n > 2y/(y^2 - 2).$$

But, as we have observed above, if we are given any rational number (in particular, $2y/(y^2 - 2)$), there is always a larger natural number. Thus, B contains no minimum element.

This completes the proof that (A, B) is a gap in $\mathscr{Q}$.

EXERCISES

8. In the rational number field $\mathcal{Q}$, let

$$A = \{u : u \in Q \wedge (u < 0 \vee u^2 < 3)\} \quad \text{and} \quad B = \{u : u \in Q \wedge 0 < u \wedge u^2 > 3\}.$$

Show that (A, B) is a gap in $\mathcal{Q}$.

9. If (A, B) is a cut in an ordered field $\mathfrak{F}$, prove that it is impossible for A to contain a maximum element and B to contain a minimum element.

Now we are able to formulate in a precise way conditions that the real number system should satisfy. The real number system must be an ordered field in which there are no gaps. (Intuitively, a gap would define a "hole" on the number line which would not correspond to any element of the field.) This leads to the following new notion.

Definition By a **complete ordered field** we mean an ordered field in which there are no gaps, that is, for every cut (A, B), either A has a maximum element or B has a minimum element.

EXAMPLE

3. The ordered field $\mathcal{Q}$ of rational numbers is not a complete ordered field. Example (2) above exhibited a gap (A, B) in $\mathcal{Q}$. (Intuitively, this gap corresponded to the "hole" in the rational numbers determined by $\sqrt{2}$.)

Our task consists in showing that the precise notion of a complete ordered field completely captures our intuitive ideas about the real number system. This task will be split into two parts.

I. We shall construct a complete ordered field. There are two principal ways of doing this: Cauchy sequences and Dedekind cuts. We shall adopt the approach via Cauchy sequences (due to Georg Cantor [1845–1918])[†]; the definitions and proofs are shorter and more elegant than those necessary under the other approach. The method of Dedekind cuts will be explained in detail in Appendix F.[‡]

[†] Cantor's method was improved upon by E. Heine [1821–1881] and had been anticipated by C. Meray [1835–1911]. Still another approach was developed by K. Weierstrass [1815–1897]. For details and references, see Manheim [1964].

[‡] Another way of constructing a complete ordered field is by means of infinite decimals. This is the clumsiest and most intricate method. For sketches of this procedure, see Auslander [1969] and Lightstone [1965].

II.   It will be proved that any two complete ordered fields are isomorphic. Hence, any complete ordered field will have the same mathematical properties as the complete ordered field constructed in I. This will mean that, for all mathematical purposes, there is *essentially* only one complete ordered field. It also tells us that the completeness property is all that we need to characterize the real number system.

In order to carry out the program that we have just sketched in I and II, certain important auxiliary ideas are necessary. Some of these ideas will be introduced and studied in the next two sections.

EXERCISES

10.   If u is rational and v is irrational, prove that $u + v, u - v, uv, u/v, v/u$ are all irrational.

11.   If u and v are irrational, must $u + v$ be irrational? Must uv be irrational?

5.2 ARCHIMEDEAN ORDERED FIELDS

Definition   Let $\mathfrak{F} = (F, +, \times, <)$ be an ordered field. We say that $\mathfrak{F}$ is **Archimedean** if and only if, for any positive elements x and y in F, there exists a natural number n such that $nx > y$, that is,

$$(\forall x)(\forall y)(0 < x \wedge 0 < y \Rightarrow (\exists n)(n \in N \wedge nx > y)).$$

Intuitively, this means that, if x and y are any positive elements of F, then, if x is added to itself often enough, we eventually obtain an element greater than y.

EXAMPLES

1.   The rational number field $\mathcal{Q}$ is Archimedean. For, if r and s are positive rational numbers, then $r = j/k$ and $s = l/m$ where j, k, l, m are natural numbers. Let $n = k(l + 1)$. Then

$$nr = k(l + 1)(j/k) = (l + 1)j \geq l + 1 > l \geq l/m = s.$$

(For example, if $r = \frac{2}{3}$ and $s = \frac{32}{5}$, then $99 \cdot \frac{2}{3} > \frac{32}{5}$.)

2.   Consider the quotient field $\mathfrak{F}$ of the ordered integral domain of polynomials with integral coefficients (see Example 3, p. 176). Remember that the elements of this field are representable as quotients

$$\frac{a_n x^n + \cdots + a_1 x + a_0}{b_k x^k + \cdots + b_1 x + b_0}$$

of polynomials with integral coefficients a_i $(i = 0, \ldots, n)$, b_j $(j = 0, \ldots, k)$, $b_k \neq 0$. Such a quotient is positive if and only if a_n and b_k have the same sign, that is, $a_n \times b_k > 0$. In particular, a polynomial $a_n x^n + \cdots + a_1 x + a_0$, being equal to

$$\frac{a_n x^n + \cdots + a_1 x + a_0}{1},$$

is positive if and only if $a_n > 0$. Thus, any monic polynomial $x^n + a_{n-1} x^{n-1} + \cdots + a_1 x + a_0$ is positive. This ordered field $\mathfrak{F}$ is non-Archimedean. To see this, consider the elements 1 and x. For any natural number n, $n1 < x$, since $n1 = n$ and $x - n$ is positive.

EXERCISES

1.$^\text{C}$ If $\mathfrak{F}$ and $\mathscr{G}$ are isomorphic ordered fields and $\mathfrak{F}$ is Archimedean, prove that $\mathscr{G}$ is also Archimedean.

2.   Prove that $\mathfrak{F}$ is Archimedean if and only if, for any positive x and y in F, there is a natural number k in $N_{\mathfrak{F}}$ such that $k \times x > y$. (*Hint*:   Every natural number in $N_{\mathfrak{F}}$ is of the form $n1$, where n is a natural number. Observe that $nx = (n1) \times x$.)

In order to get a clearer understanding of the Archimedean property, we must introduce a few ideas that will turn out to be important in their own right.

Definitions   Let $\mathfrak{F} = (F, +, \times, <)$ be an ordered field. Let $A \subseteq F$ and $z \in F$. Then

i. z is an **upper bound** of A if and only if $(\forall x)(x \in A \Rightarrow x \leq z)$.

ii. z is a **lower bound** of A if and only if $(\forall x)(x \in A \Rightarrow z \leq x)$.

Upper or lower bounds of A may or may not belong to A. For example, if $A = \{x : x \leq 1\}$, then the upper bound 1 of A belongs to A, while the upper bound 2 of A does not belong to A. If $B = \{x : x < 1\}$, then clearly no upper bound of B belongs to B. It is obvious that any element greater than an upper bound of any set A is also an upper bound of A, and any element less than a lower bound of A is also a lower bound of A.

iii. A is **bounded above** if and only if there exists some z such that z is an upper bound of A.

iv. A is **bounded below** if and only if there exists some z such that z is a lower bound of A.

v. A is **bounded** if and only if A is bounded above and below.

A set may be bounded above, but not below, for example, $A = \{x : x \leq 1\}$. Likewise, a set may be bounded below but not above, for example, $A = \{x : x \geq 1\}$. Of course, there are sets which are bounded neither above nor below, for example, the whole field F.

vi. z is a **maximum element** of A if and only if

$$z \in A \wedge (\forall x)(x \in A \Rightarrow x \leq z).$$

vii. z is a **minimum element** of A if and only if

$$z \in A \wedge (\forall x)(x \in A \Rightarrow z \leq x).$$

Observe that a set need not have a maximum element. As an example, take $A = \{x : x < 1\}$. Then, for any z in A, $(z + 1)/2$ is a larger element of A. Likewise, a set need not have a minimum element, for example, $A = \{x : x > 1\}$. However, if a maximum element of A exists, it is unique. (For, if z_1 and z_2 are maximum elements of A, then $z_1 \leq z_2$ and $z_2 \leq z_1$. Therefore, $z_1 = z_2$.) Likewise, if a minimum element of A exists, it is unique.

EXERCISES

3.^{C} Prove that z is a maximum element of A if and only if $z \in A$ and z is an upper bound of A.

4.^{C} Prove that z is a minimum element of A if and only if $z \in A$ and z is a lower bound of A.

5. Prove that any finite nonempty subset has a maximum and a minimum element.

6. Prove that every element of an ordered field is trivially an upper bound and a lower bound of the empty set.

Theorem 2.1 Let $\mathscr{F} = (F, +, \times, <)$ be an ordered field. Then $\mathscr{F}$ is Archimedean if and only if the set $N_{\mathscr{F}}$ of natural numbers of $\mathscr{F}$ is *not* bounded above.

Proof (i) Assume $\mathfrak{F}$ is Archimedean. Let us show that the supposition that the set $N_{\mathfrak{F}}$ is bounded above leads to a contradiction. Let z be an upper bound of $N_{\mathfrak{F}}$. It follows that z is positive. Applying the Archimedean property to 1 and z, we obtain a natural number n such that $n1 > z$. But $n1 \in N_{\mathfrak{F}}$, contradicting the assumption that z is an upper bound of $N_{\mathfrak{F}}$.

(ii) Conversely, assume $\mathfrak{F}$ is not Archimedean. Then there exist positive elements x and y of F such that, for every natural number n, $nx \leq y$. Thus, $(n1)x \leq y$, and, therefore, $n1 \leq y/x$. Then y/x is an upper bound of the set of natural numbers of $\mathfrak{F}$. Hence, $N_{\mathfrak{F}}$ is bounded above. ∎

EXAMPLES

3. The field of rational functions

$$\frac{a_n x^n + \cdots + a_1 x + a_0}{b_k x^k + \cdots + b_1 x + b_0}$$

(see Example 3, p. 176) is non-Archimedean, since $n < x$ for all natural numbers n.

4. The rational subfield of any ordered field $\mathfrak{F}$ is Archimedean, since $j/k < j + 1 \in N_{\mathscr{Q}}$ for every positive rational number j/k in $Q_{\mathfrak{F}}$.

Corollary 2.2 Let $\mathfrak{F} = (F, +, \times, <)$ be an ordered field. Then $\mathfrak{F}$ is Archimedean if and only if

$$(\forall z)(z \in F \wedge 0 < z \Rightarrow (\exists n)(n \in N_{\mathfrak{F}} \wedge 1/n < z)).$$

(*Note:* A positive element z of F such that $z < 1/n$ for all natural numbers n of $\mathfrak{F}$ is called an **infinitesimal**. Corollary 2.2 states that the Archimedean property is equivalent to the nonexistence of infinitesimals.)

Proof Assume $\mathfrak{F}$ is Archimedean. Assume $z \in F$ and $0 < z$. By Theorem 2.1, $N_{\mathfrak{F}}$ is not bounded above. Hence, there is some n in $N_{\mathfrak{F}}$ such that $1/z < n$. Therefore, $1/n < z$. Conversely, assume $\mathfrak{F}$ is not Archimedean. By Theorem 2.1, $N_{\mathfrak{F}}$ is bounded above. Let u be an upper bound of $N_{\mathfrak{F}}$. Then $n < u$ for all n in $N_{\mathfrak{F}}$. Hence, $1/u < 1/n$ for all n in $N_{\mathfrak{F}}$. Thus, there is some positive element z, namely $1/u$, such that there is no n in $N_{\mathfrak{F}}$ for which $1/n \leq z$. ∎

EXERCISES

7.^{C} Prove that every ordered subfield of an Archimedean ordered field must also be Archimedean. However, show that every non-Archimedean ordered field has an Archimedean subfield.

8.^{C} If $\mathfrak{F} = (F, +, \times, <)$ is an Archimedean ordered field, $x \in F$, and $x < 1/n$ for all n in $N_\mathfrak{F}$, prove that $x \le 0$.

Theorem 2.3   In an Archimedean ordered field $\mathfrak{F} = (F, +, \times, <)$,

$$(\forall x)(\exists! n)(n \in Z_\mathfrak{F} \wedge n \le x < n + 1).$$

(Every element lies between two consecutive integers.)

Proof   Assume $x \in F$. Since $\mathfrak{F}$ is Archimedean, Theorem 2.1 implies that there is some k in $N_\mathfrak{F}$ such that $x < k$. Likewise, there is some j in $N_\mathfrak{F}$ such that $-x < j$. Hence, $-j < x < k$. Thus, there exist natural numbers m such that $-j < x < -j + m$ (in particular, we can take $m = j + k$). Let h be the least natural number such that $x < -j + h$, and, finally, let $n = -j + h - 1$. Thus, $x < n + 1$. By the definition of h, we have $x \ge -j + (h - 1) = n$. We have shown that $n \le x < n + 1$.

To prove the uniqueness of n, assume that there is another integer $i \ne n$ such that $i \le x < i + 1$. Now, either $i < n$ or $n < i$. If $i < n$, then $i + 1 \le n$, and, therefore, $x < i + 1 \le n$, contradicting $n \le x$. Similarly, if $n < i$, then $n + 1 \le i$, and, therefore, $x < n + 1 \le i$, contradicting $i \le x$.   ∎

Notation   In an Archimedean ordered field, the unique integer n such that $n \le x < n + 1$ is denoted $[x]$ and is called the **greatest integer less than or equal to x.**

EXERCISES

In an Archimedean ordered field, prove:

9. $x - 1 \le [x] \le x < [x] + 1$.

10.   If k is an integer, $[k + x] = k + [x]$.

11. $[2x] = \begin{cases} 2[x] \\ \quad \text{or} \\ 2[x] + 1. \end{cases}$

12. $[x] + [y] \le [x + y] \le [x] + [y] + 1.$

13. $[x] + [-x] = \begin{cases} 0 \\ \text{or} \\ -1. \end{cases}$

14. If n is a positive integer, $x/n - [x/n] \le (n-1)/n.$

15. $[x/4] = [[x/2]/2].$

Definition Let $\mathfrak{F} = (F, +, \times, <)$ be an ordered field and let $A \subseteq F$. Then A is **dense in** $\mathfrak{F}$ if and only if

$$(\forall x)(\forall y)(x \in F \wedge y \in F \wedge x < y \Rightarrow (\exists z)(z \in A \wedge x < z < y)).$$

(Between any two elements of F, there is at least one element of A.)

EXERCISE

16. If A is dense in $\mathfrak{F}$, show that, between any two elements of F, there exist infinitely many elements of A.

Theorem 2.4 If $\mathfrak{F}$ is an Archimedean ordered field, then the subfield $Q_\mathfrak{F}$ of rational numbers is dense in $\mathfrak{F}$.

Proof Assume x and y in F with $x < y$. Then $y - x > 0$, and, therefore, $1/(y - x) > 0$. Since $\mathfrak{F}$ is Archimedean, there is some natural number n with $1/(y - x) < n$. Thus, $1/n < y - x$. Choose the least such positive integer n. Take $k = [nx + 1]$. Then $nx < k \le nx + 1$. Hence,

$$x < k/n \le x + 1/n < x + (y - x) = y.$$

Thus, the rational number k/n lies between x and y.   ∎

EXERCISE

17. If $\mathfrak{F}$ is an ordered field, prove that $\mathfrak{F}$ is Archimedean if an only if $Q_\mathfrak{F}$ is dense in $\mathfrak{F}$.

Our intuitive geometric picture of the real number system indicates that it should have the Archimedean property. The set of points on a line corresponding to the natural numbers is not bounded above, and, therefore, the Archimedean property must hold. To obtain the real number system, will it suffice simply to construct an Archimedean extension of the rational number system? A negative answer follows from the example below, which shows the existence of nonisomorphic Archimedean ordered fields $\mathfrak{F}$ and $\mathscr{G}$ neither of which is isomorphic to the rational number system.

EXAMPLE

5.   Let F be the set of all real numbers of the form $a + b\sqrt{2}$, where a and b are rational numbers. Let G be the set of all real numbers of the form $a + b\sqrt{3}$, where a and b are rational numbers. It is easy to check that both F and G are closed under subtraction and division (by nonzero elements). Hence F and G determine subfields $\mathfrak{F}$ and $\mathscr{G}$ of the real number system. $\mathfrak{F}$ is not isomorphic to $\mathcal{Q}$, since $\mathfrak{F}$ contains an element c such that $c^2 = 2$, while $\mathcal{Q}$ contains no such element. Likewise, $\mathscr{G}$ is not isomorphic to $\mathcal{Q}$, since $\mathscr{G}$ contains an element d such that $d^2 = 3$. Finally, $\mathfrak{F}$ and $\mathscr{G}$ are not isomorphic to each other, since $\mathfrak{F}$ does not contain an element c such that $c^2 = 3$. This is proved in the following way.

Assume $c^2 = 3$ with $c \in F$. Then $c = a + b\sqrt{2}$, with rational a and b. Then $3 = c^2 = a^2 + 2ab\sqrt{2} + 2b^2$.

Case 1 $a = 0$. Then $c = b\sqrt{2}$ and $3 = c^2 = 2b^2$. Let $b = m/n$, where m and n are positive integers and n is the least positive integer such that $b = l/n$ for some positive integer l. Then $3 = (2m^2)/n^2$. Hence, $3n^2 = 2m^2$. Therefore, $3n^2$ is even. Hence, n^2 is even, and so, n is even. Let $n = 2k$ for some positive integer k. Thus, $2m^2 = 12k^2$. Hence, $m^2 = 6k^2$. Thus, m^2 is even, and, therefore, so is m. Thus, $m = 2j$ for some positive integer j. Hence, $b = m/n = 2j/2k = j/k$ and $k < n$, contradicting the definition of n.

Case 2 $b = 0$. Then there is a rational number a such that $a^2 = 3$, which is impossible.

Case 3 $a \neq 0$ and $b \neq 0$. Then

$$\sqrt{2} = \frac{3 - a^2 - 2b^2}{2ab} \in Q,$$

which is impossible.

EXERCISES

18.   If $\mathfrak{F}$ and $\mathscr{G}$ are isomorphic fields, k and n are positive integers, and $\mathfrak{F}$ contains an element c such that $c^k = n1$, prove that $\mathscr{G}$ contains an element u such that $u^k = n1$.

19.   If m and n are integers such that m and n are relatively prime, and if k is a positive integer, then if there is a rational number c such that $c^k = m/n$, prove that there are integers i, j such that $c = i/j$, $i^k = m$, and $j^k = n$.

5.3 LEAST UPPER BOUNDS AND GREATEST LOWER BOUNDS

Definition Let $\mathfrak{F} = (F, +, \times, <)$ be an ordered field. Let $A \subseteq F$ and $z \in F$. We say that z is a **least upper bound** (lub)[†] of A if and only if:

i. z is an upper bound of A, and

ii. for any upper bound u of A, $z \leq u$.

Thus, z is a least upper bound of A if and only if z is the minimum element of the set of upper bounds of A.

Remarks

1. If a set A has a lub, then it has a unique lub. For, by condition (ii) of the definition, if z_1 and z_2 are lubs of A, then $z_1 \leq z_2$ and $z_2 \leq z_1$, which implies that $z_1 = z_2$. If a lub of A exists, we shall sometimes denote it by $\text{lub}(A)$.

2. If a lub z of a set A exists, then z may or may not belong to A. If $A = \{x : x < 1\}$, then 1 is the lub of A and $1 \notin A$. If $A = \{x : x \leq 1\}$, then 1 is the lub of A and $1 \in A$.

3. The empty set does not have a lub. For, every element of the field is an upper bound of the empty set, and the whole field does not have a minimum.

4. Clearly, every set which is not bounded above has no lub.

5. If A has a maximum element z, then z is the lub of A.
For, a maximum element of A is clearly an upper bound of A, and, since z is a member of A, z is less than or equal to every upper bound of A.

Lemma 3.1 If (A, B) is a cut in an ordered field $\mathfrak{F}$, then (A, B) is a gap if and only if A does not have a lub.

Proof (i) Assume (A, B) is a gap. Then A has no maximum and B has no minimum. Now, B is the set of upper bounds of A. For, on the one hand, every element of B is, by definition of a cut, an upper bound of A. On the other hand, since A has no maximum, no element of A can be an upper bound of A. (Remember that a member of A which is an upper bound of A must be a maximum element of A.) Since B is the set of upper bounds of A and B has no minimum element, A has no lub.

[†] Also called a **supremum** of A.

(ii)   Assume (A, B) is not a gap. Then either A has a maximum or B has a minimum.

Case a A has a maximum z. By Remark 5 above, z is the lub of A.

Case b B has a minimum z. Then A has no maximum. Now, B is the set of upper bounds of A. (The same argument that was given in Part i applies here.) Hence, z is a lub of A.

In either case, A has a lub.   ■

It follows from Lemma 3.1 that, in the field $\mathcal{Q}$ of rational numbers, there are sets A which are bounded above, but do not have a lub. An example of such a set is

$$A = \{u : u \in Q \wedge (u < 0 \vee u^2 < 2)\}.$$

If we let $B = \{u : u \in Q \wedge (u > 0 \wedge u^2 > 2)\}$, then (A, B) is a cut which is a gap. (This was proved on pp. 189–190.) Hence, by Lemma 3.1, A does not have a lub. It is clear that A is bounded above, since any element of B (for example, 2) is an upper bound of A.

EXERCISES

1.   Assume $w \in F$. Let $A = \{u : u \in F \wedge u < w\}$. Show that $\text{lub}(A) = w$. Similarly, if $C = \{u : u \in F \wedge u \leq w\}$, show that $\text{lub}(C) = w$.

2.   Find the lub of the following subsets of the rational number field, if such a lub exists.
   a.   The set of positive rational numbers.
   b.   The set of negative rational numbers.
   c.   The set of all elements $n/(n + 1)$, where n is any natural number.
   d.   The set of all elements $1/n$, where n is any natural number.

There is another important notion, symmetric to that of least upper bound.

Definition   Let $\mathcal{F} = (F, +, \times, <)$ be an ordered field. Let $A \subseteq F$ and $z \in F$. Then z is said to be a **greatest lower bound** (glb)† if and only if

i. z is a lower bound of A, and,

ii.   for any lower bound u of A, $u \leq z$.

Thus, z is a glb of A if and only if z is the maximum element of the set of lower bounds of A.

---

† Also called an **infimum** of A.

The following facts about glbs can be justified in the same way as the corresponding facts about lubs. Their verification is left to the reader.

1*.   If a set A has a glb, then it has a unique glb (denoted glb(A)).

2*.   If a glb z of a set A exists, z may or may not belong to A.

3*.   The empty set does not have a glb.

4*.   Every set which is not bounded below has no glb.

5*.   If A has a minimum element z, then z is a glb of A.

EXERCISES

3.   If (A, B) is a cut in an ordered field, prove that (A, B) is a gap if and only if B does not have a glb.

4.   In the rational number field, show that there is a set A which is bounded below but does not have a glb.

5.   Find the glb (if it exists) of the subsets of the rational number field given in Exercise 2 on p. 200.

Definition   Let $\mathfrak{F} = (F, +, \times, <)$ be an ordered field.

We say that $\mathfrak{F}$ has the **least upper bound (lub) property** if and only if every nonempty subset of F which is bounded above has a lub.

We say that $\mathfrak{F}$ has the **greatest lower bound (glb) property** if and only if every nonempty subset of F which is bounded below has a glb.

EXAMPLE

1.   We already have observed that the rational number field has neither the lub property nor the glb property (see p. 200).

Theorem 3.2   Let $\mathfrak{F} = (F, +, \times, <)$ be an ordered field. Then $\mathfrak{F}$ has the lub property if and only if $\mathfrak{F}$ has the glb property.

Proof   Assume $\mathfrak{F}$ has the lub property. Assume that A is a nonempty subset of F which is bounded below. We must show that A has a glb. Let $B = \{-u : u \in A\}$. (Intuitively, B is obtained by rotating A 180° around the origin. For example, if $A = \{u : u > -1\}$, then $B = \{u : u < 1\}$.) Notice that

(*)   For any x, x is an upper bound of B if and only if $-x$ is a lower bound of A.

(For, assume x is an upper bound of B. Take any element y in A. Then $-y \in B$. Hence, $-y \leq x$. Therefore, $-x \leq y$. Thus, $-x$ is a lower bound of A. Conversely, assume $-x$ is a lower bound of A. Take any element b in B. Then $-b \in A$. Hence, $-x \leq -b$. Therefore, $b \leq x$. Thus, x is an upper bound of B.)

Now, since A is bounded below, let z be a lower bound of A. Then, by (*), $-z$ is an upper bound of B. Hence, B is bounded above. Since $\mathscr{F}$ has the lub property, B has a lub w. Let us show that $-w$ is a glb of A.

(i)   Since w is an upper bound of B, (*) implies that $-w$ is a lower bound of A.

(ii)   Assume v is a lower bound of A. Then, by (*), $-v$ is an upper bound of B. Since $w = \text{lub}(B)$, $w \leq -v$. Hence, $v \leq -w$. By definition, (i) and (ii) imply that $-w$ is a lower bound of A.

We have shown that, if $\mathscr{F}$ has the lub property, then $\mathscr{F}$ has the glb property. The converse is proved in a perfectly analogous way and is left as an exercise for the reader.   ■

We now shall show that the lub property (or, by Theorem 3.2, the equivalent glb property) is equivalent to the completeness property of an ordered field. In fact, sometimes the lub property is used to define the notion of a complete ordered field.

Theorem 3.3   Let $\mathscr{F} = (F, +, \times, <)$ be an ordered field. Then $\mathscr{F}$ is complete if and only if $\mathscr{F}$ has the lub property (or, equivalently, the glb property).

Proof   (1)   Assume $\mathscr{F}$ is complete, that is, there are no gaps in $\mathscr{F}$. Now, assume that A is a nonempty subset of F which is bounded above. We must show that A has a lub. Let

$$A_1 = \{u : u \in F \wedge (\exists v)(v \in A \wedge u \leq v)\} \quad \text{and} \quad B_1 = F - A_1.$$

A_1 consists of all elements of F which are less than or equal to some element of A. It is clear that $A \subseteq A_1$ and that A and A_1 have the same upper bounds. Let us check that (A_1, B_1) is a cut in $\mathscr{F}$.

(i) $A_1 \neq \emptyset$, since $A \subseteq A_1$. In addition, $B_1 \neq \emptyset$. (For, if w is an upper bound of A, then $w + 1 \notin A_1$. Hence, $w + 1 \in B_1$.)

(ii)   By definition of B_1, $F = A_1 \cup B_1$.

(iii)   Assume $a \in A_1$ and $b \in B_1$. Since $b \in B_1$, $b \notin A_1$. Hence, $b \not< a$, and so, $a < b$.

Thus, (A_1, B_1) is a cut in $\mathfrak{F}$. Since $\mathfrak{F}$ is complete, either A_1 has a maximum or B_1 has a minimum.

Case a A_1 has a maximum z. Then z is a lub of A_1. Since A and A_1 have the same upper bounds, z is also a lub of A.

Case b B_1 has a minimum z. Therefore, A_1 has no maximum. Now, z is a lub of A. For, by (iii) above, the element z of B_1 is an upper bound of A_1, and, therefore, also an upper bound of A. Now assume w is any upper bound of A. Then w is also an upper bound of A_1. Since A_1 has no maximum, $w \notin A_1$, and, therefore, $w \in B_1$. Since z is a minimum of B_1, $z \leq w$.

In either case, A has a lub.

(2)   Assume $\mathfrak{F}$ has the lub property. Let (A, B) be any cut in $\mathfrak{F}$. To prove completeness, we must show that (A, B) is not a gap. Now, A is bounded above, since any element of B is an upper bound of A. Hence, by the lub property, A has a lub. Therefore, by Lemma 3.1, (A, B) is not a gap.   ∎

Using Theorem 3.3, we now can easily establish a relation between completeness and the Archimedean property.

Theorem 3.4   Every complete ordered field $\mathfrak{F} = (F, +, \times, <)$ is Archimedean.

Proof   Assume that $\mathfrak{F}$ is complete. By Theorem 2.1, it suffices to show that the set $N_{\mathfrak{F}}$ of natural numbers of $\mathfrak{F}$ is not bounded above. Assume, for the sake of contradiction, that $N_{\mathfrak{F}}$ is bounded above. Since $\mathfrak{F}$ is complete, we may conclude by Theorem 3.3 that $N_{\mathfrak{F}}$ has a lub z. Since $z - 1 < z$, $z - 1$ is not an upper bound of $N_{\mathfrak{F}}$. Hence, $z - 1 < n$ for some n in $N_{\mathfrak{F}}$. Then $z < n + 1$. But, since $n + 1 \in N_{\mathfrak{F}}$, this contradicts the fact that z is an upper bound of $N_{\mathfrak{F}}$.   ∎

By virtue of Theorem 3.4, all the results of Section 2 concerning Archimedean ordered fields can be applied to any complete ordered field. The converse of Theorem 3.4 is obviously false. The rational number field is Archimedean but not complete.

EXERCISES

6. Prove that, in an Archimedean ordered field $\mathfrak{F}$ which contains at least one irrational element, the set $F - Q_{\mathfrak{F}}$ of irrational elements is dense in $\mathfrak{F}$. (*Hint*: Let r denote an irrational element. If $x < y$, consider xr and yr, and use the denseness of $Q_{\mathfrak{F}}$ in $\mathfrak{F}$.)

7. In a complete ordered field $\mathfrak{F}$, prove that the set of irrational elements is dense in $\mathfrak{F}$.

8. Prove that, in a complete ordered field, there is a unique positive element z such that $z^2 = 2$.

It is convenient to derive here another fact about the Archimedean property which will be of use later in demonstrating that any two complete ordered fields are isomorphic.

Lemma 3.5 Let $\mathfrak{F} = (F, +, \times, <)$ be an Archimedean ordered field. Let $z \in F$ and $A = \{r : r \in Q_\mathfrak{F} \wedge r < z\}$. Then $z = \mathrm{lub}(A)$.

Proof (i) By definition of A, z is an upper bound of A.

(ii) Assume w is an upper bound of A. We must show that $z \leq w$. Assume, for the sake of contradiction, that $w < z$. By the density of the rational subfield in $\mathfrak{F}$ (see Theorem 2.4), there is a rational number r such that $w < r < z$. But, then, $r \in A$, and this contradicts the fact that w is an upper bound of A.

Thus, we have shown that $z = \mathrm{lub}(A)$.   ■

5.4 THE CATEGORICITY OF THE THEORY OF COMPLETE ORDERED FIELDS

A complete ordered field has been defined as an ordered field in which no cut is a gap. In Section 5.3, it was shown that the least upper bound property and the greatest lower bound property are each equivalent to completeness. In this section, we shall show that any two complete ordered fields are isomorphic, that is, that the theory of complete ordered fields is categorical.

Theorem 4.1 Any two complete ordered fields $\mathscr{K} = (K, +_K, \times_K, <_K)$ and $\mathfrak{F} = (F, +_F, \times_F, <_F)$ are isomorphic.

Proof Let Θ be an isomorphism of the rational subfield $\mathcal{Q}_{\mathscr{K}}$ of $\mathscr{K}$ with the rational subfield $\mathcal{Q}_\mathfrak{F}$ of $\mathfrak{F}$ (see Corollary 4.5.4b). Thus, $\Theta : \mathcal{Q}_{\mathscr{K}} \xrightarrow[\text{onto}]{1\text{-}1} \mathcal{Q}_\mathfrak{F}$, and

a. $\Theta(x +_K y) = \Theta(x) +_F \Theta(y)$ for all x and y in $\mathcal{Q}_{\mathscr{K}}$;
b. $\Theta(x \times_K y) = \Theta(x) \times_F \Theta(y)$ for all x and y in $\mathcal{Q}_{\mathscr{K}}$;
c. $x <_K y \Leftrightarrow \Theta(x) <_F \Theta(y)$ for all x and y in $\mathcal{Q}_{\mathscr{K}}$.

Take any z in K. Let

$$A_z = \{x : x \in Q_{\mathscr{K}} \wedge x <_K z\} \quad \text{and} \quad B_z = \{\Theta(x) : x \in A_z\}.$$

B_z is bounded above. (To see this, notice that, since $\mathcal{K}$ is complete, $\mathcal{K}$ is Archimedean by Theorem 3.4. Hence, there is a natural number n such that $n >_K z$. Now, $\Theta(n)$ is clearly an upper bound of B_z. For, if $u \in B_z$, then $u = \Theta(x)$ for some x in A_z. Hence, $x <_K z <_K n$. By (c), $u = \Theta(x) <_F \Theta(n)$.) Define $W(z)$ to be the lub of B_z in $\mathcal{F}$. Thus, $W: K \to F$. We shall divide the remainder of the proof that W is an isomorphism of $\mathcal{K}$ with $\mathcal{F}$ into a few simple steps.

(i) $z_1 <_K z_2 \Leftrightarrow W(z_1) <_F W(z_2)$.

Proof If $z_1 <_K z_2$, take r_1 in $Q_{\mathcal{K}}$ such that $z_1 <_K r_1 <_K z_2$. (We know that $Q_{\mathcal{K}}$ is dense in $\mathcal{K}$ by Theorem 2.4 and Theorem 3.4.) Similarly, take r_2 in $Q_{\mathcal{K}}$ such that $r_1 <_K r_2 <_K z_2$. Then, $\Theta(r_1)$ is an upper bound of B_{z_1} and $\Theta(r_2) \in B_{z_2}$. Hence, $W(z_1) \leq_F \Theta(r_1) <_F \Theta(r_2) \leq_F W(z_2)$.

(ii) W is one-one.

This follows immediately from (i).

(iii) W is an extension of Θ.

Proof Assume $z \in Q_{\mathcal{K}}$. Then, since $\mathcal{K}$ is Archimedean, Theorem 3.5 tells us that $z = \mathrm{lub}(A_z)$. It follows easily that $\Theta(z) = \mathrm{lub}(B_z) = W(z)$. (In particular, $W(0_K) = 0_F$ and $W(1_K) = 1_F$.)

(iv)   The range of W is F.

Proof Assume $u \in F$. Let $D_u = \{v : v \in Q_{\mathcal{F}} \wedge v <_F u\}$. Then, by Lemma 3.5, $u = \mathrm{lub}(D_u)$, since $\mathcal{F}$ is Archimedean. Let $C_u = \{\Theta^{-1}(v) : v \in D_u\} =$ the set of all elements x in $Q_{\mathcal{K}}$ such that $\Theta(x) \in D_u$. Now, C_u is bounded above in K. (If $r \in Q_{\mathcal{F}}$ and $r >_F u$, then $\Theta^{-1}(r)$ is an upper bound of C_u in K.) Let $z = \mathrm{lub}(C_u)$. Then $C_u = \{x : x \in Q_{\mathcal{K}} \wedge x <_K z\} = A_z$. (If $x \in C_u$, then $x \in Q_{\mathcal{K}}$ and $\Theta(x) \in D_u$, that is, $\Theta(x) <_F u$. Choose v in $Q_{\mathcal{F}}$ such that $\Theta(x) <_F v <_F u$. Then $x <_K \Theta^{-1}(v) \in C_u$. Hence, x is not an upper bound of C_u, and, therefore, $x <_K \mathrm{lub}(C_u) = z$. Conversely, assume $x \in Q_{\mathcal{K}}$ and $x <_K z$. Hence, x is not upper bound of C_u, that is, there is some element y in C_u such that $x <_K y$. Hence, $\Theta(x) <_F \Theta(y) <_F u$. Therefore, $x \in C_u$.) Since $C_u = A_z$, $D_u = B_z$. Therefore, $u = \mathrm{lub}(D_u) = \mathrm{lub}(B_z) = W(z)$. Thus, u is in the range of W.

(v) $W(z_1 +_K z_2) = W(z_1) +_F W(z_2)$.

Lemma 1 If $u \in Q_{\mathcal{K}}$ and $u <_K z_1 +_K z_2$, then there are r_1, r_2 in $Q_{\mathcal{K}}$ such that $r_1 <_K z_1$, $r_2 <_K z_2$, and $u = r_1 +_K r_2$.

Proof $u - z_2 <_K z_1$. Since $\mathscr{K}$ is Archimedean, we may choose r_1 in $Q_{\mathscr{K}}$ such that $u - z_2 <_K r_1 <_K z_1$. Let $r_2 = u - r_1$. Then $u = r_1 + r_2$. Since $u - z_2 <_K r_1$, it follows that $r_2 = u - r_1 <_K z_2$. $\square$

Lemma 2 $W(z_1 +_K z_2) \leq_F W(z_1) +_F W(z_2)$.

Proof   Assume $w \in B_{z_1 + z_2}$. Then $w = \Theta(u)$ for some u in $Q_{\mathscr{K}}$ such that $u <_K z_1 +_K z_2$. Choose r_1, r_2 in $Q_{\mathscr{K}}$ such that $r_1 <_K z_1$, $r_2 <_K z_2$, $u = r_1 +_K r_2$ (by Lemma 1). Hence,

$$w = \Theta(r_1 +_K r_2) = \Theta(r_1) +_F \Theta(r_2) \leq_F W(z_1) +_F W(z_2).$$

Therefore, $W(z_1) +_F W(z_2)$ is an upper bound of $B_{z_1 + z_2}$, and so, $W(z_1 +_K z_2) \leq_F W(z_1) +_F W(z_2)$. $\square$

Lemma 3 $W(z_1) +_F W(z_2) \leq_F W(z_1 +_K z_2)$.

Proof   For any n in $N_{\mathfrak{F}}$, $W(z_1) - 1/2n$ is not an upper bound of B_{z_1} and $W(z_2) - 1/2n$ is not an upper bound of B_{z_2}. Hence, there exist s_1 in B_{z_1} and s_2 in B_{z_2} such that $W(z_1) - 1/2n <_F s_1$ and $W(z_2) - 1/2n <_F s_2$. There are t_1 and t_2 in $Q_{\mathscr{K}}$ such that $\Theta(t_1) = s_1$ and $t_1 <_K z_1$, and $\Theta(t_2) = s_2$ and $t_2 <_K z_2$. Then

$$W(z_1) +_F W(z_2) <_F (s_1 +_F 1/2n) +_F (s_2 +_F 1/2n)$$
$$= \Theta(t_1) +_F \Theta(t_2) +_F 1/n$$
$$= \Theta(t_1 +_K t_2) +_F 1/n \leq_F W(z_1 +_K z_2) +_F 1/n.$$

Hence, $W(z_1) +_F W(z_2) - W(z_1 +_K z_2) \leq_F 1/n$ for all n in $N_{\mathfrak{F}}$. Therefore, $W(z_1) +_F W(z_2) -_F W(z_1 +_K z_2) \leq_F 0_F$, since $\mathfrak{F}$ is Archimedean (see Exercise 8, p. 196).

(vi) $W(-z) = -W(z)$.

Proof $W(z) +_F W(-z) = W(z +_K (-z)) = W(0_K) = 0_F$.

(vii) $W(z_1)W(z_2) = W(z_1 z_2)$.

Proof   *Case a* $z_1 = 0_K$ or $z_2 = 0_K$. Obvious.
    Case b $z_1 >_K 0_K$ and $z_2 >_K 0_K$.

Lemma 4   If $u \in Q_{\mathscr{K}}$ and $0_K <_K u <_K z_1 z_2$, then there exist r_1 and r_2 in $Q_{\mathscr{K}}$ such that $0_K <_K r_1 <_K z_1$, $0_K <_K r_2 <_K z_2$, and $u = r_1 r_2$.

Proof $0_K <_K u/z_1 <_K z_2$. Choose r_2 in $Q_{\mathscr{K}}$ such that $u/z_1 <_K r_2 <_K z_2$. Let $r_1 = u/r_2$. $\square$

Lemma 5 $W(z_1 z_2) \leq_F W(z_1) W(z_2)$.

Proof   Assume $0_F <_F w \in B_{z_1 z_2}$. There exists u in $Q_{\mathscr{H}}$ such that $0_K <_K u <_K z_1 z_2$ and $\Theta(u) = w$. By Lemma 4, there exist r_1, r_2 in $Q_{\mathscr{H}}$ such that $u = r_1 r_2$, $0_K <_K r_1 <_K z_1$, and $0_K <_K r_2 <_K z_2$. So, $w = \Theta(r_1 r_2) = \Theta(r_1) \Theta(r_2) \leq_F W(z_1) W(z_2)$. Thus, $W(z_1) W(z_2)$ is an upper bound of $B_{z_1 z_2}$. Hence, $W(z_1 z_2) \leq_F W(z_1) W(z_2)$.

Lemma 6 $W(z_1) W(z_2) \leq_F W(z_1 z_2)$.

Proof   There exists k_0 in $N_{\mathfrak{F}}$ such that $k_0 >_F W(z_1) +_F W(z_2)$. For any n in $N_{\mathfrak{F}}$, take m in $N_{\mathfrak{F}}$ such that $m >_F 2 k_0 n$ and $m >_F \max(1/W(z_1), 1/W(z_2))$. There exist s_1, s_2 such that $s_1 \in B_{z_1}$, $s_2 \in B_{z_2}$ and $0_F <_F W(z_1) - 1/m <_F s_1$, $0_F <_F W(z_2) - 1/m <_F s_2$. There exist t_1, t_2 such that $s_1 = \Theta(t_1)$, $s_2 = \Theta(t_2)$, $0_K <_K t_1 <_K z_1$, and $0_K <_K t_2 <_K z_2$. Hence,

$$W(z_1) W(z_2) <_F (\Theta(t_1) +_F 1/m)(\Theta(t_2) +_F 1/m)$$

$$= \Theta(t_1) \Theta(t_2) +_F \frac{\Theta(t_1) +_F \Theta(t_2)}{m} +_F 1/m^2$$

$$\leq \Theta(t_1 t_2) +_F \frac{W(z_1) +_F W(z_2)}{m} +_F 1/m^2$$

$$\leq W(z_1 z_2) +_F \frac{k_0}{m} +_F 1/m^2 <_F W(z_1 z_2) +_F 1/2n +_F 1/2n.$$

Therefore, $W(z_1) W(z_2) - W(z_1 z_2) <_F 1/n$ for all n in $N_{\mathfrak{F}}$. So, $W(z_1) W(z_2) - W(z_1 z_2) \leq_F 0_F$, since $\mathfrak{F}$ is Archimedean.

Case c $z_1 >_K 0_K$ and $z_2 <_K 0_K$.

Proof $W(z_1 z_2) = W(-(z_1(-z_2))) = -W(z_1(-z_2)) = -W(z_1) W(-z_2)$
$\qquad\qquad = -W(z_1)(-W(z_2)) = W(z_1) W(z_2)$.

Case d $z_1 <_K 0_K$ and $z_2 >_K 0_K$.

Proof   By symmetry from Case (c), using commutativity of multiplication.

Case e $z_1 <_K 0_K$ and $z_2 <_K 0_K$

Proof $W(z_1 z_2) = W(-(z_1)(-z_2)) = -W(z_1(-z_2)) = -W(z_1) W(-z_2)$
$\qquad\qquad = -W(z_1)(-W(z_2)) = W(z_1) W(z_2)$.   ∎

Theorem 4.2   If $\mathscr{H}$ is an Archimedean ordered field and $\mathfrak{F}$ is a complete ordered field, then $\mathscr{H}$ is isomorphic with an ordered subfield of $\mathfrak{F}$.

Proof In the proof of Theorem 4.1, the only place where the completeness of $\mathscr{K}$ was used, rather than only the Archimedean property, was in the proof in Part iv that the mapping W has range F. Thus, the proof of Theorem 4.1 shows that W is an isomorphism of the Archimedean ordered field $\mathscr{K}$ into the complete ordered field $\mathfrak{F}$.   ∎

5.5 CONVERGENT SEQUENCES AND CAUCHY SEQUENCES

In this Chapter, by a **sequence** we shall mean a denumerable sequence: $a_1, a_2, a_3, \ldots, a_n, \ldots$[†] The objects $a_1, a_2, \ldots, a_n, \ldots$ are called the **terms** of the sequence, and a_n is called the **nth term**. If A is any set, then by a **sequence in A** we mean a sequence all of whose terms belong to A. We shall often denote a sequence $a_1, a_2, \ldots, a_n, \ldots$ by $\langle a_n \rangle$.[‡]

EXAMPLES

1.   The sequence 1, 1/2, 1/3,... will be denoted $\langle 1/n \rangle$.

2.   The sequence 1, 4, 9, 16,... will be denoted $\langle n^2 \rangle$.

3.   The constant sequence 2, 2, 2,... will be denoted $\langle 2 \rangle$.

In what follows, unless something is said to the contrary,

$$\mathfrak{F} = (F, +, \times, <)$$

will designate an arbitrary ordered field.

Definition   Let $\langle a_n \rangle$ be a sequence in F, and let $b \in F$. We say that b is a **limit** of $\langle a_n \rangle$ if and only if

For every positive element ε of F, there is some positive integer n_0 such that, for any positive integer n, if $n \geq n_0$, then $|a_n - b| < \varepsilon$. (In symbols,

$$(\forall \varepsilon)(\varepsilon > 0 \wedge \varepsilon \in F \Rightarrow (\exists n_0)(n_0 \in P \wedge (\forall n)(n \in P \wedge n \geq n_0 \\ \rightarrow |a_n - b| < \varepsilon).)$$

[†] More precisely, a denumerable sequence is a function a whose domain is the set of positive integers. If a is such a function and n is any positive integer, it is traditional to write a_n instead of $a(n)$.

[‡] The standard notation for a sequence is $\{a_n\}$. However, $\{a_n\}$ already denotes the set whose only element is a_n. Therefore, we have introduced the new notation $\langle a_n \rangle$.

Intuitively, this means that, no matter how small a distance ε one takes, if one goes far enough along the sequence, from some point on all the terms of the sequence differ from b by less than ε. Pictorially (see Figure 5.3), if we draw an interval around b of radius ε, then, from some term a_{n_0} on, all terms will lie within that interval. Still another way of saying this is that we can get as close as we wish to b if we go far enough out in the sequence.

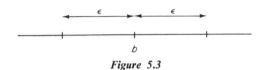

b

Figure 5.3

Lemma 5.1   A sequence $\langle a_n \rangle$ has at most one limit.

Proof   Assume that b and c are limits of the sequence $\langle a_n \rangle$. Assume $b \neq c$. Then $\varepsilon = |b - c|/2 > 0$. Hence, there exist positive integers n_1 and n_2 such that

$$n \geq n_1 \Rightarrow |a_n - b| < \varepsilon,$$
$$n \geq n_2 \Rightarrow |a_n - c| < \varepsilon.$$

Let $n_0 = \max(n_1, n_2)$. Then

$$n \geq n_0 \Rightarrow (|a_n - b| < \varepsilon \wedge |a_n - c| < \varepsilon).$$

Hence, by the Triangle Inequality,

$$|b - c| = |(b - a_{n_0}) + (a_{n_0} - c)| \leq |b - a_{n_0}| + |a_{n_0} - c|$$
$$< \varepsilon + \varepsilon = 2\varepsilon = |b - c|.$$

Thus, $|b - c| < |b - c|$, which yields a contradiction. Therefore, $b = c$.   ■

Notation   If b is a limit of the sequence $\langle a_n \rangle$, we shall write

$$b = \operatorname{Lim} a_n.$$

(In more traditional notation, this would be written: $b = \operatorname*{Lim}_{n \to \infty} a_n$.)

Definition   A sequence $\langle a_n \rangle$ is said to be **convergent** in $\mathfrak{F}$ if and only if there exists b in F such that $b = \operatorname{Lim} a_n$.

EXAMPLES

4. Consider a constant sequence $\langle c \rangle$, where $c \in F$. Then $c = \text{Lim } c$.

5. Let $\mathfrak{F}$ be Archimedean (for example, the rational number field). Then

$$0 = \text{Lim } 1/n.$$

For, assume $\varepsilon > 0$. Since $\mathfrak{F}$ is Archimedean, there exists a positive integer n_0 such that $n_0 > 1/\varepsilon$. Hence, $1/n_0 < \varepsilon$, and, therefore,

$$n \geq n_0 \Rightarrow |\, 1/n - 0\,| = 1/n \leq 1/n_0 < \varepsilon.$$

EXERCISES

1. Show that, if $\text{Lim } 1/n = 0$, then $\mathfrak{F}$ is Archimedean.

2. In an Archimedean ordered field, prove that
    a. $\text{Lim } n/(n + 1) = 1$.
    b. $\text{Lim } (n + 1)/n = 1$.
    c. $\text{Lim } n/(1 + 2n) = 1/2$.

Theorem 5.2   Assume that $\langle x_n \rangle$ and $\langle y_n \rangle$ are convergent sequences. Then so are $\langle x_n + y_n \rangle$, $\langle -y_n \rangle$, $\langle x_n - y_n \rangle$, $\langle x_n y_n \rangle$, and $\langle |\, x_n \,| \rangle$. Moreover,

i. $\text{Lim}(x_n + y_n) = \text{Lim } x_n + \text{Lim } y_n$.

ii. $\text{Lim}(-y_n) = -\text{Lim } y_n$.

iii. $\text{Lim}(x_n - y_n) = \text{Lim } x_n - \text{Lim } y_n$.

iv. $\text{Lim}(x_n y_n) = \text{Lim } x_n \cdot \text{Lim } y_n$.

v. $\text{Lim}|\, x_n \,| = |\, \text{Lim } x_n \,|$.

In addition, if $\text{Lim } y_n \neq 0$ and all $y_n \neq 0$, then

vi. $\text{Lim}(x_n/y_n) = (\text{Lim } x_n)/(\text{Lim } y_n)$.

Proof   Assume $b = \text{Lim } x_n$ and $c = \text{Lim } y_n$.

(i) We must prove that $\text{Lim}(x_n + y_n) = b + c$. Assume $\varepsilon > 0$. Then there exist positive integers n_1, n_2 such that

$$n \geq n_1 \Rightarrow |\, x_n - b\,| < \varepsilon/2 \quad \text{and} \quad n \geq n_2 \Rightarrow |\, y_n - c\,| < \varepsilon/2.$$

Let $n_0 = \max(n_1, n_2)$. Then,

$$n \geq n_0 \Rightarrow |\, (x_n + y_n) - (b + c)\,| = |\, (x_n - b) + (y_n - c)\,|$$
$$\leq |\, x_n - b\,| + |\, y_n - c\,| < \varepsilon/2 + \varepsilon/2 = \varepsilon.$$

(ii) We must prove: $\text{Lim}(-y_n) = -c$. Assume $\varepsilon > 0$. Then there exists a positive integer n_0 such that $n \geq n_0 \Rightarrow |y_n - c| < \varepsilon$. But,

$$|(-y_n) - (-c)| = |c - y_n| = |y_n - c|.$$

(iii) This follows directly from (i) and (ii), since $x_n - y_n = x_n + (-y_n)$.

(iv) We must prove: $\text{Lim}(x_n y_n) = bc$. Assume $\varepsilon > 0$. There exist positive integers n_1, n_2 such that

$$n \geq n_1 \Rightarrow |x_n - b| < \min(1, \varepsilon/3) \text{ and, if } c \neq 0, |x_n - b| < \varepsilon/3|c|;$$
$$n \geq n_2 \Rightarrow |y_n - c| < \min(1, \varepsilon/3), \text{ and, if } b \neq 0, |y_n - c| < \varepsilon/3|b|.$$

Let $n_0 = \max(n_1, n_2)$. If $n \geq n_0$, then

$$
\begin{aligned}
|x_n y_n - bc| &= |(x_n - b)(y_n - c) + b(y_n - c) + c(x_n - b)| \\
&\leq |(x_n - b)(y_n - c)| + |b(y_n - c)| + |c(x_n - b)| \\
&= |x_n - b||y_n - c| + |b||y_n - c| + |c||x_n - b| \\
&< \varepsilon/3 + \varepsilon/3 + \varepsilon/3 = \varepsilon.
\end{aligned}
$$

(v) We must prove: $\text{Lim}|x_n| = |b|$. Assume $\varepsilon > 0$. There exists a positive integer n_0 such that $n \geq n_0 \Rightarrow |x_n - b| < \varepsilon$. But,

$$||x_n| - |b|| \leq |x_n - b| \quad \text{(by Theorem 3.4.8k).}$$

(vi) It suffices to prove that, if $\text{Lim } y_n = 0$ and all $y_n \neq 0$, then $\text{Lim } 1/y_n = 1/\text{Lim } y_n$ (and then use (iv)). Thus, we assume that $\text{Lim } y_n = c \neq 0$ and all $y_n \neq 0$. Assume $\varepsilon > 0$. There exists n_1 such that $n \geq n_1 \Rightarrow |y_n - c| < (|c|^2/2)\varepsilon$. In addition, there exists n_2 such that $n \geq n_2 \Rightarrow |y_n - c| < |c|/2$. Hence,

$$n \geq n_2 \Rightarrow |y_n| = |c - (c - y_n)| \geq ||c| - |c - y_n|| \geq |c| - |c|/2 = |c|/2.$$

Let $n_0 = \max(n_1, n_2)$. Then

$$n \geq n_0 \Rightarrow |1/y_n - 1/c| = |y_n - c|/|y_n c| < \frac{\frac{1}{2}|c|^2 \varepsilon}{\frac{1}{2}|c||c|} = \varepsilon.$$

Hence, $\text{Lim } 1/y_n = 1/c$.   ■

EXERCISES

3. If $b \neq 0$, show that it is possible that $\text{Lim }|x_n| = b$, while, at the same time, $\langle x_n \rangle$ is not convergent. However, prove that, if $\text{Lim }|x_n| = 0$, then $\text{Lim } x_n = 0$.

4.^{c} a.   If $x_n \geq b$ for all n and $\langle x_n \rangle$ is convergent, prove that $\operatorname{Lim} x_n \geq b$. (Show, by a counterexample, that, if $x_n > b$ for all n and $\langle x_n \rangle$ is convergent, it is not always the case that $\operatorname{Lim} x_n > b$.)

b.   If $x_n \leq c$ for all n and $\langle x_n \rangle$ is convergent, prove that $\operatorname{Lim} x_n \leq c$.

c.   If $\langle x_n \rangle$ is a convergent sequence of nonnegative elements, prove that $\operatorname{Lim} x_n$ is nonnegative. (*Hint*: Use a.)

d.   If $\langle x_n \rangle$ is a convergent sequence of nonpositive elements, prove that $\operatorname{Lim} x_n$ is nonpositive.

5.   Prove: If $\operatorname{Lim} x_n = 0$ and there exists some n_0 such that $|y_n| < |x_n|$ for all $n \geq n_0$, then $\operatorname{Lim} y_n = 0$.

6.   In an Archimedean ordered field, find:

a. $\operatorname{Lim} \dfrac{3n}{(1 + 2n)}$.

b. $\operatorname{Lim} \dfrac{(-1)^n}{n}$.

c. $\operatorname{Lim} \dfrac{2n^2}{(1 + 2n - n^2)}$.

d. $\operatorname{Lim} \dfrac{2^n}{n!}$.

e. $\operatorname{Lim} \dfrac{n!}{n^n}$.

f. $\operatorname{Lim} \dfrac{n^2}{4^n}$.

7.   Find sequences $\langle x_n \rangle$ and $\langle y_n \rangle$ of rational numbers such that $\operatorname{Lim} x_n = \operatorname{Lim} y_n = 0$ and

a. $\operatorname{Lim} x_n/y_n = 0$;

b. $\operatorname{Lim} x_n/y_n = 1$;

c. $\operatorname{Lim} x_n/y_n$ does not exist.

Definitions   Let $\langle x_n \rangle$ be a sequence in $\mathfrak{F}$.

i. $\langle x_n \rangle$ is said to be **bounded in** $\mathfrak{F}$ if and only if there is some b in F such that $|x_n| < b$ for all n.

ii.   We say that $\langle x_n \rangle$ is a **Cauchy sequence in** $\mathfrak{F}$ if and only if, for every positive ε in F, there exists a positive integer n_0 such that

$$(n \geq n_0 \wedge k \geq n_0) \Rightarrow |x_n - x_k| < \varepsilon.$$

Thus, $\langle x_n \rangle$ is a Cauchy sequence if and only if the terms of $\langle x_n \rangle$ get arbitrarily close to each other as we move farther and farther out in the sequence.

Theorem 5.3   Every convergent sequence is a Cauchy sequence.

Proof   Let $\langle x_n \rangle$ be a convergent sequence. Let $\operatorname{Lim} x_n = b$. Assume $\varepsilon > 0$. Then there is a positive integer n_0 such that $n \geq n_0 \Rightarrow |x_n - b| < \varepsilon/2$.

Hence

$$n, k \geq n_0{}^\dagger \Rightarrow |x_n - x_k| = |(x_n - b) + (b - x_k)|$$
$$\leq |x_n - b| + |b - x_k| < \varepsilon/2 + \varepsilon/2 = \varepsilon. \qquad \blacksquare$$

Definition   An ordered field $\mathfrak{F}$ is said to be **Cauchy-complete** if and only if every Cauchy sequence in $\mathfrak{F}$ is convergent in $\mathfrak{F}$.

The converse of Theorem 5.3 is true for some ordered fields but not for others. For example, it will turn out to be true for the real number system but false for the rational number system. We shall return to this question later.

Theorem 5.4   Every Cauchy sequence is bounded.

Proof   Let $\langle x_n \rangle$ be a Cauchy sequence. There exists a positive integer n_0 such that $n, k \geq n_0 \Rightarrow |x_n - x_k| < 1$. Then

$$n \geq n_0 \Rightarrow |x_n| = |x_{n_0} + (x_n - x_{n_0})| \leq |x_{n_0}| + |x_n - x_{n_0}| < |x_{n_0}| + 1.$$

Let $b = \max(|x_0| + 1, |x_1| + 1, \ldots, |x_{n_0}| + 1)$. Then, for all positive integers n, $|x_n| < b$. $\qquad \blacksquare$

Corollary 5.5   Every convergent sequence is bounded.

Proof   This is an immediate consequence of Theorems 5.3, 5.4. $\qquad \blacksquare$

EXERCISES

8.$^\text{C}$ If $\langle y_n \rangle$ is bounded and $\operatorname{Lim} x_n = 0$, prove that $\operatorname{Lim} x_n y_n = 0$.

9. If $\langle x_n \rangle$ and $\langle y_n \rangle$ are bounded, prove that $\langle x_n + y_n \rangle$, $\langle x_n - y_n \rangle$, $\langle x_n y_n \rangle$, and $\langle |x_n| \rangle$ are also bounded.

10. Give an example of a bounded sequence which is not a Cauchy sequence (and, therefore, not a convergent sequence).

11. In an Archimedean ordered field, prove that, if $\operatorname{Lim} x_n = b$, then

$$\operatorname{Lim}(x_1 + \cdots + x_n)/n = b.$$

Theorem 5.6   If $\langle x_n \rangle$ and $\langle y_n \rangle$ are Cauchy sequences, so are $\langle x_n + y_n \rangle$, $\langle -y_n \rangle$, $\langle x_n - y_n \rangle$, $\langle x_n y_n \rangle$, and $\langle |x_n| \rangle$.

† "$n, k \geq n_0$" is an abbreviation for "$n \geq n_0$ and $k \geq n_0$."

Proof The reasoning here is quite similar to that used to prove Theorem 5.2. For that reason, we shall do only two of the five parts. The rest is left as an exercise for the reader.

(i) Let $\langle x_n \rangle$ and $\langle y_n \rangle$ be Cauchy sequences. Assume $\varepsilon > 0$. Then there exist positive integers n_1, n_2 such that

$$n, k \geq n_1 \Rightarrow |x_n - x_k| < \varepsilon/2;$$
$$n, k \geq n_2 \Rightarrow |y_n - y_k| < \varepsilon/2.$$

Let $n_0 = \max(n_1, n_2)$. Then

$$
\begin{aligned}
n, k \geq n_0 \Rightarrow |(x_n + y_n) - (x_k + y_k)| &= |(x_n - x_k) + (y_n - y_k)| \\
&\leq |x_n - x_k| + |y_n - y_k| \\
&< \varepsilon/2 + \varepsilon/2 = \varepsilon.
\end{aligned}
$$

Thus, $\langle x_n + y_n \rangle$ is a Cauchy sequence.

(iv) Let $\langle x_n \rangle$ and $\langle y_n \rangle$ be Cauchy sequences. Assume $\varepsilon > 0$. By Theorem 5.4, there exist positive elements u and v in F such that $|x_n| < u$ and $|y_n| < v$ for all n. In addition, there exist positive integers n_1, n_2 such that

$$n, k \geq n_1 \Rightarrow |x_n - x_k| < \varepsilon/2v;$$
$$n, k \geq n_2 \Rightarrow |y_n - y_k| < \varepsilon/2u.$$

Let $n_0 = \max(n_1, n_2)$. Then

$$
\begin{aligned}
n, k \geq n_0 \Rightarrow |x_n y_n - x_k y_k| &= |x_n(y_n - y_k) + y_k(x_n - x_k)| \\
&\leq |x_n(y_n - y_k)| + |y_k(x_n - x_k)| \\
&= |x_n| |y_n - y_k| + |y_k| |x_n - x_k| \\
&< (u \cdot \varepsilon/2u) + (v \cdot \varepsilon/2v) = \varepsilon/2 + \varepsilon/2 = \varepsilon.
\end{aligned}
$$

Thus, $\langle x_n y_n \rangle$ is a Cauchy sequence. ■

We shall need some properties of **subsequences** of Cauchy sequences. A subsequence is obtained from a sequence by leaving out some terms of the sequence. For example, the sequence $\langle 1/n \rangle = \{1, 1/2, 1/3, \dots\}$ has the following subsequences:

a. $\langle 1/n^2 \rangle = \{1, 1/4, 1/9, 1/16, \dots\}$.
b. $\langle 1/2n \rangle = \{1/2, 1/4, 1/6, 1/8, \dots\}$.
c. $\langle 1/(2n + 1) \rangle = \{1/3, 1/5, 1/7, 1/9, \dots\}$, etc.

More precisely, if s is a sequence and g is an increasing function from the positive integers into positive integers (that is, $g: P \to P$ and $j < k \Rightarrow$

$g(j) < g(k)$), then the sequence t, given by $t_n = s_{g(n)}$, is called a **subsequence** of s. Thus, the sequence of values of t is $s_{g(1)}, s_{g(2)}, s_{g(3)}, \ldots$. In the examples above, $g(n) = n^2$ in (a); $g(n) = 2n$ in (b); and $g(n) = 2n + 1$ in (c).

Theorem 5.7 A subsequence of a Cauchy sequence is a Cauchy sequence.

Proof Intuitively, if all the terms of a sequence get closer and closer to each other as you go far out in the sequence, then the same is true for any subsequence. For a more rigorous proof, let a subsequence t of a Cauchy sequence s be determined by a function, g, $g: P \to P$ such that $j < k \Rightarrow g(j) < g(k)$. Thus, $t_n = s_{g(n)}$ for all positive integers n. We must show that t is a Cauchy sequence. Assume $\varepsilon > 0$. Then there exists a positive integer n_0 such that

$$(*) \quad n, k \geq n_0 \Rightarrow |s_n - s_k| < \varepsilon.$$

Since g is an increasing function, it follows that $g(n) \geq n$ for all n. (The proof is by mathematical induction. Since $g(n) \in P$ for all n, $g(1) \geq 1$. If $g(n) \geq n$, then $g(n + 1) > g(n) \geq n$. Hence $g(n + 1) \geq n + 1$.) Therefore, by $(*)$,

$$n, k \geq n_0 \Rightarrow g(n), g(k) \geq n_0 \Rightarrow |s_{g(n)} - s_{g(k)}| < \varepsilon,$$

that is, $n, k \geq n_0 \Rightarrow |t_n - t_k| < \varepsilon$. ∎

Theorem 5.8 If a Cauchy sequence s has a subsequence t such that $\operatorname{Lim} t_n = b$, then $\operatorname{Lim} s_n = b$. Thus, if a subsequence of a Cauchy sequence converges, then the whole sequence converges.

Proof Intuitively, since the terms of the subsequence t get closer and closer to b, and the terms of s get closer to terms of t as you go far out in the sequence, it would follow that the terms of s also approach b. For a more rigorous proof, assume $\varepsilon > 0$. Now, $t_n = s_{g(n)}$, where g is an increasing function from P into P. Since $\operatorname{Lim} t_n = b$, there is a positive integer n_1 such that $n \geq n_1 \Rightarrow |t_n - b| < \varepsilon/2$ (that is, $n \geq n_1 \Rightarrow |s_{g(n)} - b| < \varepsilon/2$). Since s is a Cauchy sequence, there is a positive integer n_2 such that $n, k \geq n_2 \Rightarrow |s_n - s_k| < \varepsilon/2$. Let $n_0 = \max(n_1, n_2)$. Notice that, if $n \geq n_0$, then $n \geq n_2$ and $g(n) \geq n \geq n_2$. (Remember that $g(n) \geq n$; see the proof of Theorem 5.7.) Hence,

$$n \geq n_0 \Rightarrow |s_n - b| = |(s_n - s_{g(n)}) + (s_{g(n)} - b)|$$
$$\leq |s_n - s_{g(n)}| + |s_{g(n)} - b| < \varepsilon/2 + \varepsilon/2 = \varepsilon.$$

Thus, $\operatorname{Lim} s_n = b$. ∎

EXERCISE

12.^{C} Show that, if Lim $x_n = b$, then any subsequence of $\langle x_n \rangle$ also has b as a limit.

In order to formulate several additional equivalent conditions for completeness, we need a few new concepts.

I.   A sequence $\langle x_n \rangle$ is said to be

a.   **nondecreasing** if and only if $x_n \leq x_{n+1}$ for all n;

b.   **increasing** if and only if $x_n < x_{n+1}$ for all n;

c.   **nonincreasing** if and only if $x_n \geq x_{n+1}$ for all n;

d.   **decreasing** if and only if $x_n > x_{n+1}$ for all n;

e.   **monotonic** if and only if it is either nondecreasing or nonincreasing.

Let (Mon) stand for the assertion: Every bounded monotonic sequence converges.

II.   Assume $a < b$.

By the **closed interval** $[a, b]$ we mean $\{x : a \leq x \leq b\}$.

By the **open interval** (a, b) we mean $\{x : a < x < b\}$.

A sequence of closed intervals $[a_n, b_n]$ is said to be **nested** if and only if $a_n \leq a_{n+1} < b_{n+1} \leq b_n$ for all n.

Let (Nest) stand for the assertion: Every nested sequence of closed intervals has a nonempty intersection, that is, there is at least one point in all the intervals.

EXERCISES

13.   If $a = c - \varepsilon$ and $b = c + \varepsilon$, show that

$$| x - c | < \varepsilon \Leftrightarrow x \in (a, b).$$

14.   Prove: $[a, b] \subseteq [u, v] \Leftrightarrow (u \leq a \wedge b \leq v)$.

15.   If $0 < b - a < \varepsilon$, prove that

$$u, v \in [a, b] \Rightarrow | u - v | < \varepsilon.$$

16.^{D} Show that, in an ordered field, any sequence $\langle a_n \rangle$ contains a monotonic subsequence. (*Hint:* Consider the possibility that, for each term a_n, there is a later term which is the minimum of all terms after a_n.)

17.   Show that (Mon) is equivalent to the assertion that every bounded, nondecreasing sequence converges.

Theorem 5.9 For any ordered field $\mathfrak{F} = (F, +, \times, <)$, the following conditions are equivalent.

i.   completeness.

ii. (Mon).

iii. Archimedean + (Nest).

iv. Archimedean + Cauchy completeness.

Proof (i) $\Rightarrow$ (ii) Assume $\langle x_n \rangle$ is a nondecreasing bounded sequence. Let b be the lub of the set of terms of the sequence. Then, $\mathrm{Lim}\, x_n = b$. (For, assume $\varepsilon > 0$. Then $b - \varepsilon$ is not an upper bound of the set of terms of the sequence. Hence, $b - \varepsilon < x_{n_0}$ for some n_0. So, $b - \varepsilon < x_{n_0} \leq x_n$ for $n \geq n_0$. Also, $x_n \leq b$. Hence, $b - \varepsilon < x_n \leq b$ for $n \geq n_0$. Thus, $n \geq n_0 \Rightarrow |b - x_n| = b - x_n < \varepsilon$.)

(ii) $\Rightarrow$ (iii) Let us first prove (Nest). Assume $a_n \leq a_{n+1} < b_{n+1} \leq b_n$ for all n. Then $\langle a_n \rangle$ is nondecreasing and $\langle b_n \rangle$ is nonincreasing. By (ii), let $c = \mathrm{Lim}\, a_n$ and $d = \mathrm{Lim}\, b_n$. Then, $c \leq d$, since $a_n \leq b_k$ for all n and k. (Use Exercise 4a, b, p. 212). If u is any element such that $c \leq u \leq d$, then $u \in [a_n, b_n]$ for all n.

Now let us prove the Archimedean property. It suffices to prove that $N_{\mathfrak{F}}$ is not bounded above. Assume that $N_{\mathfrak{F}}$ is bounded above. Then the sequence $\langle n \rangle$, consisting of the natural numbers of $\mathfrak{F}$ in ascending order, is bounded and increasing. Hence, by (Mon), $\langle n \rangle$ is convergent. But any convergent sequence is a Cauchy sequence. Hence, there exists n_0 such that $n, k \geq n_0 \Rightarrow |n - k| < \frac{1}{2}$. In particular, when $k = n_0$ and $n = n_0 + 1$, $1 = |(n_0 + 1) - n_0| < \frac{1}{2}$, which yields a contradiction. Therefore, $\mathfrak{F}$ is Archimedean.

(iii) $\Rightarrow$ (iv) Let $\langle x_k \rangle$ be a Cauchy sequence. We must prove that $\langle x_k \rangle$ converges. We may assume:

$$(*) \quad x_k \neq x_{k+1} \quad \text{for all} \quad k.$$

(For, we have two cases.

Case 1 $\langle x_k \rangle$ is constant from some term x_{k_0} on. Then $\mathrm{Lim}\, x_k = x_{k_0}$, and we are finished.

Case 2 For any k, there exists $k' > k$ such that $x_{k'} \neq x_k$. Then obviously we can choose a subsequence $\langle x_{k_i} \rangle$ of $\langle x_k \rangle$ such that $x_{k_{i+1}} \neq x_{k_i}$ for all i. If we prove below that any Cauchy sequence satisfying $(*)$ converges, then $\langle x_{k_i} \rangle$ is convergent. Therefore, by Theorem 5.8, $\langle x_k \rangle$ would be convergent.)
Now let $\varepsilon_k = |x_{k+1} - x_k| \neq 0$. Since $\langle x_k \rangle$ is a Cauchy sequence, $\mathrm{Lim}\, \varepsilon_k = 0$.

Choose k_1 such that $j, k \geq k_1 \Rightarrow |x_k - x_j| < \varepsilon_1/2$. Let $a_1 = x_{k_1} - \varepsilon_1/2$, $b_1 = x_{k_1} + \varepsilon_1/2$. Then $x_j \in [a_1, b_1]$ for $j \geq k_1$, and $b_1 - a_1 \leq \varepsilon_1$. Assume that we have chosen points $x_{k_1}, x_{k_2}, \ldots, x_{k_{n-1}}$ and closed intervals $[a_1, b_1]$, $[a_2, b_2], \ldots, [a_{n-1}, b_{n-1}]$ such that $[a_1, b_1] \supseteq \cdots \supseteq [a_{n-1}, b_{n-1}]$, each open interval (a_j, b_j) contains all terms of the sequence from x_{k_j} on, and $b_j - a_j \leq \varepsilon_j$. There exists $k_n > k_{n-1}$ such that $j, k \geq k_n \Rightarrow |x_j - x_k| < \varepsilon_n/2$. Since $k_n > k_{n-1}$, we know that $x_{k_n} \in (a_{n-1}, b_{n-1})$. Let $a_n = \max(a_{n-1}, x_{k_n} - \varepsilon_n/2)$ and $b_n = \min(b_{n-1}, x_{k_n} + \varepsilon_n/2)$. Then $x_{k_n} \in (a_n, b_n)$ and $x_i \in (a_n, b_n)$ for all $i \geq k_n$. (Exercise for the reader.) Thus, by induction, we have constructed a subsequence $x_{k_1}, x_{k_2}, \ldots$ and a nested sequence of closed intervals $[a_n, b_n]$ such that $x_i \in [a_n, b_n]$ for all $i \geq k_n$ and $b_n - a_n \leq \varepsilon_n$. By (Nest), let $c \in [a_n, b_n]$ for all n. Then $c = \mathrm{Lim}\, x_k$. (For, assume $\varepsilon > 0$. Take n such that $\varepsilon_n < \varepsilon$. But, $c \in [a_n, b_n]$ and $k \geq k_n \Rightarrow x_k \in (a_n, b_n)$. Hence, $k \geq k_n \Rightarrow |x_k - c| < \varepsilon_n < \varepsilon$.)

(iv) $\Rightarrow$ (i)   By Theorem 3.3, it suffices to prove the lub property. Assume $A \neq \varnothing$ and A is bounded above. Let $B = \{x : (\exists y)(y \in A \wedge x \leq y)\}$. Clearly, A and B have the same upper bounds. Note that $(x \in B \wedge z \leq x) \Rightarrow z \in B$. Take any $z \in B$ and any upper bound u of B. Take any natural number n. By the Archimedean property, there is a natural number $k > n(u - z)$. Then, $z + k/n > u$ and, therefore, $z + k/n$ is an upper bound of B. Let k_n be the least natural number such that $z + k_n/n$ is an upper bound of B. Let $x_n = z + (k_n - 1)/n$ and $u_n = x_n + 1/n = z + k_n/n$. u_n is an upper bound of B and $x_n \in B$. Moreover, x_n is not an upper bound of B. Hence, for any m and n, $x_m < u_n$. Therefore,

$$x_m - x_n < u_n - x_n = u_n - (u_n - 1/n) = 1/n,$$

and, similarly, $x_n - x_m < 1/m$. Hence, $|x_m - x_n| < \max(1/n, 1/m)$. Thus, $\langle x_n \rangle$ is a Cauchy sequence. (For, assume $\varepsilon > 0$. By the Archimedean property, there exists n_0 such that $1/n_0 < \varepsilon$. Hence, if $n, m \geq n_0$, $|x_m - x_n| < \max(1/n, 1/m) \leq 1/n_0 < \varepsilon$.) By (iv), let $c = \mathrm{Lim}\, x_n$. Then

1. c is an upper bound of B. (For, assume not. Then $c < b$ for some b in B. Since $c = \mathrm{Lim}\, x_n$ and $\mathfrak{F}$ is Archimedean, there exists a natural number n such that $|x_n - c| < (b - c)/2$ and $1/n < (b - c)/2$. Hence, $x_n - c \leq |x_n - c| < (b - c)/2$, and so, $x_n < c + (b - c)/2$. Then

$$u_n = x_n + 1/n < (c + (b - c)/2) + (b - c)/2 = c + (b - c) = b,$$

contradicting the fact that u_n is an upper bound of B.)

2. c is the least upper bound of B. (For, assume $v < c$, where v is an upper bound of B. Since $c = \mathrm{Lim}\, x_n$, there is a natural number n such that

$|c - x_n| < c - v$. Hence, $c - x_n < c - v$, and so, $v < x_n$. Since $x_n \in B$, this contradicts the assumption that v is an upper bound of B.) ∎

Remark The proof of (iii) $\Rightarrow$ (iv) above actually shows that (Nest) $\Rightarrow$ Cauchy completeness.

EXERCISES

18. Prove that the ordered field of rational numbers is not Cauchy complete.

19. In a complete ordered field, show that, if $\langle [a_n, b_n] \rangle$ is a nested sequence of closed intervals such that $\mathrm{Lim}(b_n - a_n) = 0$, then there is a unique element belonging to every one of the closed intervals.

20. *Formal Laurent Series* Let $\mathfrak{F} = (F, +, \times, <)$ be an ordered field. Let L be the set of all *formal Laurent series* over $\mathfrak{F}$, that is, all objects of the form

(*) $\quad \displaystyle\sum_{n=-k_0}^{\infty} a_n x^n = a_{-k_0} x^{-k_0} + a_{-k_0+1} x^{-k_0+1} + \cdots + a_{-1} x^{-1} + a_0 + a_1 x + a_2 x^2 + \cdots,$

where k_0 is a nonnegative integer and all $a_i \in F$.

EXAMPLES

1. $3x^{-7} + 2x^{-1} + 1 + 2x^2$.

2. $-7x^{-2} + x^{-1} + x^2 + x^3 + x^4 + \cdots$

3. $4x^3 + 5x^4 + 6x^5 + 7x^6 + \cdots$.

Thus, formal Laurent series differ from polynomials in that they can involve infinitely many positive powers of x as well as finitely many negative powers of x. Formal Laurent series are added and multiplied in precisely the same way that polynomials are added and multiplied (see Appendix C). To make these operations explicit, we write each formal Laurent series (*) in the form $\sum_{n=-\infty}^{\infty} a_n x^n$, where $a_n = 0$ for all $n < -k_0$. Then

$$\sum_{n=-\infty}^{\infty} a_n x^n + \sum_{n=-\infty}^{\infty} b_n x^n = \sum_{n=-\infty}^{\infty} (a_n + b_n) x^n,$$

and

$$\sum_{n=-\infty}^{\infty} a_n x^n \times \sum_{n=-\infty}^{\infty} b_n x^n = \sum_{n=-\infty}^{\infty} c_n x^n,$$

where $c_n = \sum_{i+j=n} a_i b_j$. (The sum for c_n has a finite number of terms.) We define the set $\mathscr{P}_L$ of "positive" elements to be the set of those formal Laurent series (*) for which the first nonzero coefficient is positive in $\mathfrak{F}$. We define the order relation in the usual way in terms of $\mathscr{P}_L$ (see Theorem 3.4.3). Prove:

a. The formal Laurent series form an ordered field $\mathscr{L}_{\mathfrak{F}}$.

b. $0 < \cdots < x^3 < x^2 < x < \cdots < n < x^{-1} < x^{-2} < \cdots$.

c. $\mathscr{L}_{\mathfrak{F}}$ is non-Archimedean.

d. $\mathscr{L}_{\mathfrak{F}}$ is Cauchy complete. (*Hint:* In a Cauchy sequence in $\mathscr{L}_{\mathfrak{F}}$, show that, for any fixed integer n, the coefficients of x^n are the same from some point on.)

e.   Let $\mathscr{K}$ be the ordered subfield of $\mathscr{L}_{\mathfrak{F}}$ consisting of quotients in $\mathscr{L}_{\mathfrak{F}}$ of ordinary polynomials:

$$\frac{b_0 + \cdots + b_n x^n}{c_0 + \cdots + c_m x^m}.$$

Then (i) $\mathscr{K}$ is a proper subfield of $\mathscr{L}_{\mathfrak{F}}$. (*Hint:*   Let

$$f = \sum_{n=0}^{\infty} n!x^n = 1 + x + 2x^2 + 3!x^2 + \cdots + n!x^n + \cdots.$$

Show that $f \notin \mathscr{K}$.) (ii) $\mathscr{K}$ is non-Archimedean. (iii) $\mathscr{K}$ is not Cauchy complete.

f.   Prove that Cauchy completeness does not imply (Nest). (*Hint:*   In $\mathscr{L}_{\mathfrak{F}}$, consider the nested sequence of closed intervals $\langle [n, 1/nx] \rangle$.)

21.   Prove that the ordered field of "rational functions" (compare Example 3, p. 176 and Example 2, p. 193)

$$\frac{a_n x^n + \cdots + a_1 x + a_0}{b_k x^k + \cdots + b_1 x + b_0}$$

(where the a_i's and b_j's are rational) is not Cauchy complete. (*Hint:*   Use Exercise 20e.)

22.   Is there a field which can be made into an ordered field in at least two ways (that is, with at least two different order relations)?

23.   In a complete ordered field, prove that the limit of a bounded increasing sequence is the lub of the set of terms of the sequence.

24.   In a complete ordered field, let $a_1 = \sqrt{2}$ and $a_{n+1} = \sqrt{2a_n}$ for $n \geq 1$. Prove that $\langle a_n \rangle$ converges and find $\text{Lim } a_n$. (*Hint:*   Use (Mon).)

25.^{D} Prove that (Nest) does not imply completeness (see Borovskiĭ [1956]). (*Hint:*   It suffices to find a non-Archimedean ordered field satisfying (Nest).)

5.6 CAUCHY COMPLETION. THE REAL NUMBER SYSTEM

By Theorem 5.9, to prove the existence of a complete ordered field it suffices to prove the existence of an Archimedean, Cauchy-complete ordered field. This is what we shall do in this section.

Let $\mathfrak{F} = (F, +, \times, <)$ be an arbitrary ordered field. We shall keep $\mathfrak{F}$ fixed throughout this section.

Definition   Let $\langle x_n \rangle$ and $\langle y_n \rangle$ be Cauchy sequences in $\mathfrak{F}$. Then

$$\langle x_n \rangle \simeq \langle y_n \rangle \qquad \text{if and only if} \qquad \text{Lim}(x_n - y_n) = 0.$$

Thus, $\langle x_n \rangle \simeq \langle y_n \rangle$ if and only if x_n and y_n become arbitrarily close to each other as n increases.

Theorem 6.1 $\simeq$ is an equivalence relation, that is, for any Cauchy sequences $\langle x_n \rangle$, $\langle y_n \rangle$, $\langle z_n \rangle$:

a. $\langle x_n \rangle \simeq \langle x_n \rangle$ (reflexivity).
b. $\langle x_n \rangle \simeq \langle y_n \rangle \Rightarrow \langle y_n \rangle \simeq \langle x_n \rangle$ (symmetry).
c. $(\langle x_n \rangle \simeq \langle y_n \rangle \wedge \langle y_n \rangle \simeq \langle z_n \rangle) \Rightarrow \langle x_n \rangle \simeq \langle z_n \rangle$ (transitivity).

Proof (a) $\mathrm{Lim}(x_n - x_n) = \mathrm{Lim}\, 0 = 0.$
  (b) $\mathrm{Lim}(y_n - x_n) = -\mathrm{Lim}(x_n - y_n) = -0 = 0$ (by Theorem 5.2ii).
  (c) $\mathrm{Lim}(x_n - z_n) = \mathrm{Lim}((x_n - y_n) + (y_n - z_n))$
$$= \mathrm{Lim}(x_n - y_n) + \mathrm{Lim}(y_n - z_n) \quad \text{(by Theorem 5.2i)}$$
$$= 0 + 0 = 0. \quad \blacksquare$$

Definition If $\langle x_n \rangle$ is a Cauchy sequence, denote the equivalence class of $\langle x_n \rangle$ with respect to $\simeq$ by $[\langle x_n \rangle]$. Thus,

$$[\langle x_n \rangle] = \{\langle z_n \rangle : \langle x_n \rangle \simeq \langle z_n \rangle\}.$$

We denote by $C_{\mathfrak{F}}$ the set of all equivalence classes with respect to $\simeq$. The following lemma will enable us to define arithmetic operations in $C_{\mathfrak{F}}$.

Lemma 6.2 Let $\langle x_n \rangle$, $\langle u_n \rangle$, $\langle y_n \rangle$, $\langle v_n \rangle$ be Cauchy sequences such that $\langle x_n \rangle \simeq \langle u_n \rangle$ and $\langle y_n \rangle \simeq \langle v_n \rangle$. Then

a. $\langle x_n + y_n \rangle \simeq \langle u_n + v_n \rangle$.
b. $\langle -y_n \rangle \simeq \langle -v_n \rangle$.
c. $\langle x_n - y_n \rangle \simeq \langle u_n - v_n \rangle$.
d. $\langle x_n y_n \rangle \simeq \langle u_n v_n \rangle$.

Proof (a)

$$\mathrm{Lim}((x_n + y_n) - (u_n + v_n)) = \mathrm{Lim}((x_n - u_n) + (y_n - v_n))$$
$$= \mathrm{Lim}(x_n - u_n) + \mathrm{Lim}(y_n - v_n) = 0 + 0 = 0.$$

Hence, $\langle x_n + y_n \rangle \simeq \langle u_n + v_n \rangle$.

  (b) $\mathrm{Lim}(-y_n - (-v_n)) = \mathrm{Lim}(-(y_n - v_n))$
$$= -\mathrm{Lim}(y_n - v_n) = -0 = 0.$$

(c)   This follows from (a) and (b).

(d) $\text{Lim}(x_n y_n - u_n v_n) = \text{Lim}(x_n(y_n - v_n) + v_n(x_n - u_n))$. Since $\langle x_n \rangle$ and $\langle v_n \rangle$ are bounded and $\text{Lim}(y_n - v_n) = \text{Lim}(x_n - u_n) = 0$, it follows by Exercise 8 on p. 213 that $\text{Lim } x_n(y_n - v_n) = 0$ and $\text{Lim } v_n(x_n - u_n) = 0$. Hence,

$$\text{Lim}(x_n(y_n - v_n) + v_n(x_n - u_n)) = \text{Lim } x_n(y_n - v_n) + \text{Lim } v_n(x_n - u_n)$$
$$= 0 + 0 = 0. \quad \blacksquare$$

Definitions   Let s and t be equivalence classes in $C_{\mathcal{F}}$. Choose any $\langle x_n \rangle$ in s and any $\langle y_n \rangle$ in t. Define

$$s \oplus t = [\langle x_n + y_n \rangle].$$
$$s \otimes t = [\langle x_n y_n \rangle].$$
$$-t = [\langle -y_n \rangle].$$
$$0_C = [\langle 0 \rangle].$$
$$1_C = [\langle 1 \rangle].$$

Notice that the definitions of $\oplus$, $\otimes$, $\ominus$ depend upon Lemma 6.2a,d,b.

Lemma 6.3

a. $\langle x_n \rangle \in 0_C \Leftrightarrow \text{Lim } x_n = 0$ (that is 0_C consists of all sequences which have 0 as a limit).

b. $\langle x_n \rangle \in 1_C \Leftrightarrow \text{Lim } x_n = 1$ (that is, 1_C consists of all sequences which have 1 as a limit).

Proof   (a) $\langle x_n \rangle \in 0_C \Leftrightarrow \langle x_n \rangle \simeq \langle 0 \rangle \Leftrightarrow \text{Lim}(x_n - 0) = 0$
$$\Leftrightarrow \text{Lim } x_n = 0.$$

(b) $\langle x_n \rangle \in 1_C \Leftrightarrow \langle x_n \rangle \simeq \langle 1 \rangle \Leftrightarrow \text{Lim}(x_n - 1) = 0$. But, if $\text{Lim}(x_n - 1) = 0$, then

$$\text{Lim } x_n = \text{Lim } 1 + \text{Lim}(x_n - 1) = 1 + 0 = 1.$$

Conversely, if $\text{Lim } x_n = 1$, then

$$\text{Lim}(x_n - 1) = \text{Lim } x_n - \text{Lim } 1 = 1 - 1 = 0. \quad \blacksquare$$

Lemma 6.4   If $\langle x_n \rangle$ is a Cauchy sequence such that 0 is not a limit of $\langle x_n \rangle$, then there is a Cauchy sequence $\langle y_n \rangle$ such that $\text{Lim } x_n y_n = 1$. Moreover, there is a positive integer n_1 such that $n \geq n_1 \Rightarrow y_n = 1/x_n$.

Proof Let $\langle x_n \rangle$ be a Cauchy sequence such that 0 is not a limit of $\langle x_n \rangle$. Hence, there exists some $\varepsilon_1 > 0$ such that

(*)   for any positive integer n, there exists $k \geq n$ with $|x_k| \geq \varepsilon_1$.

In addition, since $\langle x_n \rangle$ is a Cauchy sequence, there exists a positive integer n_1 such that $n, k \geq n_1 \Rightarrow |x_n - x_k| < \varepsilon_1/2$. By (*), there is some positive integer $k_0 \geq n_1$ such that $|x_{k_0}| \geq \varepsilon_1$. Thus,

$$n \geq n_1 \Rightarrow |x_n| = |x_{k_0} - (x_{k_0} - x_n)|$$
$$\geq ||x_{k_0}| - |x_{k_0} - x_n|| \qquad \text{(by Theorem 3.4.8k)}$$
$$= |x_{k_0}| - |x_{k_0} - x_n| \geq \varepsilon_1 - \varepsilon_1/2 = \varepsilon_1/2.$$

Then, (**) for $n \geq n_1$, $|x_n| \geq \varepsilon_1/2 > 0$. Hence, for $n \geq n_1$, $x_n \neq 0$. Let

$$y_n = \begin{cases} 1 & \text{if } n < n_1; \\ 1/x_n & \text{if } n \geq n_1. \end{cases}$$

First of all, $\langle y_n \rangle$ is a Cauchy sequence. (For, assume $\varepsilon > 0$. Then there is some positive integer n_2 such that

$$k, n \geq n_2 \Rightarrow |x_n - x_k| < \varepsilon_1^2 \cdot \varepsilon/4.$$

Let $n_0 = \max(n_1, n_2)$. Then

$$n, k \geq n_0 \Rightarrow |y_n - y_k| = |1/x_n - 1/x_k|$$
$$= |(x_k - x_n)/x_k x_n|$$
$$= |x_k - x_n|/|x_k||x_n|$$
$$< \varepsilon_1^2 \cdot \varepsilon/4 \cdot 4/\varepsilon_1^2 \qquad \text{(by (**))}$$
$$= \varepsilon.)$$

Second, $\operatorname{Lim} x_n y_n = 1$. (For, $x_n y_n = 1$ for all $n \geq n_1$.)   ∎

Lemma 6.5 $(C_{\mathcal{F}}, \oplus, \otimes)$ is a field.

Proof We shall verify the field axioms (see pp. 166–167). Assume r, s, t in $C_{\mathcal{F}}$, and choose $\langle x_n \rangle \in r$, $\langle y_n \rangle \in s$, $\langle z_n \rangle \in t$.

(1) $r \oplus (s \oplus t) = (r \oplus s) \oplus t$

$$r \oplus (s \oplus t) = r \oplus ([\langle y_n \rangle] \oplus [\langle z_n \rangle]) = [\langle x_n \rangle] \oplus [\langle y_n + z_n \rangle]$$
$$= [\langle x_n + (y_n + z_n) \rangle].$$

$$(r \oplus s) \oplus t = ([\langle x_n \rangle] \oplus [\langle y_n \rangle]) \oplus [\langle z_n \rangle]$$
$$= [\langle x_n + y_n \rangle] \oplus [\langle z_n \rangle] = [\langle (x_n + y_n) + z_n \rangle]$$
$$= [\langle x_n + (y_n + z_n) \rangle].$$

(2) $r \oplus s = s \oplus r$

$$r \oplus s = [\langle x_n \rangle] \oplus [\langle y_n \rangle] = [\langle x_n + y_n \rangle] = [\langle y_n + x_n \rangle]$$
$$= [\langle y_n \rangle] \oplus [\langle x_n \rangle] = s \oplus r.$$

(3) (a) $r \oplus 0_C = [\langle x_n \rangle] \oplus [\langle 0 \rangle] = [\langle x_n + 0 \rangle] = [\langle x_n \rangle] = r.$

(b) $r \oplus (-r) = [\langle x_n \rangle] \oplus [\langle -x_n \rangle] = [\langle x_n + (-x_n) \rangle]$
$$= [\langle 0 \rangle] = 0_C.$$

(4) $r \otimes (s \otimes t) = (r \otimes s) \otimes t$

$$r \otimes (s \otimes t) = [\langle x_n \rangle] \otimes ([\langle y_n \rangle] \otimes [\langle z_n \rangle])$$
$$= [\langle x_n \rangle] \otimes [\langle y_n z_n \rangle] = [\langle x_n(y_n z_n) \rangle].$$

$$(r \otimes s) \otimes t = ([\langle x_n \rangle] \otimes [\langle y_n \rangle]) \otimes [\langle z_n \rangle]$$
$$= [\langle x_n y_n \rangle] \otimes [\langle z_n \rangle] = [\langle (x_n y_n) z_n \rangle] = [\langle x_n(y_n z_n) \rangle].$$

(5) $r \otimes (s \oplus t) = (r \otimes s) \oplus (r \otimes t)$

$$r \otimes (s \oplus t) = [\langle x_n \rangle] \otimes ([\langle y_n \rangle] \oplus [\langle z_n \rangle])$$
$$= [\langle x_n \rangle] \otimes [\langle y_n + z_n \rangle] = [\langle x_n(y_n + z_n) \rangle]$$
$$= [\langle x_n y_n + x_n z_n \rangle] = [\langle x_n y_n \rangle] \oplus [\langle x_n z_n \rangle]$$
$$= ([\langle x_n \rangle] \otimes [\langle y_n \rangle]) \oplus ([\langle x_n \rangle] \otimes [\langle z_n \rangle])$$
$$= (r \otimes s) \oplus (r \otimes t).$$

(6) $r \otimes 1_C = [\langle x_n \rangle] \otimes [\langle 1 \rangle] = [\langle x_n \times 1 \rangle] = [\langle x_n \rangle] = r.$

(7) $r \otimes s = [\langle x_n \rangle] \otimes [\langle y_n \rangle] = [\langle x_n y_n \rangle] = [\langle y_n x_n \rangle]$
$$= [\langle y_n \rangle] \otimes [\langle x_n \rangle] = s \otimes r.$$

(9) (Remember that Axiom 8 was superfluous.) Assume $r \neq 0_C$. Therefore, $\langle x_n \rangle \notin 0_C$. Hence, by Lemma 5.3a, 0 is not a limit of $\langle x_n \rangle$. Therefore, by Lemma 6.4, there is a Cauchy sequence $\langle v_n \rangle$ such that Lim $x_n v_n = 1$. Hence, if we let $r^* = [\langle v_n \rangle]$,

$$r \otimes r^* = [\langle x_n \rangle] \otimes [\langle v_n \rangle] = [\langle x_n v_n \rangle] = 1_C$$

by Lemma 6.3b.

(10) $0_C \neq 1_C$, since, by Lemma 6.3b, $\langle 0 \rangle \in 0_C$ and $\langle 0 \rangle \notin 1_C$. ∎

Now that we know that $(C_{\mathfrak{F}}, \oplus, \otimes)$ is a field, our next aim is to make it into an ordered field. We know, by Theorem 3.4.3, that it suffices to define the set of positive elements.

Definition Let $\langle x_n \rangle$ be a sequence in F. We say that $\langle x_n \rangle$ is **positive** if and only if there is some positive element c in F and there is a positive integer n_0 such that $n \geq n_0 \Rightarrow x_n \geq c$.

Thus, $\langle x_n \rangle$ is positive if and only if, from some point on, all terms of $\langle x_n \rangle$ are at least as big as some fixed positive element.

EXERCISE

1. Prove that the sequence $\langle 1/n \rangle$ of reciprocals of positive integers in an ordered field $\mathfrak{F}$ is positive if and only if $\mathfrak{F}$ is non-Archimedean.

Lemma 6.6 If $\langle x_n \rangle$ and $\langle y_n \rangle$ are Cauchy sequences such that $\langle x_n \rangle \simeq \langle y_n \rangle$ and $\langle x_n \rangle$ is positive, then $\langle y_n \rangle$ is also positive.

Proof Since $\langle x_n \rangle$ is positive, there exist some positive element c in F and a positive integer n_0 such that $n \geq n_0 \Rightarrow x_n \geq c$. Since $\langle x_n \rangle \simeq \langle y_n \rangle$, that is, $\mathrm{Lim}(x_n - y_n) = 0$, there is some positive integer n_1 such that $n \geq n_1 \Rightarrow |x_n - y_n| < c/2$. Therefore,

$$n \geq n_1 \Rightarrow x_n - y_n \leq |x_n - y_n| < c/2 \Rightarrow y_n > x_n - c/2.$$

Let $n_2 = \max(n_0, n_1)$. Then,

$$n \geq n_2 \Rightarrow y_n > x_n - c/2 \geq c - c/2 = c/2.$$

This means that $\langle y_n \rangle$ is positive. ∎

Lemma 6.7 If $\langle x_n \rangle$ is a Cauchy sequence, then exactly one of the following holds:

(i) $\langle x_n \rangle$ is positive; (i) $\mathrm{Lim}\, x_n = 0$; (iii) $\langle -x_n \rangle$ is positive.

Proof Assume that it is not the case that $\mathrm{Lim}\, x_n = 0$. Hence, there exists $\varepsilon_1 > 0$ such that, for any n, there exists $k \geq n$ with $|x_k| \geq \varepsilon_1$. Since $\langle x_n \rangle$ is a Cauchy sequence, there exists n_1 such that

$$n, k \geq n_1 \Rightarrow |x_n - x_k| < \varepsilon_1/2.$$

There exists $k_1 \geq n_1$ such that $|x_{k_1}| \geq \varepsilon_1$.

Case 1 $x_{k_1} > 0$.   Then

$$n \geq n_1 \Rightarrow x_n = x_{k_1} - (x_{k_1} - x_n) \geq \varepsilon_1 - | x_{k_1} - x_n |$$
$$> \varepsilon_1 - \varepsilon_1/2 = \varepsilon_1/2.$$

Hence, $\langle x_n \rangle$ is positive.

Case 2 $x_{k_1} < 0$.   Hence $| x_{k_1} | = -x_{k_1}$. Then

$$n \geq n_1 \Rightarrow -x_n = -x_{k_1} - (x_n - x_{k_1}) \geq \varepsilon_1 - | x_n - x_{k_1} |$$
$$> \varepsilon_1 - \varepsilon_1/2 = \varepsilon_1/2.$$

Hence, $\langle -x_n \rangle$ is positive.

To see the uniqueness, assume $\langle x_n \rangle$ is positive and $\mathrm{Lim}\ x_n = 0$. Since $\langle x_n \rangle$ is positive, there exist positive c in F and a positive integer n_0 such that $n \geq n_0 \Rightarrow x_n \geq c$. But, since $\mathrm{Lim}\ x_n = 0$, there exists a positive integer n_1 such that $n \geq n_1 \Rightarrow | x_n | < c/2$. Let $n_2 = \max(n_0, n_1)$. Since $x_{n_2} \geq c > 0$, $| x_{n_2} | = x_{n_2}$. Hence, $x_{n_2} < c/2$. This implies $c < c/2$, contradicting the fact that $c > 0$. In a similar manner, one can show that it is impossible for both $\langle x_n \rangle$ and $\langle -x_n \rangle$ to be positive, and it is also impossible for $\langle -x_n \rangle$ to be positive, while $\mathrm{Lim}\ x_n = 0$.   ∎

EXERCISE

2.^{C} If $\mathrm{Lim}\ x_n = b$, prove that $\langle x_n \rangle$ is positive if and only if $b > 0$.

Lemma 6.8   If $\langle x_n \rangle$ and $\langle y_n \rangle$ are positive Cauchy sequences, then so are $\langle x_n + y_n \rangle$ and $\langle x_n y_n \rangle$.

Proof   Assume $\langle x_n \rangle$ and $\langle y_n \rangle$ are positive Cauchy sequences. Then there exist positive elements c_1, c_2 of F and positive integers n_1, n_2 such that $n \geq n_1 \Rightarrow x_n \geq c_1$ and $n \geq n_2 \Rightarrow y_n \geq c_2$. Let $n_0 = \max(n_1, n_2)$. Then, if $n \geq n_0$, $x_n + y_n \geq c_1 + c_2$ and $x_n y_n \geq c_1 c_2$. Thus, $\langle x_n + y_n \rangle$ and $\langle x_n y_n \rangle$ are positive Cauchy sequences.   ∎

Definition   Let $r \in C_{\mathfrak{F}}$. Choose any $\langle x_n \rangle$ in r. Then we say that r is **positive** if and only if $\langle x_n \rangle$ is a positive sequence. (Notice that Lemma 6.6 tells us that it does not make any difference which sequence $\langle x_n \rangle$ we choose from r. $\langle x_n \rangle$ will be positive if and only if any other sequence of r is positive.) We shall denote the set of positive elements of $C_{\mathfrak{F}}$ by $\mathscr{P}_C$.

Definition   Assume r and s are in $C_{\mathfrak{F}}$. Then

$$r \oslash s \Leftrightarrow s - r \in \mathscr{P}_C.$$

Lemma 6.9 $\mathscr{C}_{\mathfrak{F}} = (C_{\mathfrak{F}}, \oplus, \otimes, \ominus)$ is an ordered field.

Proof   Since we already know that $(C_{\mathfrak{F}}, \oplus, \otimes)$ is a field, it suffices to verify properties (i)–(iv) of Theorem 3.4.3 (p. 106).

(i) $0_C \notin \mathscr{P}_C$. For, let us assume that $\langle x_n \rangle \in 0_C$. Then $\operatorname{Lim} x_n = 0$. By the uniqueness part of Lemma 6.7, $\langle x_n \rangle$ cannot be a positive Cauchy sequence. Hence, 0_C is not positive.

(ii)   For any r in $C_{\mathfrak{F}}$, $r \in \mathscr{P}_C$ or $r = 0_C$ or $-r \in \mathscr{P}_C$. This is a direct consequence of Lemma 6.7.

(iii), (iv)   If r and s are in $\mathscr{P}_C$, so are $r \oplus s$ and $r \otimes s$. For, take $\langle x_n \rangle$ in r and $\langle y_n \rangle$ in s. Then $\langle x_n \rangle$ and $\langle y_n \rangle$ are positive Cauchy sequences. By Lemma 6.8, $\langle x_n + y_n \rangle$ and $\langle x_n y_n \rangle$ are positive Cauchy sequences. But, $\langle x_n + y_n \rangle \in r \oplus s$ and $\langle x_n y_n \rangle \in r \otimes s$. Thus, $r \oplus s$ and $r \otimes s$ are positive.   ∎

We shall need the following two simple facts about the order relation in $\mathscr{C}_{\mathfrak{F}}$.

Lemma 6.10   Let $\langle x_n \rangle$ and $\langle y_n \rangle$ be Cauchy sequences in $\mathfrak{F}$ such that $x_n \leq y_n$ for all $n \geq n_0$, where n_0 is some fixed positive integer. Then, $[\langle x_n \rangle] \ominus [\langle y_n \rangle]$.

Proof   Since $y_n - x_n \geq 0$ for all $n \geq n_0$, we know that $\langle -(y_n - x_n) \rangle$ is not a positive sequence. Hence, by Lemma 6.7, either $\langle y_n - x_n \rangle$ is a positive sequence or $\operatorname{Lim}(y_n - x_n) = 0$. In the first case, $[\langle x_n \rangle] \ominus [\langle y_n \rangle]$. (For, $[\langle x_n \rangle] \ominus [\langle y_n \rangle]$ means that $[\langle y_n \rangle] \ominus [\langle x_n \rangle] \in \mathscr{P}_C$, that is, $\langle y_n - x_n \rangle$ is a positive sequence.) In the second case, $\langle x_n \rangle \simeq \langle y_n \rangle$, that is, $[\langle x_n \rangle] = [\langle y_n \rangle]$.   ∎

Lemma 6.11   Let $\langle x_n \rangle$ be a Cauchy sequence in $\mathfrak{F}$. Then,

$$| [\langle x_n \rangle] | = [\langle | x_n | \rangle].$$

Proof   By Lemma 6.7, we have the following three cases.

Case 1 $\langle x_n \rangle$ is positive.   Then $[\langle x_n \rangle]$ is positive in $\mathscr{C}_{\mathfrak{F}}$, and $| [\langle x_n \rangle] | = [\langle x_n \rangle]$. Also, $x_n = | x_n |$ for all $n \geq n_0$, where n_0 is some positive integer. Hence, $\langle x_n \rangle \simeq \langle | x_n | \rangle$, that is, $[\langle x_n \rangle] = [\langle | x_n | \rangle]$.

Case 2 $\operatorname{Lim} x_n = 0$.   Then $[\langle x_n \rangle] = 0_C$. Also, $\operatorname{Lim} | x_n | = | 0 | = 0$. Hence, $[\langle | x_n | \rangle] = 0_C$.

Case 3 $\langle -x_n \rangle$ is positive.   Then $[\langle x_n \rangle] \otimes 0_C$.   Hence

$$|[\langle x_n \rangle]| = -[\langle x_n \rangle] = [\langle -x_n \rangle].$$

But, $|x_n| = -x_n$ for $n \geq n_1$, where n_1 is some positive integer. Therefore, $\langle |x_n| \rangle \simeq \langle -x_n \rangle$, that is, $[\langle |x_n| \rangle] = [\langle -x_n \rangle]$.   ∎

Our original ordered field $\mathcal{F}$ is "isomorphically imbeddable" in the ordered field $\mathscr{C}_{\mathcal{F}}$ in the following sense. Let

$$\Delta(u) = [\langle u \rangle] \qquad \text{for any } u \text{ in } F.$$

Remember that $\langle u \rangle$ represents the constant sequence all terms of which are equal to u.

Theorem 6.12   The function Δ is an isomorphism of $\mathcal{F}$ into $\mathscr{C}_{\mathcal{F}}$. (The image of $\mathcal{F}$ under Δ will be denoted by $\mathcal{F}^*$. Thus, $\mathcal{F}$ is isomorphic with $\mathcal{F}^*$, and $\mathcal{F}^*$ is an ordered subfield[†] of $\mathscr{C}_{\mathcal{F}}$.)

Proof   Clearly, $\Delta: F \rightarrow C_{\mathcal{F}}$.

(i) Δ is order preserving, that is, $u < v \Leftrightarrow \Delta(u) \otimes \Delta(v)$.

$$\begin{aligned}
\Delta(u) \otimes \Delta(v) &\Leftrightarrow [\langle u \rangle] \otimes [\langle v \rangle] \\
&\Leftrightarrow [\langle v \rangle] \ominus [\langle v \rangle] \in \mathscr{P}_C \\
&\Leftrightarrow [\langle v - u \rangle] \in \mathscr{P}_C \\
&\Leftrightarrow v - u > 0 \qquad \text{(see Exercise 2, p. 226)} \\
&\Leftrightarrow u < v.
\end{aligned}$$

(ii) Δ is one-one.   This is an immediate consequence of (i).

(iii) $\Delta(u + v) = \Delta(u) \oplus \Delta(v)$

$$\Delta(u + v) = [\langle u + v \rangle] = [\langle u \rangle] \oplus [\langle v \rangle] = \Delta(u) \oplus \Delta(v).$$

(iv) $\Delta(u \times v) = \Delta(u) \otimes \Delta(v)$

$$\Delta(u \times v) = [\langle u \times v \rangle] = [\langle u \rangle] \otimes [\langle v \rangle] = \Delta(u) \otimes \Delta(v).$$ ∎

Lemma 6.13

a. $\mathcal{F}^*$ is dense in $\mathscr{C}_{\mathcal{F}}$, that is, between any two elements of $C_{\mathcal{F}}$, there is some element of $\mathcal{F}^*$.

[†] To avoid extra notation, I am taking the liberty of letting $\mathcal{F}^*$ denote both the range of Δ and the ordered subfield of $\mathscr{C}_{\mathcal{F}}$ detemined by the range of Δ.

b. If r is a positive element of $C_{\mathfrak{F}}$, there exists some u in F such that $0_C \ominus \Delta(u) \ominus r$.

Proof (a) Assume $[\langle x_n \rangle] \ominus [\langle y_n \rangle]$. Then $[\langle y_n - x_n \rangle]$ is positive, that is, there exists positive c in F such that $y_n - x_n > c$ for $n \geq n_1$, where n_1 is some positive integer. Choose $n_0 \geq n_1$ such that $k, n \geq n_0 \Rightarrow | x_n - x_k |$ $< c/4$ and $| y_n - y_k | < c/4$. Let $b = (x_{n_0} + y_{n_0})/2$. Then $[\langle b \rangle] = \Delta(b) \in \mathfrak{F}^*$, and, in addition, $[\langle x_n \rangle] \ominus [\langle b \rangle] \ominus [\langle y_n \rangle]$. (To see this, assume $n \geq n_0$. First, $y_{n_0} - x_{n_0} > c > 0$. Therefore, $x_{n_0} < b < y_{n_0}$, since the average of two numbers is greater than the smaller of the two and less than the larger of the two. Note also that $b - x_{n_0} = (y_{n_0} - x_{n_0})/2 > c/2$. Hence, $x_{n_0} < b - c/2$. In addition, since $| x_n - x_{n_0} | < c/4$,

$$x_n < x_{n_0} + c/4 < (b - (c/2)) + c/4 = b - (c/4).$$

Hence, $b - x_n > c/4$ for all $n \geq n_0$. This means that $\langle b - x_n \rangle$ is positive, and, therefore, $[\langle x_n \rangle] \ominus [\langle b \rangle]$. The analogous proof that $[\langle b \rangle] \ominus [\langle y_n \rangle]$ is left as an exercise for the reader.)

(b) Consider the two elements 0_C and r. By Part a, there must be a point z of $\mathfrak{F}^*$ between them. Then $z = \Delta(u)$ for some u in F. Since z is positive and Δ preserves order, u must be positive in $\mathfrak{F}$. ∎

EXERCISE

3. Show that, if $\langle x_n \rangle$ is a sequence in $\mathfrak{F}$, then $\langle x_n \rangle$ is a Cauchy sequence in $\mathfrak{F}$ if and only if $\langle \Delta(x_n) \rangle$ is a Cauchy sequence in $\mathscr{C}_{\mathfrak{F}}$. (*Hint*: Lemma 6.13b.)

Lemma 6.14 Let $\langle y_n \rangle$ be a Cauchy sequence in $\mathfrak{F}$ and let $r = [\langle y_n \rangle]$. Then $r = \text{Lim } \Delta(y_n)$.

Proof Let ε be a positive element of $C_{\mathfrak{F}}$. By Lemma 6.13b, there exists some positive element ε_1 in F such that $\Delta(\varepsilon_1) \ominus \varepsilon$. Since $\langle y_n \rangle$ is a Cauchy sequence in $\mathfrak{F}$, there exists a positive integer n_0 such that $k, n \geq n_0$ $\Rightarrow | y_n - y_k | < \varepsilon_1$. Take any fixed integer $j \geq n_0$. Then

$$| r \ominus \Delta(y_j) | = | [\langle y_n \rangle] \ominus [\langle y_j \rangle] | = | [\langle y_n - y_j \rangle] |$$
$$= [\langle | y_n - y_j | \rangle] \qquad \text{(by Lemma 6.11)}$$
$$\ominus \Delta(\varepsilon_1) \qquad \text{(by Lemma 6.10, since } | y_n - y_j | < \varepsilon_1 \text{ for } n \geq n_0)$$
$$\ominus \varepsilon.$$

Then, for any $j \geq n_0$, $| r \ominus \Delta(y_j) | \ominus \varepsilon$. Therefore, $r = \text{Lim } \Delta(y_n)$. ∎

EXERCISE

4. Prove that every element of $C_{\mathfrak{F}}$ is a limit of a sequence of elements of $\mathfrak{F}^*$. (*Hint*: Lemma 6.14.)

Theorem 6.15 $\mathscr{C}_{\mathfrak{F}} = (C_{\mathfrak{F}}, \oplus, \otimes, \ominus)$ is Cauchy complete.

Proof   Let $\langle r_k \rangle$ be a Cauchy sequence in $\mathscr{C}_{\mathfrak{F}}$. We may assume that $r_{k+1} \neq r_k$ for all k. (For, if there exists k_0 such that $r_{k+1} = r_k$ for all $k \geq k_0$, then Lim $r_k = r_{k_0}$. In the opposite case, we may choose a subsequence of $\langle r_k \rangle$ satisfying our assumption. (Why?) This subsequence is a Cauchy sequence by Theorem 5.7, and, if we prove the theorem under our assumption, the subsequence converges. But then, by Theorem 5.8, the original sequence converges.)

Let $\varepsilon_k = | r_{k+1} \ominus r_k |$. Then Lim $\varepsilon_k = 0_C$, since $\langle r_k \rangle$ is a Cauchy sequence. Let $r_k = [\langle x_{k,n} \rangle]$, where $\langle x_{k,n} \rangle$ is a Cauchy sequence in $\mathfrak{F}$. (*Note*: k is fixed.) By Lemma 6.14, there exists n_k such that $| r_k \ominus \varDelta(x_{k,n_k}) | \ominus \varepsilon_k$. Let $y_k = x_{k,n_k}$. Then $\langle y_k \rangle$ is a Cauchy sequence in $\mathfrak{F}$. (*Proof*:   Assume ε is a positive element of F. Let $\varepsilon^* = \varDelta(\varepsilon)$. There exists l such that

$$ p, q \geq l \Rightarrow | r_p \ominus r_q | \ominus \varepsilon^*/3. $$

In addition, there exists m such that $k \geq m \Rightarrow \varepsilon_k \ominus \varepsilon^*/3$. Let $j = \max(l, m)$. Then, for $p, q \geq j$,

$$ | \varDelta(y_p) \ominus \varDelta(y_q) | \ominus | \varDelta(y_p) \ominus r_p | \oplus | r_p \ominus r_q | \oplus | r_q \ominus \varDelta(y_q) | $$
$$ \ominus \varepsilon_p \oplus \varepsilon^*/3 \oplus \varepsilon_q \ominus \varepsilon^*/3 \oplus \varepsilon^*/3 \oplus \varepsilon^*/3 = \varepsilon^* = \varDelta(\varepsilon). $$

Hence, $| \varDelta(y_p - y_q) | \ominus \varDelta(\varepsilon)$. Then, by Lemma 6.11, with $\langle x_n \rangle$ as the constant sequence $\langle y_p - y_q \rangle$, $\varDelta(| y_p - y_q |) \ominus \varDelta(\varepsilon)$, and, therefore, $| y_p - y_q | < \varepsilon$. We also have proved that $\langle \varDelta(y_k) \rangle$ is a Cauchy sequence in $\mathscr{C}_{\mathfrak{F}}$.) Let $s = [\langle y_k \rangle]$. Then Lim $r_k = s$. (*Proof*: $s = $ Lim $\varDelta(y_k)$ by Lemma 6.14. We know that Lim $\varepsilon_k = 0_C$ and $| r_k \ominus \varDelta(y_k) | \ominus \varepsilon_k$. Hence, Lim$(r_k \ominus \varDelta(y_k)) = 0_C$. But,

$$ r_k = \varDelta(y_k) \oplus (r_k \ominus \varDelta(y_k)). $$

Hence, Lim $r_k = s$.)   ∎

Terminology   The ordered field $\mathscr{C}_{\mathfrak{F}}$ is called the **Cauchy completion** of $\mathfrak{F}$.

Now we shall turn to the problem of determining when $\mathscr{C}_{\mathfrak{F}}$ is Archimedean.

Lemma 6.16 If an ordered field $\mathfrak{F}_1$ is isomorphic with an ordered field $\mathfrak{F}_2$, and $\mathfrak{F}_1$ is Archimedean, then so is $\mathfrak{F}_2$.

Proof If ψ is the isomorphism, then ψ carries $N_{\mathfrak{F}_1}$ onto $N_{\mathfrak{F}_2}$. (For, it is easy to see that $\psi[N_{\mathfrak{F}_1}] = \{\psi(n) : n \in N_{\mathfrak{F}_1}\}$ is inductive and is a subset of every inductive set.) Hence, if z were an upper bound of $N_{\mathfrak{F}_2}$, then $\psi^{-1}(z)$ would be an upper bound of $N_{\mathfrak{F}_1}$. Therefore, by Theorem 2.1, if $\mathfrak{F}_2$ were non-Archimedean, $\mathfrak{F}_1$ would be non-Archimedean. ∎

Theorem 6.17 $\mathscr{C}_{\mathfrak{F}}$ is Archimedean if and only if $\mathfrak{F}$ is Archimedean.

Proof Since $\mathfrak{F}$ is isomorphic with $\mathfrak{F}^*$, it suffices to prove: $\mathscr{C}_{\mathfrak{F}}$ is Archimedean $\Leftrightarrow$ $\mathfrak{F}^*$ is Archimedean. Since $\mathfrak{F}^*$ is a subfield of $\mathscr{C}_{\mathfrak{F}}$, $N_{\mathfrak{F}^*} = N_{\mathscr{C}_{\mathfrak{F}}}$ (see Exercise 8a, p. 145). Clearly, if $N_{\mathfrak{F}^*}$ is bounded above in $\mathfrak{F}^*$, then $N_{\mathscr{C}_{\mathfrak{F}}}$ is bounded above in $\mathscr{C}_{\mathfrak{F}}$. Conversely, if $N_{\mathscr{C}_{\mathfrak{F}}}$ is bounded above in $\mathscr{C}_{\mathfrak{F}}$, then $N_{\mathfrak{F}^*}$ is bounded above in $\mathfrak{F}^*$. (For, if $N_{\mathscr{C}_{\mathfrak{F}}}$ has r as an upper bound in $\mathscr{C}_{\mathfrak{F}}$, then $r \ominus r \oplus 1_C$, and, therefore, by Lemma 6.13a, there must be some u in $\mathfrak{F}^*$ such that $r \ominus u \ominus r \oplus 1_C$. Thus, u is an upper bound of $N_{\mathfrak{F}^*}$ in $\mathfrak{F}^*$.) Hence, $N_{\mathfrak{F}^*}$ is bounded above in $\mathfrak{F}^*$ if and only if $N_{\mathscr{C}_{\mathfrak{F}}}$ is bounded above in $\mathscr{C}_{\mathfrak{F}}$. By Theorem 2.1, $\mathfrak{F}^*$ is non-Archimedean if and only if $\mathscr{C}_{\mathfrak{F}}$ is non-Archimedean. ∎

Theorem 6.18 The Cauchy completion $\mathscr{C}_{\mathcal{Q}}$ of the ordered field $\mathcal{Q}$ of rational numbers is a complete ordered field.

Proof By Theorem 6.15, $\mathscr{C}_{\mathcal{Q}}$ is Cauchy complete. Since $\mathcal{Q}$ is Archimedean (see Example 1, p. 192), it follows by Theorem 6.17 that $\mathscr{C}_{\mathcal{Q}}$ is Archimedean. Hence, by Theorem 5.9, $\mathscr{C}_{\mathcal{Q}}$ is a complete ordered field. ∎

Terminology and notation The complete ordered field $\mathscr{C}_{\mathcal{Q}} = (C_{\mathcal{Q}}, \oplus, \otimes, \ominus)$ will be called the **system of real numbers**, and the elements of $C_{\mathcal{Q}}$ will be called **real numbers**. We shall denote $(C_{\mathcal{Q}}, \oplus, \times, \ominus)$ by $(R, +_R, \times_R, <_R)$, and we shall write $\mathscr{R} = (R, +_R, \times_R, <_R)$.

In Appendix F, we give an alternative construction of a complete ordered field, using the method of Dedekind cuts. By Theorem 4.1, all complete ordered fields are isomorphic. Therefore, from a mathematical point of view, it does not make any difference which of these complete ordered fields is called the system of real numbers. This is similar to what happened with the number systems we studied earlier. For example, there are many Peano systems, but all of them are isomorphic. Hence, it does not matter which of them is taken to be the system of natural numbers.

Our definition of the real number system $\mathscr{R}$ has the following intuitive motivation. The rational number field had certain gaps in it, for example, the gap corresponding to $\sqrt{2}$. Now, there are sequences of rational numbers which get closer and closer to $\sqrt{2}$. Obviously, such a sequence is a Cauchy sequence. If we decided to "represent" the number $\sqrt{2}$ by such a Cauchy sequence, we would encounter the unpleasant fact that there are many sequences of rational numbers which have $\sqrt{2}$ as a limit. However, any two such sequences s and t are in the relation $\simeq$, that is, $\mathrm{Lim}(s_n - t_n) = 0$. Thus, if we take the equivalence classes of all Cauchy sequences with $\sqrt{2}$ as a limit, we can think of this class as a surrogate for the number $\sqrt{2}$. This is precisely what we have done above when we formed the Cauchy completion $\mathscr{C}_{\mathscr{Q}}$.[†]

EXERCISES

5. Prove that every Archimedean ordered field is isomorphic to an ordered subfield of the real number system $\mathscr{R}$. (*Hint*: Theorem 4.2 on p. 207.)

6. Prove that, if $\mathfrak{F} = \mathscr{Q}$, the subfield $\mathfrak{F}^*$ of $\mathscr{R}$ which is isomorphic to $\mathfrak{F}$ (see Theorem 6.12, p. 228) consists of the rational numbers of $\mathscr{R}$. (More generally, if $\mathscr{K}_1$ is a subfield of a field $\mathscr{K}$ such that $\mathscr{K}_1$ is isomorphic with the rational number field $\mathscr{Q}$, prove that $\mathscr{K}_1$ consists of the rational numbers of $\mathscr{K}$.)

7. Prove that every real number is the limit of a sequence of rational numbers.

8. Prove that $<_R$ is the only order relation which makes the system $(R, +_R, \times_R)$ into an ordered field.

9. a. Prove that no ordered subfield of $\mathscr{R}$ (except $\mathscr{R}$ itself) is complete.

    b. Prove that $\mathscr{R}$ is not isomorphic with any ordered subfield of itself (except $\mathscr{R}$).

    c. Prove that an ordered field is a maximal Archimedean ordered field (that is, is not included in any other Archimedean ordered field) if and only if it is complete.

5.7 ELEMENTARY TOPOLOGY OF THE REAL NUMBER SYSTEM

The definitions and many of the results in this section will be stated in terms of an arbitrary ordered field $\mathfrak{F} = (F, +, \times, <)$. Wherever $\mathfrak{F}$ is assumed to be a complete ordered field (that is, the real number system), this will be explicitly stated.

[†] This method of defining real numbers in terms of Cauchy sequences was originally proposed by G. Cantor [1872]. Cantor's theory was set forth in greater detail by Heine [1872].

Definition Let $A \subseteq F$. Then A is said to be an **open** set if and only if every member of A belongs to some open interval included in A.

EXAMPLES

1. Every open interval (a, b) is open. For, if $x \in (a, b)$ and if $c = \min(x - a, b - x)$, then the open interval $(x - c, x + c)$ contains x and is included in (a, b).

2. $(a, \infty) = \{u : a < u\}$ and $(-\infty, a) = \{u : u < a\}$ are open.

3. The whole set F is open.

4. The empty set $\varnothing$ is trivially open.

5. The closed interval $[a, b]$ is not open. There is no open interval containing a and entirely included in $[a, b]$.

6. A singleton $\{a\}$ is not open.

Theorem 7.1

a. The union of any collection $\mathscr{A}$ of open sets is open.
b. The intersection of a finite number of open sets is open.

Proof (a) Let $\mathscr{A}$ be a collection of open sets. Let $B = \bigcup_{A \in \mathscr{A}} A$. Assume $x \in B$. Then $x \in A$ for some A in $\mathscr{A}$. Since A is open, there is an open interval I containing x and such that $I \subseteq A$. Then $I \subseteq B$. Hence, B is open.

(b) Let B_1, B_2 be open sets. Assume $x \in B_1 \cap B_2$. Since B_1 is open, there is an open interval (a, b) containing x and included in B_1. Since B_2 is open, there is an open interval (c, d) containing x and included in B_2. Let $u = \max(a, c)$ and $v = \min(b, d)$. Then (u, v) is an open interval containing x and included in $B_1 \cap B_2$. Hence, $B_1 \cap B_2$ is open. The theorem for any finite number of open sets can be proved easily from this special case by means of mathematical induction. ∎

EXERCISE

1. Show by an example that the intersection of infinitely many open sets need not be open.

Definition Let $A \subseteq F$ and $z \in A$. We say that z is an **accumulation point** of A if and only if every open interval containing z also contains a point of $A - \{z\}$.

EXAMPLES

7.   0 is an accumulation point of the open interval $(0, 1)$. Since $0 \notin (0, 1)$, this shows that an accumulation point of a set need not belong to that set.

8.   0 is an accumulation point of the closed interval $[0, 1]$. (What are the other accumulation points of $[0, 1]$?)

9.   0 is not an accumulation point of the set $\{0\} \cup [1, 2]$. Thus, a member of a set need not be an accumulation point of that set.

Theorem 7.2

a.   If z is an accumulation point of A, then every open interval containing z contains infinitely many points of A.

b.   A finite set has no accumulation points.

Proof   (a)   Let (a, b) be an open interval I containing z. Since z is an accumulation point of A, I contains a point a_1 of $A - \{z\}$. Let

$$c_1 = \min(|a_1 - z|, |a \quad z|, |b - z|)$$

and let $I_1 = (z - c_1, z + c_1)$. Clearly, $a_1 \notin I_1$ and $I_1 \subseteq I$. Since $z \in I_1$, I_1 contains a point a_2 of $A - \{z\}$. Let $c_2 = |a_2 - z|$, and let $I_2 = (z - c_2, z + c_2)$. Then $a_2 \notin I_2$ and $I_2 \subseteq I_1 \subseteq I$. Continuing in this way, we obtain a sequence $a_1, a_2, \ldots$ of distinct points of A, all of which are in the original interval I.

(b)   This follows immediately from Part a.   ∎

EXERCISES

2.   If there is a sequence $\langle a_n \rangle$ of distinct points of A such that Lim $a_n = z$, prove that z is an accumulation point of A.

3.   We say that the ordered field $\mathfrak{F} = (F, |, \times, <)$ satisfies the **axiom of countability** if and only if there is a decreasing sequence $\langle u_n \rangle$ of positive elements such that Lim $u_n = 0$. (i)   Prove that any Archimedean ordered field satisfies the axiom of countability. (ii)   Give an example of a non-Archimedean ordered field which satisfies the axiom of countability.[†] (iii)   Assume that $\mathfrak{F} = (F, +, \times, <)$ satisfies the axiom of countability. Let $A \subseteq F$ and $z \in F$. Prove that, if z is an accumulation point of A, then there is a sequence $\langle a_n \rangle$ of distinct points of A such that Lim $a_n = z$.[‡]

[†] For ordered fields not satisfying the axiom of countability, see Sikorski [1948] and Hauschild [1967].
[‡] The proof of this result seems straightforward. However, the sophisticated reader will observe that, in addition to the Iteration Theorem, there is a subtle use of the so-called Axiom of Choice (see Appendix D, p. 297).

4. If z is a lub of A and $z \notin A$, prove that z is an accumulation point of A.

5. Show that z is an accumulation point of A if and only if, for any positive δ, there is some x such that $x \neq z$ and $|x - z| < \delta$.

6. Find sets A having: (a) exactly one accumulation point; (b) exactly two accumulation points; (c) denumerably many accumulation points.

7. Prove that every real number is an accumulation point of the set Q of rational numbers.

8. What are the accumulation points of the following sets.
   a. an open interval (a, b);
   b. a closed interval $[a, b]$;
   c. a half-open interval $[a, b)$;
   d. the set of all integers;
   e. $\{1, 1/2, 1/3, \ldots, 1/n, \ldots\}$;

Definition A is **closed** if and only if every accumulation point of A belongs to A.

EXAMPLES

10. Every closed interval $[a, b]$ is closed. The set of accumulation points of $[a, b]$ is $[a, b]$ itself.

11. Any finite set is trivially closed, since a finite set has no accumulation points. In particular, $\varnothing$ is closed.

12. $[a, \infty) = \{u : a \leq u\}$ and $(-\infty, a] = \{u : u \leq a\}$ are closed.

13. (a, b) is not closed, since a and b are accumulation points of (a, b). Likewise, $(a, b]$ and $[a, b)$ are not closed.

EXERCISES

9. Show that $N_{\mathfrak{F}}$ is closed.

10. Prove that $[a, \infty)$ and $(-\infty, a]$ are closed.

11. Show that the whole field F is closed.

Theorem 7.3

a. A is open $\Leftrightarrow F - A$ is closed.
b. A is closed $\Leftrightarrow F - A$ is open.

Thus, a set is open if and only if its complement is closed.

Proof (a) Assume A open. Assume that z is an accumulation point of $F - A$. We must prove that $z \in F - A$. Let us assume for the sake of contradiction that $z \in A$. Since A is open, there is an open interval I containing z such that $I \subseteq A$. Since z is an accumulation point of $F - A$, the interval I would have to contain some point of $(F - A) - \{z\}$, which is impossible. Therefore, $F - A$ is closed.

Conversely, assume that $F - A$ is closed. Assume $z \in A$. To prove that A is open, we must show that there is an open interval I containing z such that $I \subseteq A$. Since $z \notin F - A$ and $F - A$ is closed, z cannot be an accumulation point of $F - A$. Therefore, there is an open interval I containing z such that I contains no points of $(F - A) - \{z\}$. Hence, since z itself is a member of A, $I \subseteq A$.

(b) Substitute $F - A$ for A in Part a, and remember that

$$F - (F - A) = A. \qquad \blacksquare$$

Corollary 7.4

a. The intersection of any collection of closed sets is closed.

b. The union of any finite number of closed sets is closed.

Proof (a) Let $\mathscr{C}$ be a collection of closed sets. Let $W = \bigcap_{C \in \mathscr{C}} C$. Then $F - W = F - \bigcap_{C \in \mathscr{C}} C = \bigcup_{C \in \mathscr{C}} (F - C)$. For any $C \in \mathscr{C}$, $F - C$ is open, by Theorem 7.3b. Hence, $\bigcup_{C \in \mathscr{C}} (F - C)$ is open, by Theorem 7.1a. Therefore, W is closed, by Theorem 7.3b.

(b) Let $C_1, \ldots, C_n$ be closed. Then

$$F - (C_1 \cup \cdots \cup C_n) = (F - C_1) \cap \cdots \cap (F - C_n).$$

But each $F - C_i$ is open, by Theorem 7.3b. Hence, $F - (C_1 \cup \cdots \cup C_n)$ is open, by Theorem 7.1b, and, therefore, $C_1 \cup \cdots \cup C_n$ is closed, by Theorem 7.3b. $\blacksquare$

Definition For any set A, let the *closure* $\bar{A}$ of A be defined as the union of A with the set of accumulation points of A.

Theorem 7.5

a. $A \subseteq \bar{A}$.

b. A is closed if and only if $\bar{A} = A$.

c. $z \in \bar{A}$ if and only if every open interval containing z contains a point of A.

d. $\bar{A}$ is closed.

e. $A \subseteq B \Rightarrow \bar{A} \subseteq \bar{B}$.

f. $\overline{A \cup B} = \bar{A} \cup \bar{B}$.

g. Any least upper bound or greatest lower bound of A belongs to $\bar{A}$.

Proof (a) By definition of $\bar{A}$.

(b) Assume A closed. Then every accumulation point of A belongs to A, and, therefore, $\bar{A} = A$. Conversely, assume $\bar{A} = A$. All accumulation points of A are in $\bar{A}$, and, therefore, in A. Hence, A is closed.

(c) Assume $z \in \bar{A}$. Let I be an open interval containing z.

Case 1 $z \in A$. Then I contains a point of A, namely z.

Case 2 z is an accumulation point of A. Then I contains a point of $A - \{z\}$.

In either case, I contains a point of A. Conversely, assume every open interval I containing z contains a point of A.

Case 1 $z \in A$. Then $z \in \bar{A}$.

Case 2 $z \notin A$. Then every open interval I containing z contains a point of $A - \{z\}$. Hence, z is an accumulation point of A, and, therefore, $z \in \bar{A}$.

(d) Assume z is an accumulation point of $\bar{A}$. Let I be an open interval containing z. Then I contains a point of $\bar{A} - \{z\}$, say w. Since I is an open interval containing w and $w \in \bar{A}$, it follows by Part c that I contains a point of A. Hence, every open interval I containing z contains a point of A. By Part c, $z \in \bar{A}$. Thus, $\bar{A}$ is closed.

(e) This follows immediately from the definitions.

(f) $A \subseteq \bar{A}$ and $B \subseteq \bar{B}$. Hence, $A \cup B \subseteq \bar{A} \cup \bar{B}$. By Part (d), $\bar{A}$ and $\bar{B}$ are closed. Hence, by Corollary 7.4b, $\bar{A} \cup \bar{B}$ is closed. By Part (b), $\overline{\bar{A} \cup \bar{B}} = \bar{A} \cup \bar{B}$. Since $A \cup B \subseteq \bar{A} \cup \bar{B}$, it follows by Part (e) that $\overline{A \cup B} \subseteq \overline{\bar{A} \cup \bar{B}} = \bar{A} \cup \bar{B}$. Conversely, assume $z \notin \overline{A \cup B}$. Then, by Part c, there is an open interval I containing z such that $I \cap (A \cup B) = \varnothing$. Hence, $I \cap A = \varnothing$ and $I \cap B = \varnothing$. Therefore, $z \notin \bar{A}$ and $z \notin \bar{B}$, and so, $z \notin \bar{A} \cup \bar{B}$.

(g) Assume $z = \mathrm{lub}(A)$. If $z \in A$, then $z \in \bar{A}$. Hence, we may assume that $z \notin A$. We must show that z is an accumulation point of A. Let (a, b) be an open interval containing z. Since $a < z$, a is not an upper bound of A. Therefore, there is a point u in A such that $a < u$. Since $u \in A$, $u < z$. Hence, $u \in (a, b) - \{z\}$. Thus, z is an accumulation point of A, and, therefore, $z \in \bar{A}$. A similar proof holds for greatest lower bounds. ∎

EXERCISES

12. In any Archimedean ordered field $\mathcal{F} = (F, +, \times, <)$, show that $\overline{Q_{\mathcal{F}}} = F$.

13. Let A^{a} denote the set of accumulation points of A. Prove:
    a. $A \subseteq B \Rightarrow A^{\mathrm{a}} \subseteq B^{\mathrm{a}}$.
    b. $(A \cup B)^{\mathrm{a}} = A^{\mathrm{a}} \cup B^{\mathrm{a}}$.
    c. A^{a} is closed.
    d. $(\bar{A})^{\mathrm{a}} = A^{\mathrm{a}}$.

14. Show by an example that the union of infinitely many closed sets need not be closed.

15. If $\mathrm{Lim}\, x_n = b$ and each $x_n \in A$, prove that $b \in \bar{A}$.

16. Give an example of a subset of the real number field which is neither open nor closed.

17. Show that $\bar{A}$ is the smallest closed set including A, that is, if $A \subseteq B$ and B is closed, then $\bar{A} \subseteq B$.

18. Verify that $\bar{\varnothing} = \varnothing$ and $\bar{F} = F$.

19. For each of the following sets A, determine whether it is open, closed, or neither.
    a. $A = (0, 1) \cup [2, 3]$.
    b. $A = \{0\}$.
    c. $A = \{1, 3, 4\}$.
    d. A is the set of rational numbers $k/2^n$, where k and n are integers.

Definitions   An **open covering** of a set A is a collection $\mathcal{B}$ of open sets such that $A \subseteq \bigcup_{B \in \mathcal{B}} B$, that is, every member of A belongs to at least one set in $\mathcal{B}$. If $\mathcal{B}$ is an open covering of A, we say that A **is covered by** $\mathcal{B}$.

We say that $\mathcal{F}$ has the **Heine–Borel property** if and only if every open covering of a closed interval has a finite subcovering of that interval (that is, there is a finite subset of the covering which is itself a covering).

We say that $\mathcal{F}$ has the **Bolzano–Weierstrass property** if and only if every bounded infinite subset of F has an accumulation point.

Theorem 7.6   The following three properties of an ordered field $\mathcal{F}$ are equivalent.

a. $\mathcal{F}$ is complete.
b. $\mathcal{F}$ has the Heine–Borel property.
c. $\mathcal{F}$ has the Bolzano–Weierstrass property.

In particular, the real number system possesses the Heine–Borel and Bolzano–Weierstrass properties.

Proof (a) $\Rightarrow$ (b)   Assume $\mathfrak{F}$ is complete. Let $\mathscr{B}$ be an open covering of a closed interval $[c_0, d_0]$. We must show that there is a finite subset of $\mathscr{B}$ which is also an open covering of $[c_0, d_0]$. Assume that this is false. Divide $[c_0, d_0]$ in half: into $[c_0, (c_0 + d_0)/2]$ and $[(c_0 + d_0)/2, d_0]$. At least one of these closed intervals is not covered by a finite subcovering of $\mathscr{B}$. (For, if both were covered by finite subcoverings of $\mathscr{B}$, their union $[c_0, d_0]$ would also be covered by a finite subcovering of $\mathscr{B}$.) Let $[c_1, d_1]$ be one of these closed intervals not covered by a finite subcovering of $\mathscr{B}$. (To be specific, take $[c_1, d_1]$ to be the left interval if that is not covered by a finite subcovering of $\mathscr{B}$; otherwise, let $[c_1, d_1]$ be the right interval.) Divide $[c_1, d_1]$ in half and proceed as before, obtaining a new closed interval $[c_2, d_2]$ which is contained in $[c_1, d_1]$ and not covered by a finite subcovering of $\mathscr{B}$. We thus obtain[†] a nested sequence of closed intervals $[c_1, d_1]$, $[c_2, d_2], \ldots$ such that none of them is covered by a finite subcovering of $\mathscr{B}$, and the length $d_i - c_i$ of $[c_i, d_i]$ is $(d_0 - c_0)/2^i$. Therefore, $\mathrm{Lim}(d_i - c_i) = 0$. (For, the completeness of $\mathfrak{F}$ implies the Archimedean property. Hence, for any $\varepsilon > 0$, there is some natural number i_0 such that $i_0 > (d_0 - c_0)/\varepsilon$.

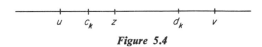

Figure 5.4

Therefore, if $i \geq i_0$, then $2^i \geq 2^{i_0} > i_0 > (d_0 - c_0)/\varepsilon$, and so, $(d_0 - c_0)/2^i < \varepsilon$.) Since $\mathfrak{F}$ is complete, $\mathfrak{F}$ satisfies the Nested Sequence Property (Nest), by Theorem 5.9. Hence, there is a point z belonging to every closed interval $[c_i, d_i]$ (see Figure 5.4). Since $z \in [c_0, d_0]$ and $\mathscr{B}$ is an open covering of $[c_0, d_0]$, z belongs to some open set B in $\mathscr{B}$. Since B is open, there is an open interval (u, v) containing z and included in B. Let $\varepsilon_0 = \min(z - u, v - z)$. Since $\mathrm{Lim}(d_i - c_i) = 0$, there exists a natural number k such that $d_k - c_k < \varepsilon_0$. Now, z belongs to each $[c_i, d_i]$ and, therefore, in particular, $z \in [c_k, d_k]$, that is, $c_k \leq z \leq d_k$. In addition, $u < c_k$. (For, $d_k - c_k < \varepsilon_0 \leq z - u \leq d_k - u$.) Moreover, $d_k < v$. (For, $d_k - c_k < \varepsilon_0 \leq v - z \leq v - c_k$.) Thus, $[c_k, d_k] \subseteq (u, v) \subseteq B$. Hence, the singleton $\{B\}$ is a finite subcovering of $\mathscr{B}$ which covers $[c_k, d_k]$, contradicting the fact that $[c_k, d_k]$ is not covered by a finite subcovering of $\mathscr{B}$.

[†] The definition of this sequence is based upon an application of the Iteration Theorem. Because of the tedious complexity required, we have omitted the details. From past experience in this book, whenever the $(n + 1)$st step of a procedure depends in a specified way upon the result of the nth step, the reader should have confidence in his ability to define the resulting sequence by a suitable use of the Iteration Theorem.

(b) $\Rightarrow$ (c)   Assume that $\mathfrak{F}$ has the Heine–Borel property. Assume that A is a bounded infinite set. We must prove that A has an accumulation point. Since A is bounded, there is a closed interval $[c, d]$ such that $A \subseteq [c, d]$. Assume that A has no accumulation point. Then, for each z in $[c, d]$, z is not an accumulation point of A, and, therefore, there is an open interval I containing z and containing no points of $A - \{z\}$, that is, the only point of I which might belong to A is z itself. Therefore, if we define $\mathscr{B}$ to be the collection of all open intervals which contain a point of $[c, d]$ but contain at most one element of A, then $\mathscr{B}$ is an open covering of A. By the Heine–Borel property, there is a finite subcovering $\{B_1, \ldots, B_n\}$ of $\mathscr{B}$ such that $\{B_1, \ldots, B_n\}$ covers A, that is, $A \subseteq B_1 \cup \cdots \cup B_n$. However, since each of the intervals $B_1, \ldots, B_n$ contains at most one point of A, A must be finite, contradicting the fact that A is infinite.

(c) $\Rightarrow$ (a)   Assume that $\mathfrak{F}$ possesses the Bolzano–Weierstrass property. To prove the completeness of $\mathfrak{F}$, it suffices by Theorem 5.9 to show that $\mathfrak{F}$ is Archimedean and Cauchy complete.

(i)   To prove that $\mathfrak{F}$ is Archimedean, it suffices by Theorem 2.1 to show that the set $N_{\mathfrak{F}}$ of natural numbers of $\mathfrak{F}$ is not bounded above. Assume that $N_{\mathfrak{F}}$ is bounded above. Since $N_{\mathfrak{F}}$ is bounded below by 1, $N_{\mathfrak{F}}$ is a bounded infinite subset of F. ($N_{\mathfrak{F}}$ is infinite in any field of characteristic zero, by Theorem 3.8.4.) By the Bolzano–Weierstrass property, there is at least one accumulation point z of $N_{\mathfrak{F}}$. Then the interval $(z - \frac{1}{2}, z + \frac{1}{2})$ must contain a point n of $N_{\mathfrak{F}} - \{z\}$. Let $c = |n - z|$. Then the interval $(z - c, z + c)$ contains a point k of $N_{\mathfrak{F}} - \{z\}$. Since $n \notin (z - c, z + c)$, $k \neq n$. Since $c < \frac{1}{2}$, $(z - c, z + c) \subseteq (z - \frac{1}{2}, z + \frac{1}{2})$. Hence, $k \in (z - \frac{1}{2}, z + \frac{1}{2})$. Since k and n both belong to $(z - \frac{1}{2}, z + \frac{1}{2})$,

$$|k - n| = |(k - z) + (z - n)| \leq |k - z| + |z - n| < \tfrac{1}{2} + \tfrac{1}{2} = 1.$$

But if k and n are integers and $|k - n| < 1$, then $|k - n| = 0$, that is, $k = n$, which contradicts the fact that $k \neq n$. Therefore, $\mathfrak{F}$ is Archimedean.

(ii)   Let $\langle x_n \rangle$ be a Cauchy sequence in $\mathfrak{F}$. We must show that $\langle x_n \rangle$ converges. By Theorem 5.4, $\langle x_n \rangle$ is a bounded sequence. Let A be the set of terms of the sequence $\langle x_n \rangle$. Then A is bounded.

Case 1 A is finite.   Then there must be a constant subsequence of $\langle x_n \rangle$. Now, any constant sequence converges. By Theorem 5.8, if any subsequence of a Cauchy sequence converges, the whole sequence converges. Hence, $\langle x_n \rangle$ converges.

Case 2 A is infinite.   By the Bolzano–Weierstrass property, A has an accumulation point z. Let us prove that $\mathrm{Lim}\ x_n = z$. Let ε be any positive element of F. Since $\langle x_n \rangle$ is a Cauchy sequence, there is some natural number n_0 such that $k, n \geq n_0 \Rightarrow | x_n - x_k | < \varepsilon/2$. Now, let c be the least positive element among $\varepsilon/2, | z - x_1 |, | z - x_2 |, \ldots, | z - x_{n_0-1} |$. (Some of the numbers $z - x_i$ may be zero.) Since z is an accumulation point of A, the open interval $(z - c, z + c)$ contains a point w of $A - \{z\}$. Since $| w - z | < c$, w is different from $x_1, x_2, \ldots, x_{n_0-1}$. Hence, $w = x_k$ for some $k \geq n_0$. Therefore, for any $n \geq n_0$,

$$| x_n - z | = | (x_n - x_k) + (x_k - z) | \leq | x_n - x_k | + | x_k - z |$$
$$< \varepsilon/2 + c \leq \varepsilon/2 + \varepsilon/2 = \varepsilon.$$

Therefore, $\mathrm{Lim}\, x_n = z$. ∎

EXERCISES

20. Prove that an ordered field $\mathfrak{F}$ is complete if and only if every bounded sequence in $\mathfrak{F}$ has a convergent subsequence.

21. In a complete ordered field $\mathfrak{F}$, if every two convergent subsequences of a given bounded sequence s have the same limit, prove that the sequence converges.

22. In a complete ordered field $\mathfrak{F}$, if the set of terms of a given bounded sequence s has a unique accumulation point b, prove that $\mathrm{Lim}\, s_n = b$.

Definition   Let $A \subseteq F$. A is said to be **compact** if and only if every open covering of A has a finite subcovering of A.

EXAMPLES

14.   The Heine–Borel property states that every closed interval is compact. Thus, by Theorem 7.6, every closed interval in the real number system is compact.

15.   The whole set F is not compact. To see this, let $\mathscr{B}$ consist of all open intervals of length 1. $\mathscr{B}$ is an open covering of F, but no finite subcovering of $\mathscr{B}$ covers F. (Why?)

EXERCISE

23. Show that the open interval $(0, 1)$ is not compact.

Theorem 7.7   The ordered field $\mathfrak{F}$ is complete if and only if the compact subsets of F are precisely the closed, bounded subsets of F.

Proof (i)   Assume $\mathfrak{F}$ is complete. Let $A \subseteq F$. We must show that A is compact if and only if A is closed and bounded. Assume A closed and bounded. Let $\mathscr{B}$ be any open covering of A. We must prove that A is covered by a finite subset of $\mathscr{B}$. Since A is bounded, there is a closed interval $[c, d]$ such that $A \subseteq [c, d]$. By Theorem 7.3b, $F - A$ is open. Then, $\mathscr{C} = \mathscr{B} \cup \{F - A\}$ is an open covering of $[c, d]$. Since $\mathfrak{F}$ is complete, the Heine–Borel property holds, by Theorem 7.6. Hence, $[c, d]$ is covered by a finite subcovering $\mathscr{C}^* \subseteq \mathscr{C}$. Then A is covered by the finite subcovering $\mathscr{B}^* = \mathscr{C}^* - \{F - A\} \subseteq \mathscr{B}$.

Conversely, assume A is compact. First, let us show that A is bounded. Since $\mathfrak{F}$ is complete, $\mathfrak{F}$ is Archimedean. Hence, $\bigcup_{n \in N_{\mathfrak{F}}} (-n, n) = F$. Therefore, the set $\mathscr{B} = \{(-n, n) : n \in N_{\mathfrak{F}}\}$ is an open covering of A. By the Heine–Borel property, A is covered by a finite subcovering of $\mathscr{B}$, say $\{(-n_1, n_1), \ldots, (-n_k, n_k).\}$ Let $m = \max(n_1, \ldots, n_k)$. Then $A \subseteq (-m, m)$. Thus, A is bounded. Second, let us prove that A is closed. It suffices by Theorem 7.3 to show that $F - A$ is open. Assume $z \in F - A$. We must prove that there is an open interval containing z and included in $F - A$. For each w in A, $|w - z| > 0$. Therefore, since $\mathfrak{F}$ is Archimedean, $|w - z| > 1/n$ for some n in $N_{\mathfrak{F}}$. The set $I_n = \{w : |w - z| > 1/n\}$ is an open set. (For, if $u \in I_n$, then $|u - z| > 1/n$. Let $c = |u - z| - 1/n$. Then $(u - c, u + c)$ is an open interval containing u and included in I_n.) Let $\mathscr{B} = \{I_n : n \in N_{\mathfrak{F}}\}$. Then $\mathscr{B}$ is an open covering of A. Since A is compact, A can be covered by a finite subset $\{I_{n_1}, \ldots, I_{n_k}\}$ of $\mathscr{B}$. Let $m = \max(n_1, \ldots, n_k)$. Now, for any w in A, $w \in I_{n_k}$ for some k. Hence, $|w - z| > 1/n_k \geq 1/m$. Therefore, if $|w - z| < 1/m$, then $w \in F - A$. Hence, the open interval $(z - 1/m, z + 1/m)$ contains z and is included in $F - A$.

(ii)   Assume that the compact subsets of F are precisely the closed, bounded subsets of F. In particular, since any closed interval is closed and bounded, any closed interval is compact, that is, the Heine–Borel property holds. Hence, by Theorem 7.6, $\mathfrak{F}$ is complete. ∎

EXERCISE

24.   In an ordered field $\mathfrak{F}$, prove that a nested sequence of nonempty compact sets B_n has a nonempty intersection. (*Hint*:   We have $B_1 \supseteq B_2 \supseteq \cdots$. If $\bigcap_{n \in N_{\mathfrak{F}}} B_n$ is empty, then $\bigcup_{n \in N_{\mathfrak{F}}} (F - B_n) = F$, and, therefore, $\mathscr{C} = \{F - B_n : n \in N_{\mathfrak{F}}\}$ is an open covering of B_1.)

Definition   Assume $A \subseteq F$. A is said to be **connected** if and only if, whenever $A = B \cup C$, where $B \neq \varnothing$, $C \neq \varnothing$, and $B \cap C = \varnothing$, either B

contains an accumulation point of C or C contains an accumulation point of B.

Thus, A is connected if and only if it cannot be divided into two nonempty parts neither of which contains an accumulation point of the other.

EXAMPLES

16. $\varnothing$ and any singleton set $\{a\}$ are trivially connected.

Theorem 7.8   An ordered field is complete if and only if every closed interval $[u, v]$ is connected.

Proof   Assume $\mathfrak{F}$ complete. Let $[u, v] = B \cup C$, where $B \neq \varnothing$, $C \neq \varnothing$, $B \cap C = \varnothing$. We must show that either B contains an accumulation point of C or C contains an accumulation point of B. Assume that this is not so. We may assume that $u \in B$. Since u is not an accumulation point of C, there is an open interval (w, z) containing u such that (w, z) does not intersect C. Thus, the set $[u, z) = \{x : u \leq x < z\}$ is a subset of B. Let $W = \{y : [u, y) \subseteq B\}$. $W \neq \varnothing$, since $z \in W$. In addition, W is bounded above by v. Since $\mathfrak{F}$ is complete, $\mathfrak{F}$ has the lub property, by Theorem 3.3. Hence, W has a lub d. Since v is an upper bound of W, $d \in [u, v]$. Notice that $[u, d) \subseteq B$. (For, if $u \leq x < d$, then x is not an upper bound of W. Hence, there is some element y of W such that $x < y$. Since $y \in W$, $[u, y) \subseteq B$. Therefore, $x \in B$.) It follows that $d \notin B$. (For, if $d \in B$, then d cannot be an accumulation point of C. Hence, there is an open interval (w_1, z_1) containing d and not containing any points of C. Therefore, $(d, z_1) \subseteq B$, and so, $[u, z_1) \subseteq B$, that is, $z_1 \in W$, contradicting the fact that d is an upper bound of W.) Therefore, $d \in C$. Then d cannot be an accumulation point of B. Hence, there is an open interval (w_2, z_2) containing d and not containing any points of B. But this contradicts the fact that (w_2, z_2) must intersect $[u, d) \subseteq B$.

Conversely, assume that every closed interval is connected. Let (A_1, A_2) be any cut in $\mathfrak{F}$. Take $u \in A_1$ and $v \in A_2$. We may assume that u is not a maximum of A_1 and v is not a minimum of A_2. (Otherwise, there is nothing more to prove.) Let $B = A_1 \cap [u, v]$ and $C = A_2 \cap [u, v]$. Then $[u, v] = B \cup C$, $B \neq \varnothing$, $C \neq \varnothing$, and $B \cap C = \varnothing$. Since $[u, v]$ is connected, there is some point z such that either ($z \in B$ and z is an accumulation point of C) or ($z \in C$ and z is an accumulation point of B).

Case 1 $z \in B$ and z is an accumulation point of C. Then z is a maximum element of A_1. (If not, there exists x such that $z < x$ and $x \in A_1$. Since (A_1, A_2) is a cut, (u, x) is an interval containing z and included in B. Thus, (u, x) contains no points of C, which contradicts the assumption that z is an accumulation point of C.)

Case 2 $z \in C$ and z is an accumulation point of B. Then z is a minimum element of A_2. (If not, there exists x such that $x < z$ and $x \in A_2$. Since (A_1, A_2) is a cut, (x, v) is an interval containing z and included in C. Hence, (x, v) contains no points of B, which contradicts the assumption that z is an accumulation point of B.)

This shows that (A_1, A_2) is not a gap in $\mathfrak{F}$. Hence, $\mathfrak{F}$ is complete.   ■

Definition   By a **generalized interval** we mean any of the following kinds of sets:

  i.   open intervals (a, b);

 ii.   closed intervals $[a, b]$;

iii.   half-open intervals

$$[a, b) = \{u : a \leq u < b\} \quad \text{and} \quad (a, b] = \{u : a < u \leq b\}.$$

 iv.   infinite half-lines:

   open half-lines $(a, \infty) = \{u : a < u\}$ and $(-\infty, a) = \{u : u < a\}$;

   closed half-lines $[a, \infty) = \{u : a \leq u\}$ and $(-\infty, a] = \{u : u \leq a\}$.

Lemma 7.9   If every closed interval is connected, then so are F itself and all generalized intervals.

Proof   Let us assume that I is either a generalized interval or F. Observe that: $x \in I \wedge y \in I \wedge x < z < y \Rightarrow z \in I$. Assume that $I = B \cup C$, where $B \neq \varnothing$, $C \neq \varnothing$, $B \cap C = \varnothing$. Take $b \in B$ and $c \in C$. We may assume $b < c$. Let $B_1 = B \cap [b, c]$ and $C_1 = C \cap [b, c]$. Then $[b, c] = B_1 \cup C_1$, $B_1 \neq \varnothing$, $C_1 \neq \varnothing$, $B_1 \cap C_1 = \varnothing$. Since $[b, c]$ is connected, there is a point z such that either ($z \in B_1$ and z is an accumulation point of C_1) or ($z \in C_1$ and z is an accumulation point of B_1). But $B_1 \subseteq B$ and $C_1 \subseteq C$. Hence, either ($z \in B$ and z is an accumulation point of C) or ($z \in C$ and z is an accumulation point of B). Therefore, I is connected.   ■

Corollary 7.10 $\mathfrak{F}$ is complete if and only if the connected subsets of F are precisely $\varnothing$, all singletons $\{a\}$, F, and all generalized intervals.

Proof (i)   Assume $\mathcal{F}$ is complete. Then use Theorem 7.8 and Lemma 7.9.

(ii)   Assume the connected sets are precisely $\varnothing$, all singletons $\{a\}$, F, and all generalized intervals. In particular, all closed intervals are connected. Hence, by Theorem 7.8, $\mathcal{F}$ is complete.   ∎

EXERCISE

25.   Show that, in an arbitrary ordered field $\mathcal{F}$, it is not necessarily true that a set A such that

$$(*) (\forall z)(\forall u)(\forall v)(u \in A \wedge v \in A \wedge u < z < v \Rightarrow z \in A)$$

is either $\varnothing$, a singleton $\{a\}$, F, or a generalized interval. Moreover, prove that $\mathcal{F}$ is complete if and only if the subsets A of F such that $(*)$ holds are precisely $\varnothing$, all singletons $\{a\}$, F, and all generalized intervals.

Lemma 7.11   If $\mathscr{A}$ is a collection of connected sets such that any two sets in $\mathscr{A}$ have points in common, then $\bigcup_{A \in \mathscr{A}} A$ is connected.

Proof   Let $D = \bigcup_{A \in \mathscr{A}} A$. Assume $D = B \cup C$, $B \neq \varnothing$, $C \neq \varnothing$, $B \cap C = \varnothing$. Take $b \in B$ and $c \in C$. Then $b \in A_1$ and $c \in A_2$ for some A_1, A_2 in $\mathscr{A}$. Then $A_1 \cap A_2 \neq \varnothing$. Let $w \in A_1 \cap A_2$. Then $w \in B$ or $w \in C$; say $w \in B$. Let $B_1 = B \cap A_2$, $C_1 = C \cap A_2$. Then $A_2 = B_1 \cup C_1$, $B_1 \neq \varnothing$ (since $w \in B \cap A_2$), $C_1 \neq \varnothing$ (since $c \in C \cap A_2$), and $B_1 \cap C_1 = \varnothing$. Since A_2 is connected, there is a point z such that either ($z \in B_1$ and z is an accumulation point of C_1) or ($z \in C_1$ and z is an accumulation point of B_1). Therefore, either ($z \in B$ and z is an accumulation point of C) or ($z \in C$ and z is an accumulation point of B). Hence, D is connected.   ∎

Definition   Let $A \subseteq F$. By a connected component of A we mean any nonempty set $B \subseteq A$ such that:

i. B is connected, and

ii.   If C is connected and $B \subseteq C \subseteq A$, then $B = C$.

Thus, a **connected component** of A is a maximal connected subset of A.

Theorem 7.12

a.   Two different connected components of A are disjoint.

b. A is the union of its connected components.

Proof (a)   Assume C_1 and C_2 are connected components of A. Assume $C_1 \cap C_2 \neq \varnothing$. Then, by Lemma 7.11, with $\mathscr{A} = \{C_1, C_2\}$, $C_1 \cup C_2$ is connected. Since C_1 and C_2 are maximal connected subsets of A, $C_1 = C_1 \cup C_2$ and $C_2 = C_1 \cup C_2$, that is, $C_2 \subseteq C_1$ and $C_1 \subseteq C_2$, which implies that $C_1 = C_2$.

(b) Let $x \in A$. Let $\mathscr{A}$ be the set of connected subsets of A containing x. By Lemma 7.11, $B = \bigcup_{C \in \mathscr{A}} C$ is a connected subset of A. Notice that $\{x\} \in \mathscr{A}$, and, therefore, $x \in B$. If C is connected and $B \subseteq C \subseteq A$, then $x \in C$; hence, $C \in \mathscr{A}$, and, therefore, $C \subseteq B$. Thus, $C = B$. B is a connected component of A containing x. ∎

Theorem 7.13

a. If A is connected, then its closure $\bar{A}$ is connected.

b. If A is closed, the connected components of A are closed.

c. Assume $\mathscr{F}$ is complete. Then, if A is open, the connected components of A are open. Hence, the connected components of a nonempty open set are open intervals, or F itself (when $A = F$), or generalized intervals of the form (a, ∞) or $(-\infty, a)$. In particular, a bounded nonempty open set is a union of disjoint open intervals.

Proof (a) Assume A connected. Let $\bar{A} = B \cup C$, $B \neq \varnothing$, $C \neq \varnothing$, $B \cap C = \varnothing$. Let $B_1 = B \cap A$, $C_1 = C \cap A$. Clearly, $B_1 \cap C_1 = \varnothing$, and $A = B_1 \cup C_1$.

Case 1 $B_1 \neq \varnothing$ and $C_1 \neq \varnothing$. Since A is connected, there is a point z such that either ($z \in B_1$ and z is an accumulation point of C_1) or ($z \in C_1$ and z is an accumulation point of B_1). Hence, either ($z \in B$ and z is an accumulation point of C) or ($z \in C$ and z is an accumulation point of B).

Case 2 $B_1 = \varnothing$ or $C_1 = \varnothing$. Say, $B_1 = \varnothing$. Then $A \subseteq C$. Take any $z \in B$. Then $z \in \bar{A} - A$, that is, z is an accumulation point of A. Since $A \subseteq C$, z is an accumulation point of C.

(b) Assume A is closed and C is a connected component of A. Since $C \subseteq A$, $\bar{C} \subseteq \bar{A}$ by Theorem 7.5e. Since A is closed, $\bar{A} = A$ by Theorem 7.5b. Thus, $C \subseteq \bar{C} \subseteq A$. By Part a, $\bar{C}$ is connected, and, therefore, $C = \bar{C}$, that is, C is closed, by Theorem 7.5b.

(c) Assume that $\mathscr{F}$ is complete. Let A be a nonempty open set. Let C be a connected component of A, and let $z \in C$. Since $z \in C \subseteq A$, there is an open interval I containing z and included in A. I is a connected set, by Corollary 7.10. By Theorem 7.11, $I \cup C$ is a connected set containing z. Since $C \subseteq I \cup C \subseteq A$, it follows that $C = I \cup C$, that is, $I \subseteq C$. Thus, C is open. ∎

EXERCISES

26. In the real number field $\mathscr{R}$, what are the connected components of the set $Q_{\mathscr{R}}$ of rational numbers?

27. If $\mathscr{F}$ is complete, show that the only subsets of F which are both open and closed are $\varnothing$ and F.

28. If $\mathscr{F}$ is complete, prove that an open set A is connected if and only if A is not the union of two disjoint nonempty open sets.

29. If $\mathscr{F}$ is complete, prove that every open set different from F itself is the union of a finite or denumerable collection of generalized open intervals (see Theorem 7.13c).

30. Let $A \subseteq F$. For any x and y in A, define $x \sim y$ to mean that x and y belong to some connected set included in A. Prove that $\sim$ is an equivalence relation and that the equivalence classes with respect to $\sim$ are the connected components of A.

Summary In succeeding sections, we shall be dealing not with arbitrary ordered fields, but with the real number field. The decision to restrict ourselves to complete ordered fields is based upon the overwhelming importance of the real number field. As an exercise, the reader can attempt to determine which ordered fields satisfy the results that we shall prove for the real number field.[†] We shall list below some of the more important properties of the real number system which have been proved in this section.

1. The Heine–Borel property.
2. The Bolzano–Weierstrass property.
3. The compact sets are precisely the closed, bounded sets.
4. The connected sets are precisely $\varnothing$, the singletons $\{a\}$, R, and all generalized intervals.
5. An open set different from R is the union of pairwise-disjoint generalized open intervals.

5.8 CONTINUOUS FUNCTIONS

Let $\mathscr{R} = (R, +, \times, <)$ be the real number system.

In this section we shall be dealing with functions $f\colon A \to R$, where the domain A is a subset of R.

[†] Some theorems of this kind may be found in Olmsted [1962], Appendix, §§ 3, 5.

Definition   Let $f: A \to R$ and $a, b \in R$. Then: **b is a limit of $f(x)$ as x approaches a** means:

i. a is an accumulation point of A;

ii.   For any positive ε in R, there exists a positive δ in R such that

$$(x \in A - \{a\} \wedge |x - a| < \delta) \Rightarrow |f(x) - b| < \varepsilon.$$

Intuitively, (ii) asserts that, as x gets closer and closer to a, $f(x)$ becomes closer and closer to b, or, synonymously, as x approaches a, $f(x)$ approaches b. Notice that a may or may not belong to the domain A of f. Whether b is a limit of $f(x)$ as x approaches a has nothing to do with whether $a \in A$ or with the value $f(a)$, if such a value exists.

Lemma 8.1   Assume $f: A \to R$. If both b and c are limits of $f(x)$ as x approaches a, then $b = c$.

Proof   Assume $b \neq c$. Let $\varepsilon = |b - c| > 0$. There exist positive δ_1 and δ_2 such that

$$(x \in A - \{a\} \wedge |x - a| < \delta_1) \Rightarrow |f(x) - b| < \varepsilon/2,$$

and

$$(x \in A - \{a\} \wedge |x - a| < \delta_2) \Rightarrow |f(x) - c| < \varepsilon/2.$$

Let $\delta = \min(\delta_1, \delta_2)$. Since a is an accumulation point of A, there is a point x in A such that $|x - a| < \delta$. Hence,

$$|b - c| = |(b - f(x)) + (f(x) - c)| \leq |b - f(x)| + |f(x) - c|$$
$$< \varepsilon/2 + \varepsilon/2 = \varepsilon = |b - c|,$$

which is a contradiction.   ∎

Notation   If there is a limit b of $f(x)$ as x approaches a, then, by Lemma 8.1, it is unique, and we shall write:

$$b = \lim_{x \to a} f(x)$$

as an abbreviation for "b is a limit of $f(x)$ as x approaches a."

EXAMPLE

1.   Constant functions. Let $f(x) = c$ for all x in A. Let a be any accumulation point of A. It is obvious that $\lim_{x \to a} f(x) = c$.

EXERCISE

1.^{C} Assume $f: A \rightarrow R$ and a is an accumulation point of A. Prove: $\lim_{x \to a} f(x) = b$ if and only if, for any sequence $\langle x_n \rangle$ of points of $A - \{a\}$ such that $\text{Lim } x_n = a$, $\text{Lim } f(x_n) = b$. (This relates the idea of **limit of a function** to that of **limit of a sequence**.

Theorem 8.2 Assume $f: A \rightarrow R$, $g: B \rightarrow R$, and a is an accumulation point of $A \cap B$. Assume $\lim_{x \to a} f(x) = b$ and $\lim_{x \to a} g(x) = c$. Then:

a. $\lim_{x \to a} (f + g)(x) = b + c$, where $(f + g)(x) = f(x) + g(x)$.

b. $\lim_{x \to a} (-g)(x) = -c$, where $(-g)(x) = -g(x)$.

c. $\lim_{x \to a} (f - g)(x) = b - c$, where $(f - g)(x) = f(x) - g(x)$.

d. $\lim_{x \to a} (fg)(x) = bc$, where $(fg)(x) = f(x)g(x)$.

e. $\lim_{x \to a} |f|(x) = |b|$, where $|f|(x) = |f(x)|$.

f. $\lim_{x \to a} (ug)(x) = uc$, where u is a constant and $(ug)(x) = u \cdot g(x)$.

Proof (a), (b) Assume $\varepsilon > 0$. Then there exist positive δ_1 and δ_2 such that

$$(x \varepsilon A - \{a\} \wedge |x - a| < \delta_1) \Rightarrow |f(x) - b| < \varepsilon/2,$$

and

$$(x \in B - \{a\} \wedge |x - a| < \delta_2) \Rightarrow |g(x) - c| < \varepsilon/2.$$

Let $\delta = \min(\delta_1, \delta_2)$. Then

$$\begin{aligned}
(x \in (A \cap B) - \{a\} \wedge |x - a| < \delta) \Rightarrow |(f(x) + g(x)) - (b + c)| \\
= |(f(x) - b) + (g(x) - c)| \\
\leq |f(x) - b| + |g(x) - c| \\
< \varepsilon/2 + \varepsilon/2 = \varepsilon.
\end{aligned}$$

Hence, $\lim_{x \to a} (f + g)(x) = b + c$. In addition,

$$(x \in B - \{a\} \wedge |x - a| < \delta_2) \Rightarrow |-g(x) - (-c)| = |c - g(x)| < \varepsilon/2.$$

Hence, $\lim_{x \to a} (-g)(x) = -c$.

(c)   This follows immediately from Parts a and b.

(d)   Assume $\varepsilon > 0$. Let

$$d = \begin{cases} \dfrac{\varepsilon}{2|b|} & \text{if} \quad b \neq 0 \\ 1 & \text{if} \quad b = 0. \end{cases}$$

There exists a positive δ_1 such that

$$(x \in B - \{a\} \wedge |x - a| < \delta_1) \Rightarrow |g(x) - c| < d.$$

Then

$$(x \in B - \{a\} \wedge |x - a| < \delta_1) \Rightarrow |g(x)| = |g(x) - c) + c|$$
$$\leq |g(x) - c| + |c| < d + |c|$$

There exists a positive δ_2 such that

$$(x \in A - \{a\} \wedge |x - a| < \delta_2) \Rightarrow |f(x) - b| < \varepsilon/(2(d + |c|)).$$

Let $\delta = \min(\delta_1, \delta_2)$. Hence,

$$(x \in (A \cap B) - \{a\} \wedge |x - a| < \delta) \Rightarrow |f(x)g(x) - bc|$$
$$\leq |(f(x) - b)g(x)| + |b(g(x) - c)|$$
$$= |g(x)||f(x) - b| + |b||g(x) - c|$$
$$\leq (d + |c|)\left(\frac{\varepsilon}{2(d + |c|)}\right) + \begin{cases} |b|\dfrac{\varepsilon}{2|b|} & \text{if} \quad b \neq 0 \\ 0 & \text{if} \quad b = 0 \end{cases}$$
$$< \varepsilon/2 + \begin{cases} \varepsilon/2 & \text{if} \quad b \neq 0 \\ 0 & \text{if} \quad b = 0 \end{cases}$$
$$\leq \varepsilon.$$

Therefore, $\lim_{x \to a} (fg)(x) = bc$.

(e)   Assume $\varepsilon > 0$. There is a positive δ such that

$$(x \in A - \{a\} \wedge |x - a| < \delta) \Rightarrow |f(x) - b| < \varepsilon.$$

Hence,

$$(x \in A - \{a\} \wedge |x - a| < \delta) \Rightarrow ||f(x)| - |b|| \leq |f(x) - b| < \varepsilon.$$

Therefore, $\lim_{x \to a} |f|(x) = |b|$.

(f)   Use Part d, with $f(x) = u$ for all x.   ∎

EXAMPLES

2. $\lim_{x \to a} x = a$. Here we are considering the identity function f such that $f(x) = x$ for all real numbers x. Assume $\varepsilon > 0$. Let $\delta = \varepsilon$. Then, $|x - a| < \delta \Rightarrow |f(x) - a| < \varepsilon$.

3.
$$\lim_{x \to 2} (3x^2 - 27x + 5) = \lim_{x \to 2} (3x^2) - \lim_{x \to 2} (27x) + \lim_{x \to 2} 5$$
$$= 3 \lim_{x \to 2} x^2 - 27 \lim_{x \to 2} x + 5$$
$$= 3 (\lim_{x \to 2} x \cdot \lim_{x \to 2} x) - 27 \cdot 2 + 5$$
$$= 3 (2 \cdot 2) - 27 \cdot 2 + 5 = -37.$$

4.
$$\lim_{x \to 2} \frac{x^2 - 4}{x - 2} = \lim_{x \to 2} (x + 2) = \lim_{x \to 2} x + \lim_{x \to 2} 2 = 2 + 2 = 4.$$

5. There is no number b such that $b = \lim_{x \to 0} 1/x$. For, if we take $\varepsilon = \frac{1}{2}$, then, for any positive δ, there is some $x \neq 0$ such that $|x - 0| = |x| < \delta$ and $|1/x - b| > \frac{1}{2}$; in fact, take any natural number $n > 1/\delta - b$, and let $x = 1/(b + n)$. Then $|x| = |1/(b + n)| < \delta$ and $|1/x - b| = |(b + n) - b| = n > \frac{1}{2}$. Intuitively, $1/x$ gets larger and larger as x approaches 0, and, therefore, $1/x$ cannot get closer and closer to any particular real number.

We shall now investigate the limit of the quotient of two functions.

Lemma 8.3

a. If $\lim_{x \to a} g(x) = c > 0$, then there is a positive δ such that
$$(x \in \mathscr{D}(g) - \{a\} \wedge |x - a| < \delta) \Rightarrow g(x) > c/2 > 0.$$

b. If $\lim_{x \to a} g(x) = c < 0$, then there is a positive δ such that
$$(x \in \mathscr{D}(g) - \{a\} \wedge |x - a| < \delta) \Rightarrow g(x) < c/2 < 0.$$

c. If $\lim_{x \to a} g(x) = c \neq 0$, then there is a positive δ such that
$$(x \in \mathscr{D}(g) - \{a\} \wedge |x - a| < \delta) \Rightarrow |g(x)| > |c|/2.$$

Proof (a) Since $\lim_{x \to a} g(x) = c$, there is some positive δ such that
$$(x \in \mathscr{D}(g) - \{a\} \wedge |x - a| < \delta) \Rightarrow |g(x) - c| < c/2$$
$$\Rightarrow c - g(x) \leq |g(x) - c| < c/2$$
$$\Rightarrow c/2 < g(x).$$

(b)   We are given that $c < 0$. Hence, $-c/2 > 0$. Since $\lim_{x \to a} g(x)\, c =$, there is some positive δ such that

$$(x \in \mathcal{D}(g) - \{a\} \wedge |x - a| < \delta) \Rightarrow |g(x) - c| < -c/2$$
$$\Rightarrow g(x) - c \leq |g(x) - c| < -c/2$$
$$\Rightarrow g(x) < c/2.$$

(c)   This follows immediately from (a) and (b).   ∎

Theorem 8.4

a.   If $\lim_{x \to a} g(x) = c \neq 0$, then $\lim_{x \to a} (1/g)(x) = 1/c$, where $(1/g)(x) = 1/g(x)$.

b.   If a is an accumulation point of $\mathcal{D}(g) \cap \mathcal{D}(f)$, $\lim_{x \to a} f(x) = b$, and $\lim_{x \to a} g(x) = c \neq 0$, then

$$\lim_{x \to a} (f/g)(x) = b/c, \qquad \text{where } (f/g)(x) = f(x)/g(x).$$

Proof   (a)   Assume $\varepsilon > 0$. Since $\lim_{x \to a} g(x) = c$, there is some positive δ_1 such that

$$(x \in \mathcal{D}(g) - \{a\} \wedge |x - a| < \delta_1) \Rightarrow |g(x) - c| < \varepsilon |c|^2/2.$$

By Lemma 8.3c, there is some positive δ_2 such that

$$(x \in \mathcal{D}(g) - \{a\} \wedge |x - a| < \delta_2) \Rightarrow |g(x)| > |c|/2.$$

Let $\delta = \min(\delta_1, \delta_2)$. Then

$$(x \in \mathcal{D}(g) - \{a\} \wedge |x - a| < \delta) \Rightarrow |1/g(x) - 1/c| = \frac{|c - g(x)|}{|c||g(x)|}$$
$$< \frac{|c - g(x)|}{|c|(|c|/2)}$$
$$< \frac{\varepsilon |c|^2/2}{|c|^2/2} = \varepsilon.$$

Notice that a is an accumulation point of $\mathcal{D}(1/g)$, since, whenever $x \in \mathcal{D}(g) - \{a\} \wedge |x - a| < \delta_2$, it follows that $g(x) \neq 0$, and, therefore, $x \in \mathcal{D}(1/g)$.

(b)   Since $f/g = f \cdot (1/g)$, this follows from Part a and Theorem 8.2d.

∎

Definition Assume $f: A \rightarrow R$. We say that f **is continuous at a** if and only if

  i. $a \in A$;

 ii. a is an accumulation point of A;

iii. $\lim_{x \to a} f(x) = f(a)$.

We say that f is **continuous** if and only if f is continuous at every point of A.

EXAMPLES

6.   Let $f(x) = 3x^2 - 27x + 5$ for all x. Then $f: R \rightarrow R$. f is continuous at 2, since $2 \in R$, 2 is an accumulation point of R, and $\lim_{x \to 2} f(x) = -37 = f(2)$ (compare Example 3 on p. 251).

7.   Let $f(x) = c$ for all x in R. Then $f: R \rightarrow R$ and f is continuous. For, if $a \in R$, $\lim_{x \to a} f(x) = c = f(a)$. Thus, every constant function is continuous.

8.   Let

$$f(x) = \begin{cases} -1 & \text{if} \quad x \leq 0 \\ 1 & \text{if} \quad x > 0 \end{cases}$$

Then $f: R \rightarrow R$. f is not continuous at 0. For, take any number b and let $\varepsilon = \frac{1}{2}$. Then, for any positive δ, there is some x such that $|x| < \delta$ and $|f(x) - b| \geq \varepsilon$. In fact, if $b \leq 0$, let $x = \delta/2$. Then

$$|f(x) - b| = |1 - b| \geq 1.$$

If $b > 0$, let $x = -\delta/2$. Then $|f(x) - b| = |-1 - b| > 1$. (However, show that f is continuous at every point different from 0.)

9.   Let

$$f(x) = \begin{cases} 0 & \text{if} \quad x \neq 0 \\ 1 & \text{if} \quad x = 0 \end{cases}$$

Then $\lim_{x \to 0} f(x) = 0$, while $f(0) = 1$. Hence, f is not continuous at 0.

EXERCISES

2.   Let

$$f(x) = \begin{cases} 1 & \text{if} \quad x \text{ is rational} \\ 0 & \text{if} \quad x \text{ is irrational.} \end{cases}$$

Show that f is not continuous at any point.

3. Let
$$f(x) = \begin{cases} x & \text{if } x \geq 0 \\ -x & \text{if } x < 0. \end{cases}$$

At which points is f continuous?

4. Let
$$f(x) = \begin{cases} 1/n & \text{if } x = m/n, \text{ where } m \text{ and } n \text{ are integers, } n > 0, \text{ and } m \text{ and } n \text{ are} \\ & \text{relatively prime;} \\ 0 & \text{if } x \text{ is irrational.} \end{cases}$$

At which points is f continuous?

5.$^\text{C}$ Assume $f: A \to R$, $a \in A$, and a is an accumulation point of A. Prove: f is continuous at a if and only if, for any sequence $\langle x_n \rangle$ such that $\operatorname{Lim} x_n = a$, $\operatorname{Lim} f(x_n) = f(a)$.

6. Assume $f: A \to R$ and f is continuous. Assume that $B \subseteq A$ and every point of B is an accumulation point of B. Then the restriction of f to B is continuous at each point of B.

7.$^\text{C}$ Assume f is continuous at a, and $f(a) \neq 0$. Then, for some open interval I containing a,
$$(x \in I \wedge x \in \mathscr{D}(f)) \Rightarrow f(x) \neq 0.$$

8. Assume $f: R \to R$, f is continuous, and $f(x + y) = f(x) + f(y)$ for all x and y. Let $c = f(1)$. Prove that $f(x) = cx$ for all x. (*Hint*: First prove this for positive integers, then integers, then rationals, and, finally, by continuity, for all real numbers.)

9. Assume $f: A \to R$, $g: A \to R$, and f and g are both continuous. Prove that the function h such that $h(x) = \max(f(x), g(x))$ for all x in A is also continuous.

10. If fg is continuous, must f and g be continuous?

11. a. For all x, let $f(x) = [x] = $ the greatest integer $\leq x$. At which points is f continuous?

    b. Let $g(x) = (x - 1)[x]$ for all x. At which points is g continuous?

12. Find a function $f: R \to R$ which is continuous at exactly one point.

Theorem 8.5   Assume $f: A \to R$, $g: B \to R$, $a \in A \cap B$, and a is an accumulation point of $A \cap B$. Assume also that f and g are continuous at a. Then

a. $f + g$ is continuous at a;

b. $-g$ is continuous at a;

c. $f - g$ is continuous at a;

d. fg is continuous at a (where $(fg)(x) = f(x)g(x)$ for $x \in A \cap B$);

e. $|f|$ is continuous at a;

f. cg is continuous at a, where c is a constant;

g. if h is continuous at $f(a)$, and a is an accumulation point of the domain of the composite function $h \circ f$, then $h \circ f$ is continuous at a.

Proof (a)–(f) $\lim\limits_{x \to a} f(x) = f(a)$ and $\lim\limits_{x \to a} g(x) = g(a)$.

Hence, by Theorem 8.2,

$$\lim_{x \to a} (f + g)(x) = \lim_{x \to a} f(x) + \lim_{x \to a} g(x) = f(a) + g(a) = (f + g)(a),$$

and similarly for $-g, f - g, fg, |f|, cg$.

(g) Assume $\varepsilon > 0$. Since h is continuous at $f(a)$, there exists a positive δ_1 such that

$$(u \in \mathscr{D}(h) \wedge |u - f(a)| < \delta) \Rightarrow |h(u) - h(f(a))| < \varepsilon.$$

Since f is continuous at a, there is a positive δ such that

$$(x \in \mathscr{D}(f) \wedge |x - a| < \delta) \Rightarrow |f(x) - f(a)| < \delta_1.$$

Therefore,

$$(x \in \mathscr{D}(h \circ f) \wedge |x - a| < \delta) \Rightarrow |f(x) - f(a)| < \delta_1$$
$$\Rightarrow |h(f(x)) - h(f(a))| < \varepsilon.$$

Hence, $h \circ f$ is continuous at a.   ∎

EXERCISE

13. If $|f|$ is continuous at a, is f necessarily continuous at a?

Corollary 8.6 Every polynomial determines a continuous function on R.

Proof All constants as well as the polynomial x determine continuous functions. But any polynomial is obtained from x and constants by addition and multiplication, and, therefore, determines a continuous function, by virtue of Theorem 8.3a,d.   ∎

Theorem 8.7 Assume $f: A \to R$, $g: B \to R$, $a \in A \cap B$, a is an accumulation point of $A \cap B$, $g(a) \neq 0$, and f and g are continuous at a. Then the function f/g is continuous at a, where $(f/g)(x) = f(x)/g(x)$ for all x in $A \cap B$ such that $g(x) \neq 0$.

Proof $\lim\limits_{x \to a} (f/g)(x) = f(a)/g(a)$ by virtue of Theorem 8.4b.   ■

EXERCISE

14. If f and g are polynomials, prove that f/g is continuous at all points of its domain, that is, for all x such that $g(x) \neq 0$.

Definition We say that B is **relatively open** in A if and only if there exists an open set V such that $B = A \cap V$.

EXAMPLES

10. Let $A = [0, 2]$ and $B = [0, 1)$. Then B is relatively open in A, since $B = A \cap (-1, 1)$.

11. Let A be the set of rational numbers in $(0, 2)$ and let B be the set of rational numbers in $(0, 1)$. Then B is relatively open in A, since $B = A \cap (0, 1)$.

EXERCISE

15. Let A be open. Prove: B is relatively open in A if and only if $B \subseteq A$ and B is open.

Lemma 8.8 Let $B \subseteq A$. Then B is relatively open in A if and only if, for any x in B, there is an open interval I containing x such that

$$(u \in I \wedge u \in A) \Rightarrow u \in B.$$

Proof Assume B relatively open in A. Then $B = A \cap V$ for some open set V. Assume $x \in B$. Hence $x \in V$, and, therefore, there is an open interval I containing x and included in V. Hence,

$$(u \in I \wedge u \in A) \Rightarrow (u \in V \wedge u \in A) \Rightarrow u \in B.$$

Conversely, assume that, for any x in B, there is an open interval I containing x such that $(u \in I \wedge u \in A) \Rightarrow u \in B$. Let $\mathscr{C}$ be the collection of all such open intervals for all x in B. Let $V = \bigcup_{I \in \mathscr{C}} I$. As a union of open sets, V is open. Clearly, $B \subseteq A \cap V$. Now, assume $y \in A \cap V$. Then $y \in I$ for some I in $\mathscr{C}$. Hence, there exists x in B such that I is an interval containing x and $(u \in I \wedge u \in A) \Rightarrow u \in B$. Since $y \in I \wedge y \in A$, it follows that $y \in B$. Therefore, $A \cap V \subseteq B$.   ■

Notation $f[B] = \{f(u) : u \in B\}$.

$\qquad f^{-1}[C] = \{u : f(u) \in C\}$.

$f[B]$ is called the **image of** B **under** f.

$f^{-1}[C]$ is called the **inverse image of** C **under** f.

Theorem 8.9   Assume $f: A \to R$. Then f is continuous if and only if, for any open set W, $f^{-1}[W]$ is relatively open in A.

Proof   Assume f is continuous. Let W be an open set, and assume $x \in f^{-1}[W]$, that is, $x \in A \wedge f(x) \in W$. Since W is open and $f(x) \in W$, there is an open interval (a, b) containing x and included in W. Let

$$\varepsilon = \min(f(x) - a,\ b - f(x)).$$

Since f is continuous, there exists a positive δ such that

$$(u \in A \wedge |u - x| < \delta) \Rightarrow |f(u) - f(x)| < \varepsilon \Rightarrow f(u) \in W.$$

Let $I = (x - \delta, x + \delta)$. Clearly, if $u \in I \wedge u \in A$, then $u \in f^{-1}[W]$. Thus, by Lemma 8.8, $f^{-1}[W]$ is relatively open in A. Conversely, assume that, for any open set $W, f^{-1}[W]$ is relatively open in A. Take any x in A. Assume $\varepsilon > 0$. Let $W = (f(x) - \varepsilon, f(x) + \varepsilon)$. Then $f^{-1}[W]$ is relatively open in A. But $x \in f^{-1}[W]$. Hence, by Lemma 8.8, there is an open interval (a, b) containing x such that

$$(u \in (a, b) \wedge u \in A) \Rightarrow u \in f^{-1}[W] \Rightarrow f(u) \in W \Rightarrow |f(u) - f(x)| < \varepsilon.$$

Let $\delta = \min(x - a,\ b - x)$. Then

$$(u \in A \wedge |u - x| < \delta) \Rightarrow |f(u) - f(x)| < \varepsilon.$$

Hence, f is continuous at x.   ∎

Theorem 8.10   Assume $f: C \to R$ and f is continuous. If C is compact, then $f[C]$ is compact.

Proof   Let $\mathscr{B}$ be an open covering of $f[C]$. We must show that $f[C]$ can be covered by a finite subset of $\mathscr{B}$. If $B \in \mathscr{B}$, then, by Theorem 8.9, $f^{-1}[B]$ is relatively open in C. Therefore, there exists open sets V_B such that $f^{-1}[B] = V_B \cap C$. Let $\mathscr{B}^*$ be the collection of all such open sets V_B for all B in $\mathscr{B}$. Clearly, $\mathscr{B}^*$ is an open covering of C. (For, if $c \in C$, then $f(c) \in f[C]$. Hence, $f(x) \in B$ for some B in $\mathscr{B}$. Then, for every V_B in $\mathscr{B}^*$, $c \in f^{-1}[B] \subseteq V_B$.) Since C is compact, C is covered by a finite subcovering

$\mathscr{B}^{**}$ of $\mathscr{B}^*$. Let $\mathscr{B}^{**} = \{V_{B_1}, V_{B_2}, \ldots, V_{B_k}\}$, where $B_1, B_2, \ldots, B_k \in \mathscr{B}$ and $V_{B_i} \cap C = f^{-1}[B_i]$. We wish to prove that $\{B_1, B_2, \ldots, B_k\}$ covers $f[C]$. Assume $z \in f[C]$. Then $z = f(c)$ for some c in C. Since $\mathscr{B}^{**}$ covers C, $c \in V_{B_i}$ for some i. Then $c \in f^{-1}[B_i]$. Therefore, $z = f(c) \in B_i$. Thus, $\{B_1, B_2, \ldots, B_k\}$ covers $f[C]$. ∎

Remark   The classical version of Theorem 8.10 reads: If $f: C \to R$, f is continuous, and C is closed and bounded, then $f[C]$ is closed and bounded. The classical proof runs as follows:

(a)   Assume $f[C]$ not bounded. Then, for each n in $N_{\mathfrak{I}}$, there exists c_n in C such that $|f(c_n)| > n$. Then $\{c_1, c_2, \ldots\}$ is an infinite subset of C. By the Bolzano–Weierstrass property, there is an accumulation point z of the set $\{c_1, c_2, \ldots\}$. Since C is closed and $\{c_1, c_2, \ldots\} \subseteq C$, $z \in C$. There is a subsequence $c_{i_1}, c_{i_2}, \ldots$ of $\langle c_n \rangle$ such that $\operatorname{Lim} c_{i_k} = z$. Since f is continuous, $\operatorname{Lim} f(c_{i_k}) = f(z)$. Since $\langle f(c_{i_k}) \rangle$ is convergent, it is bounded, contradicting the fact that $|f(c_{i_k})| > i_k$. Hence, $f[C]$ is bounded.

(b)   Assume w is an accumulation point of $f[C]$. Then there is a sequence $\langle y_n \rangle$ of points of $f[C]$ such that $\operatorname{Lim} y_n = w$. Let $y_n = f(c_n)$ for each n. Then $\langle c_n \rangle$ is a sequence in C. Since C is bounded, $\langle c_n \rangle$ has a convergent subsequence $\langle c_{i_k} \rangle$. Let $\operatorname{Lim} c_{i_k} = u$. Then $u \in \bar{C} = C$. Since f is continuous, $\operatorname{Lim} f(c_{i_k}) = f(u)$, that is, $\operatorname{Lim} y_{i_k} = f(u)$. Since $\langle y_{i_k} \rangle$ is a subsequence of $\langle y_n \rangle$, $\operatorname{Lim} y_{i_k} = \operatorname{Lim} y_n = w$. Hence, $w = f(u) \in f[C]$. Therefore, $f[C]$ contains all its accumulation points, that is, $f[C]$ is closed.

Corollary 8.11   If $f: C \to R$, f is continuous, and C is compact, then f assumes maximum and minimum values in C.

Proof   By Theorem 8.10, $f[C]$ is compact, and, therefore, closed and bounded. Let $M = \operatorname{lub}(f[C])$ and $m = \operatorname{glb}(f[C])$. Then, by Theorem 7.5g, $M \in \overline{f[C]}$ and $m \in \overline{f[C]}$. Since $f[C]$ is closed, $M \in f[C]$ and $m \in f[C]$. Hence, $M = f(c_1)$ for some c_1 in C and $m = f(c_2)$ for some c_2 in C. Therefore, M and m are the maximum and minimum values, respectively, of f in C. ∎

Let us turn now to the effect of a continuous function on a connected set.

Lemma 8.12 A is connected if and only if there do not exist sets B, C such that $A = B \cup C$, $B \neq \varnothing$, $C \neq \varnothing$, $B \cap C = \varnothing$, and B and C are relatively open in A.

Proof Assume A is connected. Assume $A = B \cup C$, $B \neq \emptyset$, $C \neq \emptyset$, $B \cap C = \emptyset$, and B and C are relatively open in A. Then $B = A \cap V_1$ and $C = A \cap V_2$, where V_1 and V_2 are open. Since A is connected, there is a point z such that either ($z \in B$ and z is an accumulation point of C) or ($z \in C$ and z is an accumulation point of B). We may assume that $z \in B$ and z is an accumulation point of C. (The proof is similar in the other case.) Then $z \in A \cap V_1$. Since V_1 is open, there is an open interval I containing z and included in V_1. Since z is an accumulation point of C, there exists some point w in $(I - \{z\}) \cap C$. Since $w \in C \subseteq A$ and $w \in I \subseteq V_1$, it follows that $w \in A \cap V_1 = B$, contradicting $B \cap C = \emptyset$.

Conversely, assume A is not connected. Then there exist sets B, C such that $A = B \cup C$, $B \neq \emptyset$, $C \neq \emptyset$, $B \cap C = \emptyset$, and there is no point z such that either ($z \in B$ and z is an accumulation point of C) or ($z \in C$ and z is an accumulation point of B). Hence, for each b in B, there is an interval I_b containing b and containing no points of C. If we take the union V_1 of all such intervals I_b for all b in B, V_1 is an open set and $V_1 \cap A = B$. Hence, B is relatively open in A. Likewise, there is an open set V_2 such that $V_2 \cap A = C$, and so, C is relatively open in A. ∎

Theorem 8.13 Assume $f: C \rightarrow R$. If f is continuous and C is connected, then $f[C]$ is connected.

Proof Assume $f[C]$ is not connected. Then there exist sets B_1, C_1 such that $f[C] = B_1 \cup C_1$, $B_1 \neq \emptyset$, $C_1 \neq \emptyset$, $B_1 \cap C_1 = \emptyset$, and B_1 and C_1 are relatively open in $f[C]$ (see Lemma 8.12). Then $B_1 = f[C] \cap V_1$ and $C_1 = f[C] \cap V_2$, where V_1 and V_2 are certain open sets. By Theorem 8.9, $f^{-1}[V_1]$ and $f^{-1}[V_2]$ are relatively open in C. Now, $C = f^{-1}[V_1] \cup f^{-1}[V_2]$. (Clearly, $f^{-1}[V_1] \subseteq C$ and $f^{-1}[V_2] \subseteq C$. In addition, assume $c \in C$. Then $f(c) \in f[C] = B_1 \cup C_1$; say, $f(c) \in B_1$. Then $f(c) \in V_1$. Hence $c \in f^{-1}[V_1]$.) In addition, $f^{-1}[V_1] \neq \emptyset$, $f^{-1}[V_2] \neq \emptyset$, and $f^{-1}[V_1] \cap f^{-1}[V_2] = \emptyset$. (For, $f^{-1}[V_1] \subseteq B_1$, $f^{-1}[V_2] \subseteq C_1$, and $B_1 \cap C_1 = \emptyset$.) Therefore, by Lemma 8.12, C is not connected. ∎

EXERCISE

16. Prove that, if n is a positive integer, then every nonnegative real number u has an nth root $\sqrt[n]{u}$. (*Hint:* The function x^n is continuous on the set of nonnegative real numbers. Use Theorem 8.13.)

Theorem 8.14 Assume $f: A \rightarrow R$ and f is continuous. If $[a, b] \subseteq A$ and f is not constant on $[a, b]$, then $f[[a, b]]$ is a closed interval, that is, nonconstant continuous functions transform closed intervals into closed intervals.

Proof Each point of $[a, b]$ is an accumulation point of $[a, b]$. Therefore, since f is continuous on A, f is continuous at each point of $[a, b]$. Thus, if we let g be the restriction of f to the domain $[a, b]$, then g is continuous. $[a, b]$ is both compact and connected. Hence, by Theorems 8.10 and 8.13, $g[[a, b]]$ is compact and connected. Since the connected sets are $\varnothing$, R, singletons $\{u\}$, and generalized intervals, while compact sets are closed and bounded, $g[[a, b]]$ is either a singleton $\{u\}$ or a closed interval. Since f is not constant on $[a, b]$, $g[[a, b]]$ is a closed interval. But $g[[a, b]]$ $= f[[a, b]]$. ∎

EXERCISE

17.^{D} If f is a continuous one-one function with domain $[a, b]$, prove that f^{-1} is continuous.

Corollary 8.15 *Intermediate Value Theorem.* Assume $f: A \to R$ and f is continuous. If $[a, b] \subseteq A$ and $f(a) \neq f(b)$, then, for any value d between $f(a)$ and $f(b)$, there exists c between a and b such that $f(c) = d$. In particular, if $f(a)$ and $f(b)$ have different signs, then there exists c between a and b such that $f(c) = 0$.

Proof Immediate from Theorem 8.14. ∎

EXERCISES

18. Prove that the polynomial $x^3/3 - x^2/2 - 2x + 1$ has roots between -2 and -3 between 0 and 1, and between 3 and 4.

19. Let $F = (F, +, \times, <)$ be any ordered field.
   a. Prove that $|u| < 1 \Rightarrow 1 + u > 0$.
   b. If $f(x) = x^n + a_{n-1}x^{n-1} + \cdots + a_1 x + a_0$, where each $a_i \in F$, and $M =$, $\max(1, |a_1| + \cdots + |a_n|)$, show that:
      i. $x > M \Rightarrow f(x) > 0$;
      ii. $x < -M \Rightarrow (-1)^n f(x) > 0$. (Hence, if n is odd, $x < -M \Rightarrow f(x) < 0$.)

20. Any polynomial with real coefficients and of odd degree has a real root. (*Hint:* We may assume that the polynomial is of the form $x^n + a_{n-1}x^{n-1} + \cdots + a_1 x + a_0$. Now, use Exercise 19b above and the Intermediate Value Theorem.)

21. Locate the roots of $x^3 - 2x^2 + x - 1$ between consecutive integers (as in Exercise 18), and then approximate the roots with an error of less than .1.

22. If f is defined on an interval and f is continuous and one-one, prove that f is either increasing or decreasing.

23. Assume that $f: [0, 1] \to [0, 1]$ and f is continuous. Prove that $f(c) = c$ for at least one c in $[0, 1]$.

24. If $f: A \to R$ and f is continuous,
    a. does the fact that A is open imply that $f[A]$ is open?
    b. does the fact that A is closed imply that $f[A]$ is closed?
    c. does the fact that A is bounded imply that $f[A]$ is bounded?

A somewhat stronger condition than continuity is often useful.

Definition Let $f: A \to R$. We say that f is **uniformly continuous** on A if and only if every point of A is an accumulation point of A and, for every $\varepsilon > 0$, there exists $\delta > 0$ such that

$$(x_1 \in A \wedge x_2 \in A \wedge |x_1 - x_2| < \delta) \Rightarrow |f(x_1) - f(x_2)| < \varepsilon.$$

Notice that uniform continuity implies continuity. The converse is not true.

EXAMPLE

12. Let $f(x) = 1/x$ for $0 < x < 1$. f is continuous. Assume f is uniformly continuous on $(0, 1)$. Take $\varepsilon = \frac{1}{2}$. Then there should be some $\delta > 0$ such that

$$(0 < x_1 < 1 \wedge 0 < x_2 < 1 \wedge |x_1 - x_2| < \delta) \Rightarrow |1/x_1 - 1/x_2| < \frac{1}{2}.$$

Since $\text{Lim } 1/n = 0$, there exists n such that $1/n < \delta$. Hence,

$$0 < 1/n - 1/(n + 1) < \delta.$$

Now, let $x_1 = 1/n$ and $x_2 = 1/n + 1$. Then $|x_1 - x_2| < \delta$. But

$$|f(x_1) - f(x_2)| = |n - (n + 1)| = 1,$$

which is a contradiction.

The following theorem shows that, for compact domains, continuity *does* imply uniform continuity.

Theorem 8.16 If $f: A \to R$, A is compact, and f is continuous, then f is uniformly continuous on A.

Proof Let $\varepsilon > 0$. Assume $z \in A$. By continuity of f at z, there exists $\delta > 0$ such that

$$(u \in A \wedge |u - z| < \delta) \Rightarrow |f(u) - f(z)| < \varepsilon/2.$$

Let $I_z = (z - \delta/2, z + \delta/2)$. Let $\mathscr{B}$ be the collection of all such open intervals I_z for all z in A. Then $\mathscr{B}$ is an open covering of A. Hence, by the compactness of A, A is covered by a finite subcollection of $\mathscr{B}$:

$$\{(z_1 - \delta_1/2, z_1 + \delta_1/2), \ldots, (z_k - \delta_k/2, z_k + \delta_k/2)\}.$$

Let $\delta^* = \min(\delta_1/2, \ldots, \delta_k/2)$. Assume $x_1 \in A$, $x_2 \in A$, and $|x_1 - x_2| < \delta^*$. Now, $x_1 \in (z_j - \delta_j/2, z_j + \delta_j/2)$ for some j. Then $|x_1 - z_j| < \delta_j/2 < \delta_j$. Hence, $|f(x_1) - f(z_j)| < \varepsilon/2$. Also,

$$|x_2 - z_j| = |(x_2 - x_1) + (x_1 - z_j)| \leq |x_2 - x_1| + |x_1 - z_j|$$
$$< \delta^* + \delta_j/2 \leq \delta_j/2 + \delta_j/2 = \delta_j.$$

Therefore, $|f(x_2) - f(z_j)| < \delta/2$. Finally,

$$|f(x_1) - f(x_2)| = |(f(x_1) - f(z_j)) + (f(z_j) - f(x_2))|$$
$$\leq |f(x_1) - f(z_j)| + |f(z_j) - f(x_2)|$$
$$< \varepsilon/2 + \varepsilon/2 = \varepsilon.$$

Therefore, f is uniformly continuous on A. ∎

EXERCISES

25. If f is uniformly continuous on a bounded set B, prove that $f[B]$ is bounded.

26. In each of the following cases, determine whether f is uniformly continuous on A.

   a. $f(x) = x^2$ for all x in $A = (0, 1)$.

   b. $f(x) = x^2$ for all x in the set A of positive real numbers.

   c. $f(x) = 1/x$ for all x in the set $A = \{u : u \geq 1\}$.

For some more information about the significance of uniform continuity, see Anderson and Hall [1963], pp. 121–135.

5.9 INFINITE SERIES

Let $\langle a_n \rangle$ be a sequence of real numbers. By the **infinite series** $\sum a_n$ we mean the sequence $\langle S_n \rangle$, where

$$S_n = a_1 + a_2 + \cdots + a_n = \sum_{i=1}^{n} a_i.$$

S_n is called the **nth partial sum** of ther series. To say that the series $\sum a_n$ converges is the same thing as saying that the sequence $\langle S_n \rangle$ converges. To say that $\sum a_n$ is **divergent** means that it is not convergent. If $\sum a_n$ is convergent, we let $\sum_{n=1}^{\infty} a_n$ denote $\operatorname{Lim} S_n$. $\sum_{n=1}^{\infty} a_n$ is called the **sum** of the series $\sum a_n$. Sometimes one writes $a_1 + a_2 + \cdots + a_n + \cdots$ instead of $\sum_{n=1}^{\infty} a_n$. The numbers a_n are called the **terms** of the series $\sum a_n$.

EXAMPLES

1. Let $a_n = 1/n - 1/(n+1)$. Thus, $a_1 = 1 - \frac{1}{2}$, $a_2 = \frac{1}{2} - \frac{1}{3}$, $a_3 = \frac{1}{3} - \frac{1}{4}$, etc. Then $S_n = a_1 + a_2 + \cdots + a_n = 1 - 1/(n+1)$, and, therefore, $\operatorname{Lim} S_n = \operatorname{Lim}(1 - 1/(n+1)) = 1$. Thus, $\sum a_n$ is convergent, and $\sum_{n=1}^{\infty} a_n = 1$.

2. Let $a_n = 1/(n(n+1))$. Then $\sum a_n$ is convergent and $\sum_{n=1}^{\infty} a_n = \frac{1}{2} + \frac{1}{2} \cdot 3 + \frac{1}{3} \cdot 4 + \cdots = 1$. For, $1/(n(n+1)) = 1/n - 1/(n+1)$. Hence,

$$S_n = \left(1 - \frac{1}{2}\right) + \left(\frac{1}{2} - \frac{1}{3}\right) + \left(\frac{1}{3} - \frac{1}{4}\right) + \cdots + \left(\frac{1}{n} - \frac{1}{n+1}\right)$$

$$= 1 - \frac{1}{n+1}.$$

But, $\operatorname{Lim}\left(1 - \dfrac{1}{n+1}\right) = 1$.

EXERCISES

1. Prove that

$$\sum_{n=1}^{\infty} \frac{1}{n(n+1)(n+2)} = \frac{1}{4}.$$

(*Hint*: Consider

$$\frac{1}{n(n+1)} - \frac{1}{(n+1)(n+2)}.)$$

2. Evaluate

$$\frac{1}{1 \cdot 3} + \frac{1}{2 \cdot 4} + \cdots + \frac{1}{n(n+2)} + \cdots.$$

Theorem 9.1 *Cauchy Criterion.* $\sum a_n$ converges if and only if, for any $\varepsilon > 0$, there exists a positive integer n_0 such that

$$(n \geq n_0 \wedge j > 0) \Rightarrow |a_{n+1} + a_{n+2} + \cdots + a_{n+j}| < \varepsilon.$$

Proof Let $S_n = a_1 + a_2 + \cdots + a_n$. Then $\sum a_n$ converges if and only if $\langle S_n \rangle$ converges. But $\langle S_n \rangle$ converges if and only if $\langle S_n \rangle$ is a Cauchy sequence (by the Cauchy completeness of the real number system and Theorem 5.3). Hence, $\langle S_n \rangle$ converges if and only if, for any $\varepsilon > 0$, there is a positive integer n_0 such that $n, k \geq n_0 \Rightarrow |S_k - S_n| < \varepsilon$. Therefore,

$$k > n \geq n_0 \Rightarrow |S_k - S_n| = |a_{n+1} + \cdots + a_k| < \varepsilon.$$

If we let $j = k - n$, we obtain the desired result.   ■

Notice that Theorem 9.1 implies that whether $\sum a_n$ converges depends upon what happens far out in the series; changing some terms at the beginning of the series does not affect convergence or divergence.

Corollary 9.2 If $\sum a_n$ converges, then $\mathrm{Lim}\, a_n = 0$.

Proof Assume $\sum a_n$ converges. Assume $\varepsilon > 0$. Then, by Theorem 9.1, there is a positive integer n_0 such that $n > n_0 \Rightarrow |a_n| < \varepsilon$. (We take $j = 1$ in Theorem 9.1.) Hence, $\mathrm{Lim}\, a_n = 0$.   ■

EXERCISE

3. Prove that the series $\sum \dfrac{n}{1 + n}$ is divergent.

Corollary 9.3 *Comparison Test* If $\sum a_n$ is a convergent series of non-negative terms, and $|b_n| \leq a_n$ for all n, then $\sum b_n$ converges.

Proof

$$|b_{n+1} + b_{n+2} + \cdots + b_{n+j}| \leq |b_{n+1}| + |b_{n+2}| + \cdots + |b_{n+j}|$$
$$\leq a_{n+1} + a_{n+2} + \cdots + a_{n+j}.$$

Hence, the Cauchy criterion (Theorem 9.1) for $\sum a_n$ yields the Cauchy criterion for $\sum b_n$.   ■

EXAMPLES

3. Let $b_n = 1/(n+1)^2$ and $a_n = 1/n(n+1)$. Then $b_n \leq a_n$ for all n. By Example 2 on p. 263, $\sum a_n$ converges. Hence, by Corollary 9.3, $\sum b_n$ converges.

4. $\sum 1/n^2$ converges, since it is obtained from $\sum 1/(n+1)^2$ by adding an extra term.

EXERCISES

4. If $\sum a_n$ is a convergent series of nonnegative terms, and $|b_n| \leq a_n$ for all $n \geq n_0$, prove that $\sum b_n$ converges.

5. Let k be an integer greater than or equal to 2. Prove that $\sum 1/n^k$ converges.

6. If $\sum a_n$ and $\sum b_n$ converge, prove that $\sum (a_n + b_n)$ and $\sum ca_n$ also converge.

7. a. If $a_n > 0$, $b_n > 0$, and $\lim a_n/b_n = c > 0$, prove that $\sum a_n$ converges if and only if $\sum b_n$ converges.

   b. Does $\sum 1/(3n^2 - 2n + 1)$ converge?

Lemma 9.4

a. If n is any natural number, then

$$\delta > -1 \Rightarrow (1 + \delta)^n \geq 1 + n\delta.$$

b. $|u| > 1 \Rightarrow |u|^n$ is an increasing sequence which is not bounded above.

c. $|r| < 1 \Rightarrow \lim r^n = 0$.

Proof (a) The proof proceeds by mathematical induction. When $n = 1$, $(1 + \delta)^1 = 1 + \delta$. Now, assume $(1 + \delta)^n \geq 1 + n\delta$. Then

$$(1 + \delta)^{n+1} = (1 + \delta)^n(1 + \delta) \geq (1 + n\delta)(1 + \delta)$$
$$= 1 + (n + 1)\delta + n\delta^2 \geq 1 + (n + 1)\delta.$$

(b) If $|u| > 1$, then $|u|^{n+1} = |u|^n |u| > |u|^n$. Thus, $\langle |u|^n \rangle$ is increasing. Let $\delta = |u| - 1 > 0$. By the Archimedean property, for any positive element v, there exists a natural number n such that $n\delta > v$. Hence, by Part a,

$$|u|^n = (1 + \delta)^n \geq 1 + n\delta > 1 + v.$$

Thus, $\langle |u|^n \rangle$ is not bounded above.

(c)   Assume $0 < |r| < 1$. Let $u = 1/r$. By Part b, $\langle |u|^n \rangle$ is increasing and not bounded above. Assume $\varepsilon > 0$. Then there exists n_0 such that

$$n \geq n_0 \Rightarrow |u|^n > 1/\varepsilon$$

$$\Rightarrow \frac{1}{|r|^n} > 1/\varepsilon$$

$$\Rightarrow |r^n| < \varepsilon.$$

Therefore, $\operatorname{Lim} r^n = 0$.   ∎

EXERCISE

8.   If $|u| > 1$ and k is a positive integer, prove that $\operatorname{Lim} n^k/u^n = 0$.

EXAMPLES

5.   Assume $|r| < 1$. Let $b \in R$ and $a_n = br^{n-1}$. Thus, $a_1 = b$, $a_2 = br$, $a_3 = br^2, \ldots.$ The series $\sum a_n$ has the partial sums

$$(*) \quad S_n = b + br + \cdots + br^{n-1}.$$

Such a series is called a **geometric series with ratio** r. Now,

$$(**) \quad rS_n = br + br^2 + \cdots + br^{n-1} + br^n.$$

Subtracting $(**)$ from $(*)$, we obtain

$$(1 - r)S_n = b - br^n = b(1 - r^n).$$

Therefore,

$$S_n = \frac{b(1 - r^n)}{1 - r}.$$

Hence, by Lemma 9.4c,

$$\sum_{n=1}^{\infty} a_n = \operatorname{Lim} S_n = \frac{b}{1-r} \operatorname{Lim}(1 - r^n) = \frac{b}{1-r}(\operatorname{Lim} 1 - \operatorname{Lim} r^n) - \frac{b}{1-r}.$$

Thus, $b + br + \cdots + br^{n-1} + \cdots = b/(1 - r)$. For example, when $b = 1$ and $r = \frac{1}{2}$,

$$1 + \frac{1}{2} + \cdots + \frac{1}{2^{n-1}} + \cdots = \frac{1}{1 - \frac{1}{2}} = 2.$$

EXERCISES

9. a.   Calculate $1 + 1/10 + 1/100 + \cdots + 1/10^n + \cdots$.

    b.   If $|r| < 1$, evaluate

$$\sum_{n=0}^{\infty} (n + 1)r^n = 1 + 2r + 3r^2 + \cdots.$$

    (*Hint*:   Let $T_n = 1 + 2r + 3r^2 + \cdots + (n + 1)r^n$. Subtract rT_n from T.)

10.   Assume $|r| \geq 1$. Let $b \in R$ and $a_n = br^{n-1}$. As in Example 1 above, the partial sum

$$S_n = b + br + \cdots + br^{n-1} = \frac{b(1 - r^n)}{1 - r},$$

and, therefore,

$$r^n = 1 - \frac{1 - r}{b} S_n.$$

By Lemma 9.4b, $|r|^n$ increases without bound. Therefore, $\text{Lim } S_n$ does not exist, and $\sum a_n$ is divergent.

11.   Evaluate $\sum_{n=0}^{\infty} (2/3)^n$.

12.   Evaluate $1 + 9/10 + (9/10)^2 + (9/10)^3 + \cdots$.

Theorem 9.5   Assume $a_n \geq 0$ for all n. Then $\sum a_n$ converges if and only if the sequence of partial sums $S_n = a_1 + a_2 + \ldots + a_n$ is bounded above.

Proof   Assume $\sum a_n$ converges. Then, by definition, $\langle S_n \rangle$ converges. By Theorem 5.5, $\langle S_n \rangle$ is bounded. Conversely, assume $\langle S_n \rangle$ is bounded above. Since each $a_n \geq 0$, $\langle S_n \rangle$ is nondecreasing. Hence, by (Mon) (see Theorem 5.9), $\langle S_n \rangle$ converges, that is, $\sum a_n$ converges.

EXAMPLE

6.   The series $\sum 1/n$ is called the **harmonic series**. It is not convergent, because $\langle S_n \rangle$ is not bounded above. To see this, it is only necessary to observe that

$S_2 = 1 + \frac{1}{2}$,

$S_4 = 1 + \frac{1}{2} + \frac{1}{3} + \frac{1}{4} \geq 1 + \frac{1}{2} + \frac{1}{4} + \frac{1}{4} = 1 + \frac{2}{2}$.

$S_8 = S_4 + \frac{1}{5} + \frac{1}{6} + \frac{1}{7} + \frac{1}{8} \geq S_4 + \frac{1}{8} + \frac{1}{8} + \frac{1}{8} + \frac{1}{8} = S_4 + \frac{1}{2} \geq 1 + \frac{3}{2}$.

$S_{16} = S_8 + \frac{1}{9} + \cdots + \frac{1}{16} \geq S_8 + \frac{1}{16} + \cdots + \frac{1}{16} = S_8 + \frac{8}{16} \geq 1 + \frac{4}{2}$.

$\vdots$

$S_{2^k} \geq 1 + \dfrac{k}{2}$.

This example shows that the fact that $\text{Lim } a_n = 0$ does not necessarily imply that $\sum a_n$ converges.

EXERCISES

13. Show that $\sum \dfrac{n}{1 + n^2}$ is divergent.

14. Show that $\sum \dfrac{n}{1 + n^3}$ is convergent.

15. Show that $\sum \dfrac{1}{2n - 1}$ is divergent.

16. Show that $\sum \dfrac{1}{\sqrt{n}}$ is divergent.

Definition   We say that $\sum a_n$ **converges absolutely** if and only if $\sum |a_n|$ converges.

Theorem 9.6   If $\sum a_n$ converges absolutely, then $\sum a_n$ converges.

Proof   Assume $\sum |a_n|$ converges. Then, by Corollary 9.3, $\sum a_n$ converges.   ∎

Theorem 9.7   *Alternating Series*   If the terms of $\sum a_n$ alternate in sign (say, $a_n > 0$ for n odd and $a_n < 0$ for n even), and if $\text{Lim } a_n = 0$ and $\langle |a_n| \rangle$ is nonincreasing (that is, $|a_1| \geq |a_2| \geq |a_3| \geq \cdots$), then $\sum a_n$ converges.

Proof   Let i and j be positive integers, and $R_{i,j} = a_i + a_{i+1} + \cdots + a_{i+j}$. Assume $\varepsilon > 0$. Since $\text{Lim } a_n = 0$, there is a positive integer n_0 such that $n \geq n_0 \Rightarrow |a_n| < \varepsilon$. But,

$$R_{i,j} = (-1)^{i+1} |a_i| + (-1)^{i+2} |a_{i+1}| + \cdots + (-1)^{i+j} |a_{i+j}|.$$

Hence,

$$R_{i,j} = \begin{cases} (-1)^{i+1}[(|a_i| - |a_{i+1}|) + (|a_{i+2}| - |a_{i+3}|) + \cdots \\ \quad + (|a_{i+j-1}| - |a_{i+j}|)] \qquad \text{if } j \text{ is odd,} \\[2mm] (-1)^{i+1}[(|a_i| - |a_{i+1}|) + (|a_{i+2}| - |a_{i+3}|) + \cdots \\ \quad + (|a_{i+j-2}| - |a_{i+j-1}|) + |a_{i+j}|] \qquad \text{if } j \text{ is even.} \end{cases}$$

In either case, $R_{i,j}(-1)^{i+1} \geq 0$. Also,

$$R_{i,j}(-1)^{i+1} = \begin{cases} |a_i| - (|a_{i+1}| - |a_{i+2}|) - \cdots - (|a_{i+j-2}| - |a_{i+j-1}|) \\ \quad - |a_{i+j}| \quad \text{if } j \text{ is odd,} \\ \\ |a_i| - (|a_{i+1}| - |a_{i+2}|) - \cdots - (|a_{i+j-1}| - |a_{i+j}|) \\ \quad \text{if } j \text{ is even} \end{cases}$$

Hence, $R_{i,j}(-1)^{i+1} \leq |a_i|$. Therefore, $|R_{i,j}| \leq |a_i|$. Hence, if $i \geq n_0$, $|R_{i,j}| < \varepsilon$. Therefore, by the Cauchy criterion, $\sum a_n$ converges. ∎

Let $\sum a_n$ be a series, and let $i \geq 0$. Then $\sum a_{n+i}$ is a new series whose partial series are obtained from those of $\sum a_n$ by omitting the first i terms. If $\sum a_n$ converges, then, by the Cauchy criterion, $\sum a_{n+i}$ also converges. We denote the sum of the series $\sum a_{n+i}$ by $\sum_{n=i+1}^{\infty} a_n$ or R_i. R_i is called the **remainder of the series $\sum a_n$ after the ith term**. It follows from the Cauchy criterion that $\text{Lim } R_i = 0$. (Why?) In the case of an alternating series $\sum a_n$ with $\text{Lim } a_n = 0$ and $\langle a_n \rangle$ nonincreasing, we have, in the course of the proof of Theorem 9.7, derived the inequality $|R_{i,j}| \leq |a_i|$. Hence, $|R_{i-1}| \leq |a_i|$ (see Exercise 4b, p. 212). Thus, $|R_i| \leq |a_{i+1}|$. In other words, the remainder after the ith term is no bigger in magnitude than the first term neglected.

EXAMPLE

7. Consider the alternating series $\sum (-1)^{n+1}/n$. Here, $a_n = (-1)^{n+1}/n$. Its partial sums are $1 - \frac{1}{2} + \frac{1}{3} - \cdots + (-1)^{n+1}(1/n)$. Notice that $\langle |a_n| \rangle$ is decreasing. Hence, by Theorem 9.7, $\sum (-1)^{n+1}/n$ converges. However, $\sum |a_n| = \sum 1/n$ does not converge (see Example 6, p. 267). Thus, $\sum (-1)^{n+1}/n$ is convergent but not absolutely convergent.

Theorem 9.8 *Ratio Test* Assume $a_n \neq 0$ for all n. Then $\sum a_n$ converges absolutely if $\text{Lim } |a_{n+1}/a_n| < 1$, and diverges if $\text{Lim } |a_{n+1}/a_n| > 1$ or if $\text{Lim } |a_{n+1}/a_n| = \infty$.†

Proof (i) Assume $\text{Lim } |a_{n+1}/a_n| = b < 1$. Then there exists a positive integer n_0 such that

$$n \geq n_0 \Rightarrow \left| \left| \frac{a_{n+1}}{a_n} \right| - b \right| < \frac{1-b}{2}.$$

---

† To say that $\text{Lim } u_n = \infty$ means that, for any z, there is a positive integer n_0 such that $n \geq n_0 \Rightarrow u_n \geq z$.

Hence,

$$n \geq n_0 \Rightarrow \left| \frac{a_{n+1}}{a_n} \right| \leq b + \frac{1-b}{2} = \frac{1+b}{2} = c < 1.$$

Therefore, $n \geq n_0 \Rightarrow |a_{n+1}| \leq c \, |a_n|$, where $0 < c < 1$. Then

$$|a_{n_0}| + |a_{n_0+1}| + \cdots + |a_{n_0+j}| \leq |a_{n_0}| + c \, |a_{n_0}| + c^2 \, |a_{n_0}| + \cdots + c^j \, |a_{n_0}|$$

$$= |a_{n_0}|(1 + c + c^2 + \cdots + c^j)$$

$$\leq |a_{n_0}| \, \frac{1}{1-c}.$$

Therefore, every partial sum

$$S_n = |a_1| + \cdots + |a_{n_0-1}| + |a_{n_0}| + \cdots + |a_{n_0+j}|$$

$$\leq |a_1| + \cdots + |a_{n_0-1}| + |a_{n_0}| \, \frac{1}{1-c}.$$

Hence, by Theorem 9.5, $\sum |a_n|$ converges.

(ii)   Assume $\text{Lim} \, |a_{n+1}/a_n| = b > 1$. Then there exists a positive integer n_0 such that

$$n \geq n_0 \Rightarrow \left| \left| \frac{a_{n+1}}{a_n} \right| - b \right| < \frac{b-1}{2}.$$

Hence,

$$n \geq n_0 \Rightarrow b = |b| = \left| \left(b - \left| \frac{a_{n+1}}{a_n} \right| \right) + \left| \frac{a_{n+1}}{a_n} \right| \right|$$

$$\leq \left| b - \left| \frac{a_{n+1}}{a_n} \right| \right| + \left| \frac{a_{n+1}}{a_n} \right| < \frac{b-1}{2} + \left| \frac{a_{n+1}}{a_n} \right|$$

$$\Rightarrow \left| \frac{a_{n+1}}{a_n} \right| > b - \left(\frac{b-1}{2} \right) = \frac{b+1}{2} = c, \qquad \text{where } 1 < c.$$

Then

$$|a_{n_0+1}| \geq |a_n| \, c \geq |a_{n_0}|, \quad |a_{n_0+2}| \geq |a_{n_0+1}| \, c \geq |a_{n_0}| \, c^2 \geq |a_{n_0}|,$$

and, in general, $|a_{n_0+j}| \geq |a_{n_0}| \, c^j \geq |a_{n_0}|$. Therefore, since $|a_{n_0}| \neq 0$, $\text{Lim} \, a_n$ cannot be 0, and, by Corollary 9.2, $\sum a_n$ does not converge.

(iii)   Assume $\text{Lim} \, |a_{n+1}/a_n| = \infty$. Then there exists a positive integer n_0 such that $n \geq n_0 \Rightarrow |a_{n+1}/a_n| > 2$. The rest of the proof proceeds as in Case ii.   ∎

EXERCISES

17. Prove: $\text{Lim} \dfrac{u^n}{n!} = 0.$

18. By a **rearrangement** of a series $\sum a_n$ we mean a series $\sum a_{\sigma(n)}$, where σ is a permutation of the positive integers, that is, $\sigma: P \xrightarrow[\text{onto}]{1-1} P$. If $\sum a_n$ converges absolutely, prove that any rearrangement of $\sum a_n$ also converges and has the same sum as $\sum a_n$. (*Hint*: Use the Cauchy criterion for $\sum a_n$.)

Example. The geometric series $1 - \tfrac{1}{2} + \tfrac{1}{4} - \tfrac{1}{8} + \cdots$ converges absolutely, and its sum is $\tfrac{2}{3}$. Hence the series $1 + \tfrac{1}{4} - \tfrac{1}{2} - \tfrac{1}{8} + \tfrac{1}{16} + \tfrac{1}{32} - \tfrac{1}{64} - \tfrac{1}{128} + \cdots$ also converges and has sum $\tfrac{2}{3}$.

19. a. If a series $\sum a_n$ converges but does not converge absolutely, prove that, for every real number r, there is a rearrangement of $\sum a_n$ whose sum is r. (*Hint*: The series must have infinitely many positive and infinitely many negative terms. (Why?) Show that the series based upon the positive terms diverges and the series based upon the negative terms diverges. Add positive terms until we have a sum $> r$; then add negative terms until we have a sum $< r$, and then again add positive terms to obtain a sum $> r$, etc.)

   b. Compare the sum of the series $1 - \tfrac{1}{2} + \tfrac{1}{3} - \tfrac{1}{4} + \tfrac{1}{5} - \cdots$ with the sum of the series $1 - \tfrac{1}{2} - \tfrac{1}{4} + \tfrac{1}{3} - \tfrac{1}{6} - \tfrac{1}{8} + \tfrac{1}{5} - \tfrac{1}{10} - \tfrac{1}{12} + \cdots$.

   c. Show how the series $1 - \tfrac{1}{2} + \tfrac{1}{3} - \tfrac{1}{4} + \tfrac{1}{5} - \cdots$ can be rearranged to obtain a series with sum 0.

20. If $\sum a_n$ and $\sum b_n$ are convergent, is $\sum a_n b_n$ necessarily convergent?

21. Show that

$$\frac{1}{1} + \frac{1\cdot 3}{1\cdot 4} + \frac{1\cdot 3\cdot 5}{1\cdot 4\cdot 7} + \cdots$$

is convergent.

22. For what values of x is

$$\frac{x}{1} + \frac{x^2}{2} + \frac{x^3}{3} + \cdots$$

convergent?

Let us turn now to the study of infinite decimals, which are really special kinds of infinite series.

By an **infinite decimal** we mean an infinite sequence $a_0 \cdot a_1 a_2 a_3 \cdots$, where a_0 is an arbitrary integer and, for $i > 0$, a_i is an integer between 0 and 9, that is, $0 \leq a_i \leq 9$. The dot between a_0 and a_1 is called a **decimal point**.

Examples of decimals. (i) $121.01341777\cdots$.

(ii) $-98.1212121212\cdots$.

Theorem 9.9   The series

$$\sum \frac{a_n}{10^n} = a_0 + \frac{a_1}{10} + \frac{a_2}{10^2} + \cdots + \frac{a_n}{10^n} + \cdots,$$

determined by an infinite decimal $a_0 \cdot a_1 a_2 a_3 \cdots$, converges. (The sum of the series is said to be the number **represented** by the infinite decimal.)

Proof

$$\frac{a_1}{10} < \frac{10}{10} = 1, \qquad \frac{a_2}{10^2} < \frac{10}{10^2} = \frac{1}{10},$$

$$\frac{a_3}{10^3} < \frac{10}{10^3} = \frac{1}{10^2}, \qquad \cdots, \qquad \frac{a_n}{10^n} < \frac{10}{10^n} = \frac{1}{10^{n-1}}.$$

Hence, by comparison with the geometric series

$$1 + 1/10 + 1/10^2 + 1/10^3 + \cdots,$$

the series $a_1/10 + a_2/10^2 + a_3/10^3 + \cdots$ converges (see Corollary 9.3). Therefore, $a_0 + a_1/10 + a_2/10^2 + \cdots$ converges, since it differs from the previous series in having one additional term.   ∎

EXAMPLES

8. $7.3333\cdots$ represents

$$7 + \frac{3}{10} + \frac{3}{10^2} + \frac{3}{10^3} + \cdots = 7 + \frac{3}{10}\left(1 + \frac{1}{10} + \frac{1}{10^2} + \cdots\right)$$

$$= 7 + \frac{3}{10}\left(\frac{1}{1 - \frac{1}{10}}\right) = 7 + \frac{3}{9} = \frac{22}{3}.$$

9. $.14141414\cdots$ represents

$$\frac{1}{10} + \frac{4}{10^2} + \frac{1}{10^3} + \frac{4}{10^4} + \frac{1}{10^5} + \frac{4}{10^6} + \cdots$$

$$= \frac{14}{10^2} + \frac{14}{10^4} + \frac{14}{10^6} + \cdots = \frac{14}{10^2}\left(1 + \frac{1}{10^2} + \frac{1}{10^4} + \frac{1}{10^6} + \cdots\right)$$

$$= \frac{14}{10^2}\left(\frac{1}{1 - 1/10^2}\right) = \frac{14}{99}.$$

10.   By a periodic decimal we mean one which, after a certain point, repeats a certain finite sequence of digits:

$$a_0 \cdot a_1 a_2 \cdots a_k b_1 \cdots b_m b_1 \cdots b_m b_1 \cdots b_m \cdots.$$

By the **period** of the decimal we mean the length m of the repeated finite sequence. Such a decimal represents

$$a_0 + \frac{a_1}{10} + \cdots + \frac{a_k}{10^k} + \frac{b_1}{10^{k+1}} + \cdots + \frac{b_m}{10^{k+m}}$$

$$+ \frac{b_1}{10^{k+m+1}} + \cdots + \frac{b_m}{10^{k+2m}} + \cdots$$

$$= \left(a_0 + \frac{a_1}{10} + \cdots + \frac{a_k}{10^k} \right) + \frac{b_1 \cdots b_m{}^{\dagger}}{10^{k+m}} + \frac{b_1 \cdots b_m}{10^{k+2m}} + \cdots$$

$$= \left(a_0 + \frac{a_1}{10} + \cdots + \frac{a_k}{10^k} \right) + \frac{b_1 \cdots b_m}{10^{k+m}} \left(1 + \frac{1}{10^m} + \frac{1}{10^{2m}} + \cdots \right)$$

$$= \left(a_0 + \frac{a_1}{10} + \cdots + \frac{a_k}{10^k} \right) + \frac{b_1 \cdots b_m}{10^{k+m}} \left(\frac{1}{1 - 1/10^m} \right)$$

$$= \left(a_0 + \frac{a_1}{10} + \cdots + \frac{a_k}{10^k} \right) + \frac{b_1 \cdots b_m}{10^k} \frac{1}{10^m - 1}.$$

Thus, a periodic decimal represents a rational number.

EXERCISES

Find the rational numbers represented by:

23. $0.23172172172172 \cdots$

24. $0.625000000 \cdots$

25. $4.37373737 \cdots.$

Lemma 9.10 Every real number such that $0 \le x < 1$ is represented by an infinite decimal $.a_1 a_2 a_3 \cdots$.

Proof We shall define the digits a_i recursively, as follows: Let $a_1 = [10x]$. (Remember that $[y]$ is the greatest integer $\le y$.) Since $0 \le x < 1$, $0 \le 10x < 10$. Hence, $0 \le a_1 \le 9$. Since $a_1 \le 10x < a_1 + 1$, we have $a_1/10 \le x < a_1/10 + 1/10$. Now, let us assume that we have defined decimal digits $a_1, a_2, \ldots, a_n$ such that

$$\frac{a_1}{10} + \frac{a_2}{10^2} + \cdots + \frac{a_n}{10^n} \le x < \frac{a_1}{10} + \frac{a_2}{10^2} + \cdots + \frac{a_n}{10^n} + \frac{1}{10^n}.$$

Let

$$A_n = \frac{a_1}{10} + \frac{a_2}{10^2} + \cdots + \frac{a_n}{10^n}.$$

† $b_1 \cdots b_m$ is decimal notation. It represents $b_1 10^{m-1} + \cdots + b_{m-1} 10 + b_m$.

Thus, $A_n \leq x < A_n + 1/10^n$. Hence, $0 \leq x - A_n < 1/10^n$. Therefore, $0 \leq 10^{n+1}(x - A_n) < 10$. Let $a_{n+1} = [10^{n+1}(x - A_n)]$. Then, $0 \leq a_{n+1} \leq 9$ and

$$a_{n+1} \leq 10^{n+1}(x - A_n) < a_{n+1} + 1.$$

Thus,

$$\frac{a_{n+1}}{10^{n+1}} \leq x - A_n < \frac{a_{n+1}}{10^{n+1}} + \frac{1}{10^{n+1}},$$

and, therefore,

$$\frac{a_1}{10} + \cdots + \frac{a_n}{10^n} + \frac{a_{n+1}}{10^{n+1}} \leq x < \frac{a_1}{10} + \cdots + \frac{a_n}{10^n} + \frac{a_{n+1}}{10^{n+1}} + \frac{1}{10^{n+1}}.$$

For every positive integer n,

$$0 \leq x - \left(\frac{a_1}{10} + \cdots + \frac{a_n}{10^n} \right) < \frac{1}{10^n}.$$

Hence, $\mathrm{Lim}(x - (a_1/10 + \cdots + a_n/10^n)) = 0$, and, therefore,

$$x - \mathrm{Lim}\left(\sum_{i=1}^{n} \frac{a_i}{10^i} \right) = 0.$$

This means that $x = \sum_{i=0}^{\infty} a_i/10^i$, which shows that x is represented by the decimal $.a_1 a_2 a_3 \cdots$.   ∎

Theorem 9.11 Every nonnegative real number is represented by a unique infinite decimal, except that a number represented by a finite decimal $b_0 \cdot b_1 \cdots b_{k-1} b_k 000 \cdots$ (with $b_k > 0$) is also represented by

$$b_0 \cdot b_1 \cdots b_{k-1}(b_k - 1)999 \cdots.$$

Proof (I) First let us show the existence of an infinite decimal representation. The case $0 \leq x < 1$ has been handled in Lemma 9.10. If $x \geq 1$, let $a_0 = [x]$. Then $0 \leq x - a_0 < 1$. If $.a_1 a_2 \cdots$ is a decimal representation for $x - a_0$ given by Lemma 9.10, then $a_0 \cdot a_1 a_2 \cdots$ is clearly a decimal representation for x.

(II) It clearly suffices to prove the uniqueness for the case $0 \leq x < 1$. Assume then that x has two different decimal representations: $x = .a_1 a_2 \cdots$ and $x = .b_1 b_2 \cdots$. Let k be the first integer for which $a_k \neq b_k$. Thus, $a_1 = b_1$, $a_2 = b_2, \ldots, a_{k-1} = b_{k-1}$, $a_k \neq b_k$. We may assume that $a_k < b_k$. Hence,

$1 \leq a_k + 1 \leq b_k \leq 9$. Therefore,

$$(1) \quad x = \sum_{n=1}^{k-1} \frac{a_n}{10^n} + \frac{a_k}{10^k} + \sum_{n=k+1}^{\infty} \frac{a_n}{10^n}$$

$$(2) \quad \leq \sum_{n=1}^{k-1} \frac{a_n}{10^n} + \frac{a_k}{10^k} + \sum_{n=k+1}^{\infty} \frac{9}{10^n} = \sum_{n=1}^{k-1} \frac{a_n}{10^n} + \frac{a_k}{10^k} + \frac{9}{10^{k+1}} \sum_{n=1}^{\infty} \frac{1}{10^{n-1}}$$

$$(3) \quad = \sum_{n=1}^{k-1} \frac{a_n}{10^n} + \frac{a_k}{10^k} + \frac{9}{10^{k+1}} \cdot \frac{1}{1 - \frac{1}{10}} = \sum_{n=1}^{k-1} \frac{a_n}{10^n} + \frac{a_k}{10^k} + \frac{1}{10^k}$$

$$(4) \quad = \sum_{n=1}^{k-1} \frac{b_n}{10^n} + \frac{a_k}{10^k} + \frac{1}{10^k}$$

$$(5) \quad = \sum_{n=1}^{k-1} \frac{b_n}{10^n} + \frac{a_k + 1}{10^k} \leq \sum_{n=1}^{k-1} \frac{b_n}{10^n} + \frac{b_k}{10^k}$$

$$(6) \quad \leq \sum_{n=1}^{k-1} \frac{b_n}{10^n} + \frac{b_k}{10^k} + \sum_{n=k+1}^{\infty} \frac{b_n}{10^n}$$

$$(7) \quad = \sum_{n=1}^{\infty} \frac{b_n}{10^n} = x.$$

Hence, all the inequalities must be equalities. In particular, from line (5),

$$\sum_{n=1}^{k-1} \frac{b_n}{10^n} + \frac{a_k + 1}{10^k} = \sum_{n=1}^{k-1} \frac{b_n}{10^n} + \frac{b_k}{10^k},$$

whence $a_k + 1 = b_k$.

In addition, from lines (5), (6),

$$\sum_{n=1}^{k-1} \frac{b_n}{10^n} + \frac{b_k}{10^k} = \sum_{n=1}^{k-1} \frac{b_n}{10^n} + \frac{b_k}{10^k} + \sum_{n=k+1}^{\infty} \frac{b_n}{10^n},$$

whence $\sum_{n=k+1}^{\infty} b_n/10^n = 0$. This implies that $b_n = 0$ for $n \geq k + 1$. From lines (1), (2),

$$\sum_{n=1}^{k-1} \frac{a_n}{10^n} + \frac{a_k}{10^k} + \sum_{n=k+1}^{\infty} \frac{a_n}{10^n} = \sum_{n=1}^{k-1} \frac{a_n}{10^n} + \frac{a_k}{10^k} + \sum_{n=k+1}^{\infty} \frac{9}{10^n},$$

whence $\sum_{n=k+1}^{\infty} a_n/10^n = \sum_{n=k+1}^{\infty} 9/10^n$. This implies that $a_n = 9$ for $n \geq k + 1$. (For, $\sum_{n=k+1}^{\infty} (9 - a_n)/10^n = 0$, whence $9 - a_n = 0$ for all $n \geq k + 1$.) We have proved that, if x has two distinct decimal representations, one must be of the form $.b_1 b_2 \cdots b_{k-1} b_k 000$, with $b_k > 0$, and the other of the form $.b_1 b_2 \cdots b_{k-1}(b_k - 1)999\cdots$. ∎

EXAMPLE

11. .23999⋯ represents

$$\frac{2}{10} + \frac{3}{10^2} + \frac{9}{10^3} + \frac{9}{10^4} + \cdots$$

$$= \frac{2}{10} + \frac{3}{10^2} + \frac{9}{10^3}\left(1 + \frac{1}{10} + \frac{1}{10^2} + \cdots\right)$$

$$= \frac{2}{10} + \frac{3}{10^2} + \frac{9}{10^3}\left(\frac{1}{1 - \frac{1}{10}}\right)$$

$$= \frac{2}{10} + \frac{3}{10^2} + \frac{1}{10^2} = \frac{2}{10} + \frac{4}{10^2} = \frac{24}{100} = \frac{6}{25}.$$

But 6/25 also has the decimal representation .24000⋯.

The role that ten plays in decimal representation is not essential. If m is any integer ≥ 2, we can use infinite m-ary representations for real numbers: $a_0 \cdot a_1 a_2 \cdots$, where a_0 is an arbitrary integer and, for $i > 0$, a_i is an integer between 0 and $m - 1$, that is, $0 \leq a_i \leq m - 1$. (Thus, $m - 1$ plays the same role here that nine played in decimal representations.) The m-ary representation $a_0 \cdot a_1 a_2 \cdots$ stands for the real number

$$a_0 + \frac{a_1}{m} + \frac{a_2}{m^2} + \frac{a_3}{m^3} + \cdots + \frac{a_k}{m^k} + \cdots.$$

The theorems presented above for decimal representations carry over to m-ary representations by an almost word-for-word translation (10 being replaced by m, and 9 by $m - 1$).

EXAMPLES

12. Binary representation: $m - 2$.

  a. .11010101⋯ represents

$$\frac{1}{2} + \frac{1}{2^2} + \frac{0}{2^3} + \frac{1}{2^4} + \frac{0}{2^5} + \frac{1}{2^6} + \cdots$$

$$= \frac{1}{2} + \frac{1}{2^2} + \frac{1}{2^4} + \frac{1}{2^6} + \frac{1}{2^8} + \cdots$$

$$= \frac{1}{2} + \frac{1}{2^2}\left(1 + \frac{1}{2^2} + \frac{1}{2^4} + \frac{1}{2^6} + \cdots\right)$$

$$= \frac{1}{2} + \frac{1}{2^2}\left(\frac{1}{1 - \frac{1}{4}}\right) = \frac{1}{2} + \frac{1}{3} = \frac{5}{6}.$$

b. .0101111··· represents

$$\frac{1}{2^2} + \frac{1}{2^4} + \frac{1}{2^5} + \frac{1}{2^6} + \cdots = \frac{1}{2^2} + \frac{1}{2^4}\left(1 + \frac{1}{2} + \frac{1}{2^2} + \cdots\right)$$

$$= \frac{1}{2^2} + \frac{1}{2^4}\left(\frac{1}{1 - \frac{1}{2}}\right) = \frac{1}{2^2} + \frac{1}{2^3} = \frac{3}{8}.$$

$\frac{3}{8}$ also has the binary representation .011000···.

13. Ternary representations: $m = 3$.   .0101111··· represents

$$\frac{0}{3} + \frac{1}{3^2} + \frac{0}{3^3} + \frac{1}{3^4} + \frac{1}{3^5} + \cdots = \frac{1}{3^2} + \frac{1}{3^4}\left(1 + \frac{1}{3} + \frac{1}{3^2} + \cdots\right)$$

$$= \frac{1}{3^2} + \frac{1}{3^4}\left(\frac{1}{1 - \frac{1}{3}}\right) = \frac{1}{9} + \frac{1}{3^3}\frac{1}{2} = \frac{1}{9} + \frac{1}{54} = \frac{7}{54}.$$

This number does not have any other ternary representation. Notice that the same sequence .010111··· represents different numbers in the binary and ternary systems.

We have been using only periodic representations for the sake of simplicity. Hence, we have obtained only rational numbers. Of course, by Theorem 9.11, every positive real number has an m-ary representation.

EXERCISES

26. What are the decimal representations of the numbers with the following binary representations?

   a.   .0101010101 ····.

   b.   .110110110110 ····.

27. Find the decimal representations of the numbers with the following ternary representations.

   a.   .012012012012 ····.

   b.   .0101010101 ····.

28. Find the binary representations of the following numbers.

   a. $\frac{1}{3}$.

   b. $\frac{3}{32}$.

   c. $\frac{2}{7}$.

29. a.   Assume k and n are relatively prime positive integers. Prove that k/n has a finite decimal representation if and only if n is a factor of a power of 10, that is, $n = 2^j 5^l$ for some nonnegative integers j and l.

b.   Find the decimal representation of $\frac{7}{40}$.

c.   Reformulate the result of Part a in the case where, instead of decimal representations, we consider *m*ary representations (for some integer $m > 1$).

30.   a.   Prove that a rational number has a periodic decimal representation. (*Hint*: Let k and n be relatively prime positive integers. Verify that Lemmas 9.10, 9.11 yield the following familiar division algorithm for generating the decimal representation of k/n.

$$\frac{k}{n} = a_0 + \frac{k_1}{n}, \qquad \text{where } 0 \le k_1 < n;$$

$$10\,\frac{k_1}{n} = a_1 + \frac{k_2}{n}, \qquad \text{where } 0 \le k_2 < n;$$

$$10\,\frac{k_2}{n} = a_2 + \frac{k_3}{n}, \qquad \text{where } 0 \le k_3 < n, \text{ etc.}$$

Here, $k_1, k_2, \ldots$ are integers. Clearly, the values of k_{i+1} and a_i are determined by the value of k_i. Since $0 \le k_i < n$, after at most n steps some value of k_i must be repeated, say $k_j = k_{j+l}$. Then the terms from a_j to a_{j+l-1} will be the same as the terms from a_{j+l} to a_{j+2l-1}, which, in turn, will be the same as the terms from a_{j+2l} to a_{j+3l-1}, etc.)

b.   Find the decimal representations for: (i) $\frac{2}{3}$; (ii) $\frac{3}{7}$.

c.   Prove that the number represented by the decimal

$$.12345678910111213141516\ldots,$$

obtained by writing the positive integers in ascending order, is irrational.

For further information about decimal representations, see Beck, Bleicher, and Crowe [1969], Chapter 6; Hardy and Wright [1954], Chapter 9; Sierpinski [1964], Chapter 7.

APPENDIX A

EQUALITY

In most mathematical theories, equality is taken to be a primitive unde-fined relation. In order to use the equality relation in the traditional way, it suffices to postulate the following properties.

Assumption I   For all x, $x = x$.   (Reflexivity)

Assumption II   For all x and y,

$$x = y \Rightarrow (\mathscr{A}(x, x) \Rightarrow \mathscr{A}(x, y))$$ (Substitutivity)

(Here, $\mathscr{A}(x, x)$ stands for any formula of the given theory, and $\mathscr{A}(x, y)$ results from $\mathscr{A}(x, x)$ by replacing some, but not necessarily all, occurrences of x by y. Moreover, $\mathscr{A}(x, x)$ may contain other variables in addition to x.)

All the standard properties of equality follow from these assumptions. We shall present a few illustrations of this fact.

Theorem A1   (*Symmetry*) $x = y \Rightarrow y = x$.

Proof   Let $\mathscr{A}(x, x)$ be $x = x$, and let $\mathscr{A}(x, y)$ be $y = x$. Then, by Assumption II,

$$x = y \Rightarrow (x = x \Rightarrow y = x).$$

This is equivalent to: $x = x \Rightarrow (x = y \Rightarrow y = x).$[†] Hence, by Assumption I, $x = y \Rightarrow y = x$.   ∎

[†] In general, $A \Rightarrow (B \Rightarrow C)$ is equivalent to $B \Rightarrow (A \Rightarrow C)$. Remember that $A \Rightarrow (B \Rightarrow C)$ is equivalent to $(A \wedge B) \Rightarrow C$.

Theorem A2 (*Transitivity*) $(x = y \wedge y = z) \Rightarrow x = z$.

Proof By Assumption II (interchanging x and y),

$$(*) \quad y = x \Rightarrow (y = z \Rightarrow x = z).$$

(Here, $\mathscr{A}(y, y)$ is $y = z$ and $\mathscr{A}(y, x)$ is $x = z$.) Assume $x = y \wedge y = z$. By Theorem A1, $y = x$. Hence, $(*)$ yields $y = z \Rightarrow x = z$. Since $y = z$, we obtain $x = z$. ∎

Let us give an example of an application of the substitutivity assumption in a particular case. Let us assume that we are dealing with a theory in which $+$ is a binary operation.

Theorem A3

a. $x = y \Rightarrow x + z = y + z$.
b. $x = y \Rightarrow z + x = z + y$.

Proof (a) In the substitutivity assumption (II), let $\mathscr{A}(x, x)$ be $x + z = x + z$, and let $\mathscr{A}(x, y)$ be $x + z = y + z$. We obtain

$$x = y \Rightarrow (x + z = x + z \Rightarrow x + z = y + z),$$

which is equivalent to

$$x + z = x + z \Rightarrow (x = y \Rightarrow x + z = y + z).$$

By Assumption I, $x + z = x + z$. Hence, $x = y \Rightarrow x + z = y + z$.

(b) This is proved in the same way as (a), taking $\mathscr{A}(x, x)$ to be $z + x = z + x$, and $\mathscr{A}(x, y)$ to be $z + x = z + y$. ∎

Theorem A3 is a typical application of the substitutivity assumption in mathematics. It generalizes to the case where $\tau(x, x)$ is any term and $\tau(x, y)$ is obtained from $\tau(x, x)$ by replacing some but not necessarily all occurrences of x by y. Then $x = y \Rightarrow \tau(x, x) = \tau(x, y)$ is provable.

Assumption II need not be assumed in its full generality. It suffices to consider only "atomic" formulas $\mathscr{A}(x, x)$, that is, formulas in which no logical connectives or quantifiers occur. For details, see, Mendelson [1964], pp. 75–79.

FINITE SUMS AND THE Σ NOTATION

Let f be a function from the set P of positive integers into a set A, in which there is defined an associative binary operation $+$. It is customary to abbreviate the finite sum

$$f(1) + \cdots + f(n) \qquad \text{by the expression} \qquad \sum_{i=1}^{n} f(i)$$

which is also written as $\sum_{i=1}^{n} f(i)$. The basic properties that $\sum_{i=1}^{n} f(i)$ must satisfy are

1. $\displaystyle\sum_{i=1}^{1} f(i) = f(1)$.

2. $\displaystyle\sum_{i=1}^{n+1} f(i) = \sum_{i=1}^{n} f(i) + f(n+1)$.

This can be justified on the basis of the Iteration Theorem of Chapter 2.

In the Iteration Theorem, let W be the set $P \times A$, let $c = (1, f(1))$, and let $g(n, a) = (n+1, a+f(n+1))$ for any $(n, a) \in P \times A$. Then there exists a unique function $\Phi \colon P \to P \times A$ such that

$$\Phi(1) = (1, f(1)),$$

and

$$\Phi(n+1) = g(\Phi(n)).$$

Define $\sum_{i=1}^{n} f(i)$ to be the second component of $\Phi(n)$. Thus, since $\Phi(1) = (1, f(1))$, we obtain $\sum_{i=1}^{1} f(i) = f(1)$. Since

$$\Phi(2) = g(\Phi(1)) = g(1, f(1)) = (2, f(1) + f(2)),$$

it follows that $\sum_{i=1}^{2} f(i) = f(1) + f(2)$. In general,

$$\sum_{i=1}^{n+1} f(i) = \sum_{i=1}^{n} f(i) + f(n+1).$$

To see this, first observe that each $\Phi(n)$ is of the form $(n, \sum_{i=1}^{n} f(i))$. This follows easily by induction with respect to n. Hence,

$$\Phi(n+1) = g(\Phi(n)) = g\left(n, \sum_{i=1}^{n} f(i)\right) = \left(n+1, \sum_{i=1}^{n} f(i) + f(n+1)\right).$$

Therefore,

$$\sum_{i=1}^{n+1} f(i) = \text{the second component of } \Phi(n+1)$$

$$= \sum_{i=1}^{n} f(i) + f(n+1).$$

There are other variations of the $\sum$ notation. For example, if the function f is defined for nonnegative integers, $\sum_{i=0}^{n} f(i)$ can be defined as $f(0) + \sum_{i=1}^{n} f(i)$ if $n > 0$, and as $f(0)$ if $n = 0$.

If the function f is defined for positive integers, and j and k are positive integers with $1 < j \leq k$, then $\sum_{i=j}^{k} f(i)$ is intended to denote $f(j) + f(j+1) + \cdots + f(k)$. It can be defined as

$$\sum_{i=1}^{k} f(i) - \sum_{i=1}^{j-1} f(i).$$

If a multiplication operation is also defined in A, then finite products can be treated in a manner similar to the way finite sums are handled. The standard notation for a finite product is

$$\prod_{i=1}^{n} f(i).$$

EXERCISES

1. Prove: $\sum_{i=1}^{n} cf(i) = c\sum_{i=1}^{n} f(i)$ if there is a multiplication operation in A satisfying the distributive law with respect to addition.

2. If the operation $+$ is commutative in A, prove

$$\sum_{i=1}^{n} (f(i) + g(i)) = \sum_{i=1}^{n} f(i) + \sum_{i=1}^{n} g(i).$$

3. If $1 \leq k < n$, prove

   a. $\displaystyle\sum_{i=1}^{n} f(i) = \sum_{i=1}^{k} f(i) + \sum_{i=k+1}^{n} f(i).$

   b. $\displaystyle\sum_{i=1}^{n} f(i) = \sum_{i=1}^{k} f(i) + \sum_{i=1}^{n-k} f(k + i).$

4. Using mathematical induction, prove

   a. $\displaystyle\sum_{i=1}^{n} i = \frac{n(n + 1)}{2}.$

   b. $\displaystyle\sum_{i=1}^{n} (2i - 1) = n^{2}.$

   c. $\displaystyle\sum_{i=1}^{n} i^{2} = \frac{n(n + 1)(2n + 1)}{6}.$

5. If f and g are functions defined for non negative integers and with values in A, propose a suitable definition for

   $$\sum_{i+j=n} f(i)g(j) = f(0)g(n) + f(1)g(n - 1) + \cdots + f(n)g(0).$$

   Do the same for

   $$\sum_{i+j+k=n} f(i)g(j)h(k),$$

   if h is another function defined for nonnegative integers and with values in A.

POLYNOMIALS

Let $\mathscr{D} = (D, +, \times)$ be an integral domain. By a polynomial over $\mathscr{D}$ one ordinarily means an "expression" of the form

$$a_n x^n + a_{n-1} x^{n-1} + \cdots + a_1 x + a_0,$$

where each $a_i \in D$. If one looks carefully at the definition, it becomes apparent that it is unsuitable. For one thing, the expressions $(1 + 2)x^2 + 7$ and $3x^2 + 7$ are different, but they stand for the same polynomial. Furthermore, the different expressions $4t^2 + 1$ and $4x^2 + 1$ represent the same polynomial. Thus, we must reformulate the definition of a polynomial in a rigorous manner.

Let N denote the set of nonnegative integers. By a **polynomial over** $\mathscr{D}$ we mean any function $f: N \to D$ such that there exists some n_0 in N for which $(\forall n)(n \geq n_0 \Rightarrow f(n) = 0)$, that is, the values of f are 0 from some point on. For example, the polynomial $x^3 + 3x - 2$ is the function f such that $f(0) = -2$, $f(1) = 3$, $f(2) = 0$, $f(3) = 1$, and $f(n) = 0$ for all $n \geq 4$. The polynomial $7x^4 - 4x^3 + x$ is the function g such that $g(0) = 0$, $g(1) = 1$, $g(2) = 0$, $g(3) = -4$, $g(4) = 7$, and $g(n) = 0$ for all $n > 4$.

Let $\mathscr{D}[x]$ denote the set of all polynomials over $\mathscr{D}$. By the **zero polynomial** $\square$ we mean the constant function on N which is always equal to 0. If f is a polynomial different from the zero polynomial, then the largest integer n such that $f(n) \neq 0$ is called the **degree** of the polynomial and is denoted by $\deg(f)$; $f(n)$ is called the **leading coefficient** of the polynomial. The usual way of writing polynomials can be regained by the convention of expressing a nonzero polynomial f in the following manner: $a_0 + a_1 x + \cdots + a_n x^n$, where $a_i = f(i)$ and $n = \deg(f)$. (Sometimes the polynomial is written in the reverse order: $a_n x^n + \cdots + a_1 x + a_0$.) The number a_i is called the **ith coefficient of f** or **the coefficient of x^i.**

Definition   Let $f, g \in \mathscr{D}[x]$. By $f + g$ we mean the polynomial h such that $h(n) = f(n) + g(n)$ for every n in N.[†]

By $-g$ we mean the polynomial φ such that $\varphi(n) = -(g(n))$ for every n in N.

Finally, let $f - g = f + (-g)$.

Lemma C1   Let $f, g, h \in \mathscr{D}[x]$. Then

a. $f + g = g + f$.

b. $f + (g + h) = (f + g) + h$.

c. $f + \square = f$.

d. $f + (-f) = \square$.

Proof   (a) $(f + g)(n) = f(n) + g(n) = g(n) + f(n) = (g + f)(n)$ for all n in N. Hence, $f + g = g + f$.

(b) $(f + (g + h))(n) = f(n) + (g + h)(n) = f(n) + (g(n) + h(n))$. $((f + g) + h)(n) = (f + g)(n) + h(n) = (f(n) + g(n)) + h(n)$, for all n in N. Hence, $f + (g + h) = (f + g) + h$.

(c) $(f + \square)(n) = f(n) + \square(n) = f(n) + 0 = f(n)$, for all n in N. Hence, $f + \square = f$.

(d) $(f + (-f))(n) = f(n) + (-f)(n) = f(n) + (-(f(n))) = 0 = \square(n)$ for all n in N. Hence, $f + (-f) = \square$.   ∎

In order to define multiplication of polynomials, we examine the usual intuitive picture.

$$(a_k x^k + \cdots + a_1 x + a_0)(b_m x^m + \cdots + b_1 x + b_0)$$

is evaluated by "multiplying out." As a result, each power x^n in the product, has as its coefficient $a_0 b_n + a_1 b_{n-1} + \cdots + a_{n-1} b_1 + a_n b_0$. For example, the coefficient of x^2 is $a_0 b_2 + a_1 b_1 + a_2 b_0$. This leads to the appropriate rigorous definition.

Definition   *Multiplication*   Let $f, g \in \mathscr{D}[x]$. Then fg is the polynomial h such that

$$h(n) = \sum_{i=0}^{n} f(i)g(n - i).[‡]$$

[†] We use the same symbol for addition of polynomials as for addition in $\mathscr{D}$ in order to avoid notational complexity. The meaning of $+$ will always be clear from the context.

[‡] If ψ is any function with domain N, the notation $\sum_{i=0}^{n} \psi(i)$ represents the sum $\psi(0) + \psi(1) + \cdots + \psi(n)$. See Appendix B.

Observe that, if we let $a_i = f(i)$, $b_i = g(i)$, $c_i = h(i)$, then

$$c_n = \sum_{i=0}^{n} a_i b_{n-i} = a_0 b_n + a_1 b_{n-1} + \cdots + a_{n-1} b_1 + a_n b_0.$$

In particular,

$$c_0 = a_0 b_0,$$
$$c_1 = a_0 b_1 + a_1 b_0,$$
$$c_2 = a_0 b_2 + a_1 b_1 + a_2 b_0,$$
$$c_3 = a_0 b_3 + a_1 b_2 + a_2 b_1 + a_3 b_0, \quad \text{etc.}$$

Lemma C2   Let $f, g, h \in \mathcal{D}[x]$. Then

a. $fg = gf$.

b. $f(gh) = (fg)h$.

c. $f1 = f$ (where 1 is the "constant" polynomial such that $1(0) = 1$ and $1(n) = 0$ for all $n > 0$).

d. $f(g + h) = (fg) + (fh)$.

Proof   (a) $(fg)(n) = \sum_{i=0}^{n} f(i)g(n - i) = \sum_{i=0}^{n} f(n - i)g(i) = (gf)(n).$

(b) $(f(gh))(n) = \sum_{i=0}^{n} f(i)((gh)(n - i))$

$$= \sum_{i=0}^{n} \left(f(i) \left(\sum_{j=0}^{n-i} g(j)h(n - i - j) \right) \right)$$

$$= \sum_{i=0}^{n} \left(\sum_{j=0}^{n-i} f(i)g(j)h(n - i - j) \right)$$

$$= \sum_{i+j+k=n} f(i)g(j)h(k).$$

$$((fg)h)(n) = \sum_{i=0}^{n} (fg)(i)h(n - i) = \sum_{i=0}^{n} \left(\sum_{j=0}^{i} f(j)g(i - j) \right) h(n - i)$$

$$= \sum_{i=0}^{n} \left(\sum_{j=0}^{i} f(j)g(i - j)h(n - i) \right)$$

$$= \sum_{j+l+k=n} f(j)g(l)h(k) = \sum_{i+j+k=n} f(i)g(j)h(k).$$

(c) $(f1)(n) = \sum_{i=0}^{n} f(i)1(n - i) = f(n) \cdot 1 = f(n).$

(Notice that $1(n - i) = 0$ except when $n - i = 0$, that is, except when $i = n$.)

(d) $(f(g + h))(n) = \sum_{i=0}^{n} f(i)((g + h)(n - i))$

$= \sum_{i=0}^{n} f(i)(g(n - i) + h(n - i))$

$= \sum_{i=0}^{n} (f(i)g(n - i) + f(i)h(n - i))$

$= \sum_{i=0}^{n} f(i)g(n - i) + \sum_{i=0}^{n} f(i)h(n - i)$

$= (fg)(n) + (fh)(n) = ((fg) + (fh))(n).$ ∎

Lemma C3   Let f, g be nonzero polynomials, that is, $f \neq \square$, $g \neq \square$. Then $fg \neq \square$ and $\deg(fg) = \deg(f) + \deg(g)$.

Proof   Let $\deg(f) = n_0$ and $\deg(g) = k_0$. Then

$$(fg)(n_0 + k_0) = \sum_{i=0}^{n_0+k_0} f(i)g(n_0 + k_0 - i).$$

If $i > n_0$, then $f(i) = 0$. If $i < n_0$, then $n_0 + k_0 - i > k_0$ and $g(n_0 + k_0 - i) = 0$. Hence, $(fg)(n_0 + k_0) = f(n_0)g(k_0) \neq 0$. Therefore, $fg \neq \square$. To show that $\deg(fg) = n_0 + k_0$, we must still prove that $(fg)(n) = 0$ for all $n > n_0 + k_0$. So, assume $n > n_0 + k_0$. Then $(fg)(n) = \sum_{i=0}^{n} f(i)g(n - i)$. When $i > n_0$, $f(i) = 0$. When $i \leq n_0$, then $n - i \geq n - n_0 > (n_0 + k_0) - n_0 = k_0$, and, therefore, $g(n - i) = 0$. Therefore, $(fg)(n) = 0$ for all $n > n_0 + k_0$.   ∎

Theorem C4 $(\mathcal{D}[x], +, \times)$ is an integral domain, where $+$, $\times$ represent addition and multiplication of polynomials.

Proof   The axioms for an integral domain have been verified in Lemmas C1–C3.   ∎

Notice that $(\mathcal{D}[x], +, \times)$ contains an isomorphic copy of $\mathcal{D}$. Define the function $\Phi: D \to \mathcal{D}[x]$ by specifying that, for any d in D, $\Phi(d)$ is the "constant" polynomial which assigns to 0 the value d and to any positive integer the value 0. It is easy to check that Φ is a one-one function from D

onto the set consisting of $\square$ and all polynomials of degree zero, and that

$$\Phi(d_1 + d_2) = \Phi(d_1) + \Phi(d_2);$$
$$\Phi(d_1 d_2) = \Phi(d_1)\Phi(d_2).$$

We shall follow the customary procedure of not distinguishing between d and $\Phi(d)$, that is, the "constant" polynomials are identified with the corresponding elements of D.

If the coefficients of the polynomials come from not just an integral domain but from an ordered integral domain $\mathscr{D} = (D, +, \times, <)$, then the integral domain of polynomials also can be ordered in the following way.

Definition   Let $f \in \mathscr{D}[x]$. We say that f is **positive** if and only if the leading coefficient of f is a positive element of D. (The zero polynomial $\square$ is assumed not to be positive.)

For any f, g in $\mathscr{D}[x]$, we define $f < g$ to mean that $g - f$ is positive. Notice that this means that $f \neq g$ and, if n is the largest integer for which $f(n) \neq g(n)$, then $f(n) < g(n)$.

Let $\mathscr{P}^*$ denote the set of positive polynomials. Then

i. $\square \notin \mathscr{P}^*$;
ii. for any f in $\mathscr{D}[x]$, $f \in \mathscr{P}^*$ or $f = \square$ or $-f \in \mathscr{P}^*$;
iii. for any $f, g \in \mathscr{P}^*$, $f + g \in \mathscr{P}^*$.

(For, if $n = \deg(f)$ and $k = \deg(g)$, and $m = \max(n, k)$, then $f(m) + g(m) > 0$ and $\deg(f + g) = m$. Hence, $f + g \in \mathscr{P}^*$.)

iv. for any $f, g \in \mathscr{P}^*$, $fg \in \mathscr{P}^*$.

(For, if $n = \deg(f)$ and $k = \deg(g)$, then $\deg(fg) = n + k$, and $(fg)(n + k) = f(n)g(k) > 0$.)

Hence, by Theorem 3.4.3, $(\mathscr{D}[x], +, \times, <)$ is an ordered integral domain (where $+$ and $\times$ are addition and multiplication of polynomials, and $<$ is the relation just defined.)

EXERCISE

1. Let $\mathscr{P}^\#$ be the set of nonzero polynomials f such that, if n is the **least** integer such that $f(n) \neq 0$, then $f(n) > 0$. If we define $f <^\# g$ to mean that $g - f \in \mathscr{P}^\#$, does this transform the domain of polynomials into an ordered integral domain $(\mathscr{D}[x], +, \times, <^\#)$?

Definition A **monic** polynomial is a polynomial whose leading coefficient is 1.

Theorem C5 Let $\mathscr{D} = (D, +, \times)$ be an integral domain. Let f be a fixed monic polynomial in $\mathscr{D}[x]$ of degree ≥ 1. Then, for any g in $\mathscr{D}[x]$, there exist unique polynomials q, r in $\mathscr{D}[x]$ such that:

$$g = fq + r \quad \text{and} \quad (r = \square \text{ or } \deg(r) < \deg(f)).$$

Proof Let k be the degree of f.

(I) Existence of q and r.

If $g = \square$, let $q = r = \square$. Assume now that $g \neq \square$, and let n be the degree of g. The proof will proceed by mathematical induction. If $n = 0$, let $q = \square$ and $r = g$. Now assume the result true for all polynomials of degree $\leq n$, and let g be a polynomial of degree $n + 1$. Let a_{n+1} be the leading coefficient of g.

Case 1 $n + 1 < k$. Then, let $q = \square$ and $r = g$.

Case 2 $n + 1 \geq k$. Let $h = g - a_{n+1}x^{n+1-k}f$.[†] Then h has degree $\leq n$. Hence, there exist q_1, r_1 such that

$$h = q_1 f + r_1 \quad \text{and} \quad (r_1 = \square \text{ or } \deg(r_1) < \deg(f)).$$

Hence,

$$g = h + a_{n+1}x^{n+1-k}f = q_1 f + r_1 + a_{n+1}x^{n+1-k}f$$
$$= (q_1 + a_{n+1}x^{n+1-k})f + r_1.$$

(II) Uniqueness of q and r.

Assume

$$g = q_1 f + r_1 \wedge (r_1 = \square \quad \text{or} \quad \deg(r_1) < \deg(f)),$$

and

$$g = q_2 f + r_2 \wedge (r_2 = \square \quad \text{or} \quad \deg(r_2) < \deg(f)).$$

Hence, $(q_1 - q_2)f = r_2 - r_1$. Assume, for the sake of contradiction, that $r_1 \neq r_2$. Then $r_2 - r_1 \neq \square$, and, therefore, $q_1 - q_2 \neq \square$. Hence, $\deg((q_1 - q_2)f) \geq \deg(f)$. But $r_2 - r_1$ has degree no greater than $\max(\deg(r_1), \deg(r_2)) < \deg(f)$. This contradiction shows that $r_1 = r_2$. Hence, $(q_1 - q_2)f = \square$, and, therefore, $q_1 = q_2$. ∎

† We use here the traditional notation for polynomials.

Corollary C6 If $\mathscr{D}$ is an integral domain $(D, +, \times)$ and $c \in D$, then, for any polynomial g in $\mathscr{D}[x]$, there exists a polynomial q in $\mathscr{D}[x]$ and an element r in D such that

$$g = q(x - c) + r.$$

Proof This follows from Theorem C5 by letting $f = x - c$. Since $r = \square$ or $\deg(r) < \deg(x - c) = 1$, it follows that r must be a constant polynomial, that is, an element of D. ■

The polynomials, as we have defined them, are often called **polynomial forms**. However, polynomials are usually treated in another way, as functions. In fact, if a polynomial g is $a_n x^n + \cdots + a_1 x + a_0$, then the corresponding function assigns to each element u of D the value obtained by "substituting" u for x in $a_n x^n + \cdots + a_1 x + a_0$ and then evaluating. We shall make this precise in the following way.

Let u be an element of the domain D. We shall define a corresponding function $E_u : \mathscr{D}[x] \to D$. Intuitively, if g is a polynomial which is written in traditional form as $a_n x^n + \cdots + a_1 x + a_0$, then $E_u(g) = a_n u^n + \cdots + a_1 u + a_0$, that is, the result of "substituting" u for x and "evaluating." First, let $E_u(\square) = 0$. Second, assume that we have defined $E_u(h)$ for all polynomials h of degree $< n$, and assume that a given polynomial g has degree n. Then, if $a_n = g(n)$ and $h = g - a_n x^n$, then h is either $\square$ or a polynomial of degree $< n$. Then, let $E_u(g) = E_u(h) + a_n u^n$. In particular, if g is a constant polynomial c, then $E_u(g) = c$; if g is $a_1 x + a_0$, then $E_u(g) = a_1 u + a_0$; etc. Now, by the **polynomial function** associated with a polynomial g we mean the function $\hat{g} : D \to D$ defined by

$$\hat{g}(u) = E_u(g).$$

EXERCISE

2. Prove that $E_u(f + g) = E_u(f) + E_u(g)$ and $E_u(fg) = E_u(f)E_u(g)$, that is, $\widehat{f + g}(u) = \hat{f}(u) + \hat{g}(u)$ and $\widehat{fg}(u) = \hat{f}(u)\hat{g}(u)$.

By a **root** of a polynomial f we mean an element u of D such that $\hat{f}(u) = 0$.

Theorem C7 If $g \in \mathscr{D}[x]$ and $c \in D$, then c is a root of g if and only if g is divisible by $x - c$, that is, $g = q(x - c)$ for some polynomial q in $\mathscr{D}[x]$.

Proof By Corollary C6, there is a polynomial q and a constant r such that $g = q(x - c) + r$. If c is a root of g, then

$$0 = \hat{g}(c) = \hat{q}(c) \cdot \widehat{(x - c)}(c) + \hat{r}(c) = \hat{q}(c) \cdot (c - c) + r$$
$$= \hat{q}(c) \cdot 0 + r = r.$$

Hence, $g = q(x - c)$. Conversely, if g is divisible by $x - c$, then, by the uniqueness part of Theorem C5, $g = q(x - c)$. Hence,

$$\hat{g}(c) = \hat{q}(c) \cdot \widehat{(x - c)}(c) = \hat{q}(c) \cdot (c - c) = \hat{q}(c) \cdot 0 = 0.$$

Thus, c is a root of g.   ∎

Corollary C8 If $g \in \mathscr{D}[x]$ and $\deg(g) = n$, then g has at most n distinct roots.

Proof By mathematical induction. If $\deg(g) = 0$, then g has no roots, since it is a nonzero constant. Assume that a polynomial of degree n has at most n distinct roots, and let g be a polynomial of degree $n + 1$. We must prove that g has at most $n + 1$ distinct roots. If g has no roots, we are finished. If g has at least one root, let c be such a root. By Theorem C7, $g = q(x - c)$ for some polynomial q in $\mathscr{D}[x]$. Since $\deg(g) = \deg(q) + \deg(x - c) = \deg(q) + 1$, it follows that $\deg(q) = n$. Hence, q has at most n distinct roots. Since $g = q(x - c)$, any root of g different from c must be a root of q. (For, if $\hat{g}(d) = 0$, and $d \neq c$, then

$$0 = \hat{g}(d) = \hat{q}(d) \cdot \widehat{(x - c)}(d) = \hat{q}(d) \cdot (d - c),$$

and, therefore, $\hat{q}(d) = 0$.) Therefore, g has at most n roots different from c. Thus, g has at most $n + 1$ roots.   ∎

The question suggests itself as to whether different polynomials can determine the same function.

Theorem C9 If $\mathscr{D} = (D, +, \times)$ is an **infinite** integral domain, then different polynomials in $\mathscr{D}[x]$ determine different functions.

Proof Assume $f, g \in \mathscr{D}[x], f \neq g$, and f and g determine the same function, that is $\hat{f} = \hat{g}$. Hence, for all u in D, $(\widehat{f - g})(u) = \hat{f}(u) - \hat{g}(u) = 0$ for all u in D. Thus, $f - g$ has infinitely many roots. Now, $f - g \neq \Box$, and therefore, $f - g$ has a degree. But this contradicts Corollary C8.   ∎

The situation is different when we consider finite integral domains (that is, finite fields). For example, if $\mathscr{D} = \mathscr{X}_2 = (Z_2, +_2, \times_2)$, the domain of integers modulo 2, the different polynomials x^2 and x determine the same function: $0^2 = 0$ and $1^2 = 1$.

For further information about polynomials, the reader is referred to standard books on modern algebra (for example, van der Waerden [1949]) or the theory of equations (for example, Borofsky [1950]).

FINITE, INFINITE, AND DENUMERABLE SETS. CARDINAL NUMBERS

We say that sets A and B are **equinumerous** if and only if there is a one-one correspondence between A and B. We shall write $A \approx B$ to mean that A and B are equinumerous. (Basic information about one-one correspondences may be found in Chapter 1, Section 20.)

Theorem D1

a. $A \approx A$.

b. $A \approx B \Rightarrow B \approx A$.

c. $(A \approx B \land B \approx C) \Rightarrow A \approx C$.

Proof (a) The identity function I_A on A, defined by $I_A(x) = x$ for all x in A, is a one-one correspondence between A and itself.

(b) If $F: A \xrightarrow[\text{onto}]{1-1} B$, then $F^{-1}: B \xrightarrow[\text{onto}]{1-1} A$.

(c) If $F: A \xrightarrow[\text{onto}]{1-1} B$ and $G: B \xrightarrow[\text{onto}]{1-1} C$, then $G \circ F: A \xrightarrow[\text{onto}]{1-1} C$. ∎

EXERCISES

1. If $A \approx \varnothing$, prove that $A = \varnothing$.

2. Prove: $A \approx A \times \{y\}$. (*Hint*: Let $\theta(a) = (a, y)$ for all a in A.)

Two sets which are equinumerous are said to have the same **cardinal number**.[†] We assume that each set A has an associated cardinal number $\mathscr{K}(A)$. Hence, $\mathscr{K}(A) = \mathscr{K}(B) \Leftrightarrow A \approx B$.

[†] Synonyms for **cardinal number** are **power** and **cardinality**.

Definitions $A \leq B$ means $(\exists F)(F: A \xrightarrow{1-1} B)$.

$\qquad\qquad\quad$ $A < B$ means $A \leq B \land A \not\approx B$.[†]

Intuitively, $A \leq B$ means that B has at least as many elements as A.

Theorem D2

a. $A \leq A$.

b. $A \leq B \land B \leq C \Rightarrow A \leq C$.

c. $A \not< A$.

d. $A \subseteq B \Rightarrow A \leq B$.

e. $A \leq B \Leftrightarrow (A < B \lor A \approx B)$.

f. $A \leq B \Leftrightarrow (\exists W)(W \subseteq B \land A \approx W)$.

g. $(A \leq B \land B \approx C) \Rightarrow A \leq C$.

h. $(A \leq B \land A \approx C) \Rightarrow C \leq B$.

Proof   (a) $I_A : A \xrightarrow{1-1} A$.

$\quad$ (b) If $F: A \xrightarrow{1-1} B$ and $G: B \xrightarrow{1-1} C$, then $G \circ F: A \xrightarrow{1-1} C$.

$\quad$ (c) $A \approx A$. Hence, $A \not< A$.

$\quad$ (d) If $A \subseteq B$, then $I_A : A \xrightarrow{1-1} B$.

$\quad$ (e) If $A \leq B$, and $A \not< B$, then, by definition, $A \approx B$. Conversely, if $A < B$ or $A \approx B$, then obviously $A \leq B$.

$\quad$ (f) $A \leq B \Leftrightarrow (\exists F)(F: A \xrightarrow{1-1} B)$. If $A \leq B$, take $W = \mathscr{R}(F)$. Then $W \subseteq B \land A \approx W$. Conversely, if $(\exists W)(W \subseteq B \land A \approx W)$, then there is a function $F: A \xrightarrow{1-1} W$, and, therefore, $F: A \xrightarrow{1-1} B$. Hence, $A \leq B$.

$\quad$ (g)–(h) Easy exercises for the reader. ∎

We can define an order relation $<_K$ on cardinal numbers. Let $\mathfrak{m}, \mathfrak{n}$ be cardinal numbers. Take any sets A, B having cardinal numbers $\mathfrak{m}, \mathfrak{n}$, respectively. Then, by definition, $\mathfrak{m} \leq_K \mathfrak{n}$ if and only if $A \leq B$, and $\mathfrak{m} <_K \mathfrak{n}$ if and only if $A < B$. By Theorem D2g,h, it does not make any difference in these definitions which sets A and B we choose having cardinal numbers $\mathfrak{m}$ and $\mathfrak{n}$, respectively.

One can obtain theorems about cardinal numbers analogous to those proved above. For example, the analogue of Theorem D2b is $(\mathfrak{m} \leq_K \mathfrak{n} \land \mathfrak{n} \leq_K \mathfrak{p}) \Rightarrow \mathfrak{m} \leq_K \mathfrak{p}$, for any cardinal numbers $\mathfrak{m}$, $\mathfrak{n}$, and $\mathfrak{p}$.

[†] $A \not\approx B$ means: not$(A \approx B)$.

Theorem D3 *(Cantor's Theorem)* $A < \mathscr{P}(A)$.

Proof (i) $A \leq \mathscr{P}(A)$. For, if we let $F(u) = \{u\}$, for all u in A, then $F: A \xrightarrow{\text{1-1}} \mathscr{P}(A)$.

(ii) We must show that $A \not\approx \mathscr{P}(A)$. Assume that $A \approx \mathscr{P}(A)$. Let $G: A \xrightarrow[\text{onto}]{\text{1-1}} \mathscr{P}(A)$. Let $B = \{y : y \in A \wedge y \notin G(y)\}$. Then $B \subseteq A$ and $(\forall y)(y \in A \Rightarrow (y \in B \Leftrightarrow y \notin G(y)))$. Since $B \in \mathscr{P}(A)$, $B = G(y)$ for some y in A. Then, $y \in B \Leftrightarrow y \notin G(y)$, by definition of B. But, $y \notin G(y) \Leftrightarrow y \notin B$, since $B = G(y)$. Hence, $y \in B \Leftrightarrow y \notin B$, which yields a contradiction.   ∎

The next result expresses the simple fact that, if B has at least as many elements as A and A has at least as many elements as B, then A and B are equinumerous. Surprisingly, the proof is not obvious.

Theorem D4 *(Schröder–Bernstein Theorem)*

$$(A \leq B \wedge B \leq A) \Rightarrow A \approx B.$$

Proof We may assume $A \cap B = \varnothing$. (If not, we could use $A_1 = A \times \{0\}$ and $B_1 = B \times \{1\}$ instead of A and B, respectively.) Let us assume that $F: A \xrightarrow{\text{1-1}} B$ and $G: B \xrightarrow{\text{1 1}} A$. By a **B-thread** (see Figure D.1) we mean a function f from the set P of positive integers into $A \cup B$ such that

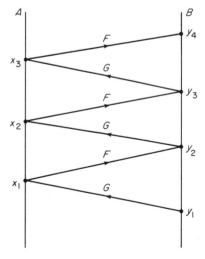

Figure D.1

$y_1 = f(1) \in B - \mathscr{R}(F), f(n) \in B \Rightarrow f(n+1) = G(f(n)); f(n) \in A \Rightarrow f(n+1)$
$= F(f(n))$. Thus, a B-thread is an infinite sequence of points starting with a point y_1 which is not in the range of F, and then using the functions G and F alternately. Let W be the set of elements of A which belong to the range of some B-thread. Then the desired one-one correspondence H between A and B is given by $H = (F \restriction (A - W)) \cup (G^{-1} \restriction W)$, that is, for points x in W, $H(x) = G^{-1}(x)$, and, for points not in W, $H(x) = F(x)$. It is easy to see that $H: A \to B$. Let us show that $\mathscr{R}(H) = B$. Assume $b \in B$.

Case 1 $G(b) \in W$. Then $H(G(b)) = G^{-1}(G(b)) = b$.

Case 2 $G(b) \in A - W$. Then there is an x in A such that $F(x) = b$. (If there were no such x, there would be a B-thread beginning at b and containing $G(b)$, contradicting the assumption that $G(b) \notin W$.) Now, $x \in A - W$. (For, if there were a B-thread containing x, the same B-thread would contain $G(b) = G(F(x))$, contradicting the assumption that $G(b) \notin W$.) Hence, $H(x) = F(x) = b$. Thus, $\mathscr{R}(H) = B$. Finally, H is one-one. For, assume $H(x) = H(u) \wedge x \neq u$. It is clear that x and u cannot both be in $A - W$, since F is one-one, and equally clear that x and u cannot both be in W, since G^{-1} is one-one. Hence, we may assume that $x \in A - W$ and $u \in W$. Then $F(x) = H(x) = H(u) = G^{-1}(u)$. Therefore, $G(F(x)) = u$. Hence, $G(F(x)) \in W$. But, since $G(F(x))$ is on a B-thread, x must also be on the same B-thread, contradicting the assumption that $x \notin W$.[†] ∎

Corollary D5

a. $A \leq B \Rightarrow B \not< A$.

b. $(A \leq B \wedge B < C) \Rightarrow A < C$.

c. $(A < B \wedge B \leq C) \Rightarrow A < C$.

d. $(A < B \wedge B < C) \Rightarrow A < C$.[‡]

Proof (a) Assume $A \leq B$ and $B < A$. Then $B \leq A$. Hence, by Theorem D4, $A \approx B$, contradicting $B < A$.

(b) Assume $A \leq B \wedge B < C$. Hence $B \leq C$, and, therefore, $A \leq C$.

[†] For another proof of Theorem C4 and further discussion, see Fraenkel [1961]. Notice that the cardinal number formulation of Theorem C4 reads: $(\mathfrak{m} \leq_K \mathfrak{n} \wedge \mathfrak{n} \leq_K \mathfrak{m}) \Rightarrow \mathfrak{m} = \mathfrak{n}$.

[‡] All of these results and others formulated in terms of $\approx$, $\leq$, $<$ can be reformulated in terms of cardinal numbers. For example, Corollary D5a becomes: $\mathfrak{m} \leq_K \mathfrak{n} \Rightarrow \mathfrak{n} \not<_K \mathfrak{m}$ for any cardinal numbers $\mathfrak{m}$ and $\mathfrak{n}$.

Assume $A \approx C$. Then $C \leq A$. Therefore, $C \leq B$. Hence, by Theorem D4, $B \approx C$, contradicting $B < C$.

(c), (d)   are left as exercises for the reader.   ∎

EXERCISE

3. Prove that there is no universal set U of all sets. (*Hint*: $\mathscr{P}(U) \leq U$, since $\mathscr{P}(U) \subseteq U$. Use Corollary D5a and Cantor's Theorem D3.)

Another apparently simple fact about sets is that, for any sets A, B, either $A \leq B$ or $B \leq A$, that is, either B has at least as many elements as A or A has at least as many elements as B.[†] This result, known as the Trichotomy Law, turns out to be equivalent to the celebrated proposition of set theory known as the Axiom of Choice. This proposition, which was once the center of a great deal of controversy as to its validity, is so useful in modern abstract mathematics that it has now achieved fairly wide acceptance. Nevertheless, since it is not essential for our purposes here, the reader is referred to any standard treatment of set theory (for example, Sierpinski [1958] or Mendelson [1964], Chapter IV) for a discussion of the Axiom of Choice and a proof of its equivalence with the Trichotomy Law.[‡]

Let us turn now to the study of finite sets.

Notation   If n is a natural number, let

$$P_n = \{k : k \text{ is a natural number and } k \leq n\}.$$

For example, $P_1 = \{1\}$, $P_2 = \{1, 2\}$, $P_3 = \{1, 2, 3\}$, etc.

Definition   We say that A is **finite** if and only if

$$A = \varnothing \quad \text{or} \quad (\exists n)(A \approx P_n).$$

Thus, the finite sets consist of the empty set plus all those sets which can be enumerated by the positive integers from 1 up to some positive integer n.[§]

[†] In terms of cardinal numbers, this reads: For any cardinal numbers $\mathfrak{m}$, $\mathfrak{n}$, either $\mathfrak{m} \leq_K \mathfrak{n}$ or $\mathfrak{n} \leq_K \mathfrak{m}$.

[‡] The Axiom of Choice reads: If A is any set of nonempty sets, then there is a function F with domain A such that $F(B) \in B$ for every set B in A.

[§] The notion of *finite set* should be distinguished from that of *finite sequence*. By a **finite sequence** we mean a function with domain P_n, for some n. The number n is called the **length** of the sequence, and the values of the function are called the **terms** of the sequence.

Definition   We say that A is **infinite** if and only if A is not finite.

It is clear that any set equinumerous with a finite set is finite, and, therefore, that any set equinumerous with an infinite set is infinite. Obviously, $\varnothing$ and all the sets P_n are finite.

Lemma D6 $(\forall A)(A \subsetneqq P_n \Rightarrow A \not\approx P_n)$, that is, $\{1, 2, \ldots, n\}$ is not equinumerous with any of its proper subsets.

Proof   We shall prove this by mathematical induction with respect to n. When $n = 1$, if $A \subsetneqq P_1 = \{1\}$, then $A = \varnothing$, and, therefore, $A \not\approx P_1$. Now assume that the result is true for n: $(\forall A)(A \subsetneqq P_n \Rightarrow A \not\approx P_n)$ (inductive hypothesis). Assume $A \subsetneqq P_{n+1}$ and $A \approx P_{n+1}$. Let φ be a one-one correspondence between P_{n+1} and A. Then, $\varphi: P_{n+1} \xrightarrow[\text{onto}]{1-1} A$. We shall derive a contradiction from these assumptions.

Case 1 $n + 1 \neq \varphi(i)$ for any $i \leq n$. Then the restriction of φ to P_n is a one-one correspondence between P_n and $A - \{\varphi(n + 1)\}$. But, $A - \{\varphi(n + 1)\} \subsetneqq P_n$, and this contradicts the inductive hypothesis.

Case 2 $n + 1 = \varphi(i)$ for a certain $i \leq n$. Define ψ as follows:

$$\psi(j) = \begin{cases} \varphi(j) & \text{for } j \neq i \wedge j \leq n \\ \varphi(n + 1) & \text{for } j = i. \end{cases}$$

Then ψ is a one-one correspondence between P_n and $A - \{n + 1\}$. Since $n + 1 \in A$ and $A \subsetneqq P_{n+1}$, it follows that $A - \{n + 1\} \subsetneqq P_n$. But this contradicts the inductive hypothesis.   ∎

Theorem D7   If B is finite, then B is not equinumerous with any of its proper subsets.

Proof   If $B = \varnothing$, then B has no proper subsets. Assume $B \approx P_n$. Let $\varphi: B \xrightarrow[\text{onto}]{1-1} P_n$. Assume $C \subsetneqq B$. We wish to show that $C \not\approx B$. Now, $\varphi[C] = \{\varphi(c) : c \in C\} \subsetneqq P_n$. Hence, by Lemma D6, $\varphi[C]$ is not equinumerous with P_n. Therefore, $C \not\approx B$. (For, if $\theta: B \xrightarrow[\text{onto}]{1-1} C$, then $\varphi \circ \theta \circ \varphi^{-1}$ would be a one-one correspondence between P_n and $\varphi[C]$.)   ∎

The property mentioned in Theorem D7 will turn out to be characteristic for finite sets. All infinite sets will be equinumerous with proper subsets of themselves.

Corollary D8

a. $n \neq k \Rightarrow P_n \not\approx P_k$.

b. If B is finite, then either $B = \varnothing$ or there is a unique positive integer n such that $B \approx P_n$.

Proof (a) Assume $n \neq k$. Say, $n < k$. Then $P_n \subsetneqq P_k$, and, by Lemma D6, $P_n \not\approx P_k$.

(b) Assume B finite. If $B \neq \varnothing$, then there exists some n such that $B \approx P_n$. If there were another integer k such that $B \approx P_k$, then $P_n \approx P_k$, contradicting Part a.   ∎

Terminology If $B \approx P_n$, we say that **B has n elements,** or that **n is the number of elements in B.**

Convention We assign the following cardinal numbers. $\mathscr{K}(\varnothing) = 0$, and $\mathscr{K}(P_n) = n$.

Thus, if B has n elements, then $\mathscr{K}(B) = n$.

Theorem D9

a. If B is finite, so is $B \cup \{y\}$.

b. If B is finite and $D \subseteq B$, then D is finite.

c. If B is finite and $D \leq B$, then D is finite.

Proof (a) If $B = \varnothing$, then $B \cup \{y\} = \{y\} \approx P_1$. If $B \approx P_n$, then, either $y \in B$ and $B \cup \{y\} = B \approx P_n$, or $y \notin B$ and $B \cup \{y\} \approx P_{n+1}$.

(b) We shall prove by mathematical induction that

$$(*)\quad (\forall D)(D \subseteq P_n \Rightarrow D \text{ is finite}).$$

If $n = 1$ and $D \subseteq P_1$, then $D = \varnothing$ or $D = \{1\}$, and, therefore, D is finite. Assume as the inductive hypothesis that $(*)$ holds for n. Assume $D \subseteq P_{n+1}$. Let $D^* = D - \{n+1\}$. Then $D^* \subseteq P_n$. Hence, D^* is finite. But either $D = D^*$ or $D = D^* \cup \{n+1\}$. Hence, by Part (a), D is finite.

Now, assume B finite and $D \subseteq B$. If $B = \varnothing$, then $D = \varnothing$ and D is finite. If $B \approx P_n$, then D is equinumerous with a subset of P_n, and, therefore, by $(*)$, D, being equinumerous with a finite set, must be finite.

(c) Assume B finite and $D \leq B$. Then D is equinumerous with a subset of B, and, therefore, by Part (b), D is finite.   ∎

EXERCISES

4. If D is infinite and $D \subseteq B$, prove that B is infinite.

5. If D is infinite and $D \leq B$, prove that B is infinite.

6. Assume B is finite, $D \subsetneqq B$, $\mathcal{K}(B) = n$, and $\mathcal{K}(D) = k$. Prove: $k < n$.

Theorem D10   The set P of natural numbers is equinumerous with a proper subset of itself.

Proof   Let $\varphi(n) = n + 1$ for all n in P. Then $\varphi: P \xrightarrow[\text{onto}]{1-1} P - \{1\}$.   ∎

Corollary D11   The set P of positive integers is infinite.

Proof   Use Theorems D7 and D10.   ∎

Definition   A set B is said to be **denumerable** if and only if $B \approx P$.

Clearly, any set equinumerous with a denumerable set is denumerable. By Theorem D10, any denumerable set is equinumerous with a proper subset of itself, and, therefore, every denumerable set is infinite.[†]

EXERCISES

7. Prove that any infinite set B contains a denumerable subset. (*Hint*:   This requires the Axiom of Choice, which provides a function ψ such that $\psi(D) \in D$ for all non-empty subsets D of B. Define, by the Iteration Theorem, a function $\theta: P \to B$ such that $\theta(1) = \psi(B)$ and $\theta(n + 1) = \psi(B - \theta(n))$. Then $\theta: P \xrightarrow{1-1} B$ and $\mathcal{R}(\theta)$ is a denumerable subset of B.)

8. Show that any infinite set B is equinumerous with a proper subset of itself. (*Hint*: By Part (a), let D be an infinite denumerable subset of B. Let φ be a one-one correspondence between D and a proper subset of D. Extend φ to a function φ^* defined on B by letting $\varphi^*(u) = u$ for any u in $B - D$. Then φ^* is a one-one correspondence between B and a proper subset of itself.)

Theorem D12   Assume that A and B are finite. Then

a. $A \cap B$ is finite.

b. $A \cup B$ is finite.

c. $A \times B$ is finite.

d. $\mathscr{P}(A)$ is finite.

[†] A denumerable set is to be distinguished from a **denumerable sequence**. The latter is simply a function with domain P. Sometimes denumerable sequences are loosely referred to as **infinite sequences**.

Proof (a) $A \cap B \subseteq A$, and any subset of a finite set is finite.

(b) If $B = \varnothing$, then $A \cup B = A$. Let $B \approx P_n$. We shall prove that $A \cup B$ is finite by mathematical induction with respect to n. If $n = 1$, then $B = \{y\}$ for some y. Then $A \cup B = A \cup \{y\}$, and we apply Theorem D9a. Now assume that $(\forall B)(B \approx P_n \Rightarrow A \cup B$ is finite$)$. Let $B \approx P_{n+1}$. We must show that $A \cup B$ is finite. Let $\theta \colon B \xrightarrow[\text{onto}]{1-1} P_{n+1}$. Let $B^* = B - \{\theta^{-1}(n + 1)\}$. Then $B^* \approx P_n$. Hence, $A \cup B^*$ is finite. But,

$$A \cup B = (A \cup B^*) \cup \{\theta^{-1}(n + 1)\},$$

and, therefore, by Theorem D9a, $A \cup B$ is finite.

(c) If $B = \varnothing$, then $A \times B = A \times \varnothing = \varnothing$. Let $B \approx P_n$. We shall prove by mathematical induction with respect to n that $A \times B$ is finite. When $n = 1$, $B = \{y\}$ for some y, and $A \times B = A \times \{y\} \approx A$. Hence, $A \times B$ is finite, since A is finite. Now assume that $(\forall B)(B \approx P_n \Rightarrow A \times B$ is finite$)$. Let $B \approx P_{n+1}$; say, $\theta \colon B \xrightarrow[\text{onto}]{1-1} P_{n+1}$. We must show that $A \times B$ is finite. Let $B^* = B - \{\theta^{-1}(n + 1)\}$. Then $B^* \approx P_n$, and therefore, by inductive hypothesis, $A \times B^*$ is finite. But, $A \times B = (A \times B^*) \cup (A \times \{(\theta^{-1}(n + 1)\})$. Now, since A is finite, $A \times \{\theta^{-1}(n + 1)\}$ is finite. Hence, by Part b, $A \times B$ is finite.

(d) If $A = \varnothing$, then $\mathscr{P}(A) = \{\varnothing\}$, which is finite. Now let $A \approx P_n$. We shall prove by mathematical induction that $\mathscr{P}(A)$ is finite. If $n = 1$, then $A = \{y\}$ for some y, and $\mathscr{P}(A) = \{\varnothing, \{y\}\}$, which is finite. Now assume that $(\forall A)(A \approx P_n \Rightarrow \mathscr{P}(A)$ is finite$)$. Assume $A \approx P_{n+1}$. We must prove that $\mathscr{P}(A)$ is finite. Let $\theta \colon A \xrightarrow[\text{onto}]{1-1} P_{n+1}$. Let $A^* = A - \{\theta^{-1}(n + 1)\}$. Hence, $A^* \approx P_n$. By inductive hypothesis, $\mathscr{P}(A^*)$ is finite. Let $B = \mathscr{P}(A) - \mathscr{P}(A^*)$. The sets in B are those subsets of A containing $\theta^{-1}(n + 1)$. Define the following function Ψ from $\mathscr{P}(A^*)$ into B. $\Psi(C) = C \cup \{\theta^{-1}(n + 1)\}$ for every C in $\mathscr{P}(A^*)$. It is easy to see that $\Psi \colon \mathscr{P}(A^*) \xrightarrow[\text{onto}]{1-1} B$. Thus, $\mathscr{P}(A^*)$ and B are equinumerous, and, since $\mathscr{P}(A^*)$ is finite, B must also be finite. But, $\mathscr{P}(A) = \mathscr{P}(A^*) \cup B$, and, therefore, by Part b, $\mathscr{P}(A)$ is finite. ∎

Definition For any sets A and B, let

$$A^B = \{f \colon (f \colon B \to A)\}.$$

Thus, A^B is the set of all functions from B into A.

Corollary D13 If A and B are finite, so is A^B.

Proof If $f \in A^B$, then $f \subseteq B \times A$, and, therefore, $f \in \mathscr{P}(B \times A)$. Thus, $A^B \subseteq \mathscr{P}(B \times A)$. By Theorem D12c, $B \times A$ is finite. Hence, by Theorem D12d, $\mathscr{P}(B \times A)$ is finite. Finally, by Theorem D9b, A^B is finite. ∎

EXERCISES

9. Assume that A is a set of sets. Prove: A is a finite set whose elements are finite sets if and only if $\bigcup_{B \in A} B$ is finite.

10. If $\mathscr{P}(A)$ is finite, prove that A is finite.

11. If A is finite and B is any set, prove the trichotomy property: $A \leq B$ or $B \leq A$.

12. If A is finite and B is infinite, prove that $A \leq B$.

13. a. Prove that a set A is finite if and only if every nonempty subset of $\mathscr{P}(A)$ has a minimal element (with respect to $\subseteq$).

    b. Prove that a set A is finite if and only if every nonempty subset of $\mathscr{P}(A)$ has a maximal element (with respect to $\subseteq$).

14. If $A \subseteq P$ and A is finite, prove that A has a greatest element.

15. Prove: (i) $A^\varnothing = \{\varnothing\}$. (ii) $A \approx A^{\{x\}}$. (*Hint*: For any a in A, let $\Phi(a) = \{(x, a)\}$. Then $\Phi \colon A \xrightarrow[\text{onto}]{1\text{-}1} A^{\{x\}}$.) (iii) $B \neq \varnothing \Rightarrow \varnothing^B = \varnothing$.

It is useful at this point to prove a few lemmas that will be important in what follows.

Lemma D14   Assume $A \approx B$ and $C \approx D$. Then

a. $A \times C \approx B \times D$.

b. $A^C \approx B^D$.

c. If $A \cap C = \varnothing$ and $B \cap D = \varnothing$, then $A \cup C \approx B \cup D$.

Proof Let $F \colon A \xrightarrow[\text{onto}]{1\text{-}1} B$ and $G \colon C \xrightarrow[\text{onto}]{1\text{-}1} D$.

(a) Define $H(a, c) = (F(a), G(c))$ for any $(a, c) \in A \times C$. Then $H \colon A \times C \xrightarrow[\text{onto}]{1\text{-}1} B \times D$.

(b) Define $\Psi(f) = F \circ f \circ G^{-1}$ for any f in A^C. Then $\Psi \colon A^C \xrightarrow[\text{onto}]{1\text{-}1} B^D$.

(c) Define

$$\Phi(u) = \begin{cases} F(u) & \text{if } u \in A \\ G(u) & \text{if } u \in C. \end{cases}$$

Then $\Phi \colon A \cup C \xrightarrow[\text{onto}]{1\text{-}1} B \cup D$. ∎

EXERCISES

16. Prove that $A \times B \approx B \times A$.

17. Prove that $A \times (B \times C) \approx (A \times B) \times C$.

Lemma D15

a. $A \cap B = \varnothing \Rightarrow C^{A \cup B} \approx C^A \times C^B$.

b. $(A^B)^C \approx A^{B \times C}$.

Proof (a) Assume $A \cap B = \varnothing$. Let $\Psi(f) = (f \restriction A, f \restriction B)$ for any f in $C^{A \cup B}$. (Remember that $f \restriction A$ denotes the restriction of f to the domain A.) Then, $\Psi\colon C^{A \cup B} \xrightarrow[\text{onto}]{1\text{-}1} C^A \times C^B$.

   (b) If f is any element of $(A^B)^C$, let $\Theta(f)$ be the function in $A^{B \times C}$ such that $(\Theta(f))(b, c) = (f(c))(b)$ for any $(b, c) \in B \times C$. Then $\Theta\colon (A^B)^C \xrightarrow[\text{onto}]{1\text{-}1} A^{B \times C}$. (To see that $\mathscr{R}(\Theta) = A^{B \times C}$, let $g \in A^{B \times C}$. Let f be the function in $(A^B)^C$ such that, for any c in C, $f(c)$ is that function in A^B for which $(f(c))(b) = g(b, c)$ for any b in B. Then $\Theta(f) = g$.)   ■

EXERCISES

Assume $A \leq B$ and $C \leq D$. Prove:

18. $B \cap D = \varnothing \Rightarrow A \cup C \leq B \cup D$.

19. $A \times C \leq B \times D$.

20. $A \cup B \cup C \neq \varnothing \Rightarrow A^C \leq B^D$.

Theorem D16 $\{0, 1\}^A \approx \mathscr{P}(A)$.

Proof For each B in $\mathscr{P}(A)$, define $\Psi(B)$ to be the function in $\{0, 1\}^A$ such that

$$(\Psi(B))(a) = \begin{cases} 0 & \text{if} \quad a \in A \\ 1 & \text{if} \quad a \notin A. \end{cases}$$

($\Psi(B)$ is called the **characteristic function of** B.) Then $\Psi\colon \mathscr{P}(A) \xrightarrow[\text{onto}]{1\text{-}1} \{0, 1\}^A$. (To see that $\mathscr{R}(\Psi) = \{0, 1\}^A$, let $g \in \{0, 1\}^A$. Define $B = \{a : a \in A \wedge g(a) = 0\}$. Then $B \in \mathscr{P}(A)$ and $\Psi(B) = g$.)   ■

Let us turn now to the study of denumerable sets.

Lemma D17 $A \subseteq P \Rightarrow A$ is finite or denumerable.

Proof   Assume $A \subseteq P$ and A infinite. We must show that A is denumerable, that is, $A \approx P$. Now, by the Iteration Theorem, there is a function f, defined on P, such that $f(1)$ is the least element of A, and $f(n + 1)$ is the least element of A greater than $f(n)$. Then $f: P \xrightarrow{1-1} A$. To prove that $\mathscr{R}(f) = A$, assume not, and let k be the least element of $A - \mathscr{R}(f)$. Hence, there is an element of A smaller than k. Therefore, $P_{k-1} \cap A$ is a nonempty finite set. Let r be its largest element. Then, by the minimality of k, $r \in \mathscr{R}(f)$, say $r = f(i)$. Hence, $f(i + 1)$ is the least element of A greater than $f(i)$, namely, k. This contradicts the assumption that $k \notin \mathscr{R}(f)$.   ■

Corollary D18   Any subset of a denumerable set is either finite or denumerable.

Proof   Exercise for the reader. (Use Lemma D17.)   ■

EXERCISE

21.   Prove that, if A is denumerable and $B < A$, then B is finite.

Theorem D19 $P \times P$ is denumerable.

Proof   (i) $P \leq P \times P$. (Let $f(n) = (n, 1)$ for all n in P. Then $f: P \xrightarrow{1-1} P \times P$.)

(ii)   Let $g(n, k) = 2^n(2k + 1)$ for any $n, k \in P$. Then, $g: P \times P \xrightarrow{1-1} P$. From (i), (ii), by Theorem D4, $P \times P \approx P$.   ■

A more graphic proof of Theorem D19 can be given by arranging the element of $P \times P$ in an infinite matrix:

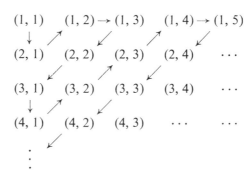

and then enumerating all these ordered pairs by proceeding up and down the consecutive diagonals, as indicated in the diagram.

Corollary D20

a. If A and B are denumerable, so is $A \times B$.

b. If A is denumerable, $A^{Pn} \approx A$.

c. If A is denumerable, B is finite, and $B \neq \varnothing$, then A^B is denumerable.

Proof (a) Assume $A \approx P$ and $B \approx P$. By Lemma D14a, $A \times B \approx P \times P$. Then, by Theorem D19, $A \times B \approx P$.

(b) Let us use mathematical induction. $A^{P_1} = A^{\{1\}} \approx A \approx P$. Now assume $A^{Pn} \approx A$. Then,

$$A^{Pn+1} = A^{Pn \cup \{n+1\}} \approx A^{Pn} \times A^{\{n+1\}} \qquad \text{by Theorem D15a,}$$

$$\approx A \times A \approx A \qquad \text{by Part a.}$$

(c) This follows directly from Part b.   ■

Lemma D21 If A has at least two elements and B has at least two elements, then $A \cup B \leq A \times B$.

Proof Let $a_1, a_2 \in A$ with $a_1 \neq a_2$. Let $b_1, b_2 \in B$ with $b_1 \neq b_2$. Define $F: A \cup B \rightarrow A \times B$ as follows:

$$F(u) = \begin{cases} (u, \, b_1) & \text{if} \quad u \in A, \\ (a_1, \, u) & \text{if} \quad u \in B - A \text{ and } u \neq b_1, \\ (a_2, \, b_2) & \text{if} \quad u = b_1 \in B - A. \end{cases}$$

Then $F: A \cup B \xrightarrow{\text{1-1}} A \times B$.   ■

EXERCISES

22. If A and B are denumerable, prove that $A \cup B$ is denumerable.

23. If A is denumerable and B is finite, prove that $A \cup B$ is denumerable.

24. If A is a nonempty finite set of denumerable sets, prove that $\bigcup_{B \in A} B$ is denumerable.

25. If A is a denumerable set of denumerable sets, prove that $\bigcup_{B \in A} B$ is denumerable.

Now let us show that certain familiar concrete sets are denumerable.

Theorem D22   The following sets are denumerable.

a.   The set Z of all integers.

b.   The set Q^+ of all positive rational numbers.

c.   The set Q of all rational numbers.

Proof   (a)   We can enumerate the integers as follows: 0, 1, -1, 2, -2, 3, -3, etc. (In Chapter 1, Example 20.19, an explicit one-one correspondence is defined.)

(b)   Clearly, $P \leq Q^+$. Conversely, for any r in Q^+, let $r = k/n$, where $k, n \in P$ and the greatest common divisor of k and n is 1, that is, k and n have no common factors other than 1. Define $\Phi(r) = (k, n)$. Then $\Phi: Q^+ \xrightarrow{\;1\text{-}1\;} P \times P$. Hence, $Q^+ \leq P \times P$. But, $P \times P \approx P$, by Theorem D19. Hence, $Q^+ \leq P$. Then, by Theorem D4, $Q^+ \approx P$.

(A more picturesque proof can be obtained by constructing the following infinite array:

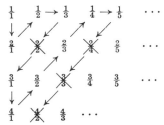

and enumerating along the diagonals as shown, except that any repetition of a fraction already enumerated is skipped.)

(c)   By Part b, let Φ be a one-one correspondence between Q^+ and P. Then, define

$$\Psi(r) = \begin{cases} \Phi(r) & \text{if } r \in Q^+, \\ 0 & \text{if } r = 0, \\ -\Phi(-r) & \text{if } -r \in Q^+. \end{cases}$$

Then $\Psi: Q \xrightarrow[\text{onto}]{1\text{-}1} Z$. But, by Part a, $Z \approx P$. Hence, $Q \approx P$.   ∎

Notation   The cardinal number $\mathcal{H}(P)$ is traditionally denoted by $\aleph_0$. Thus, all denumerable sets have cardinal number $\aleph_0$. By Exercise 21 on p. 304, the only cardinal numbers smaller than $\aleph_0$ are the finite cardinal numbers. By Exercise 7 on p. 300, $\aleph_0$ is the smallest infinite cardinal number.

By Cantor's Theorem (D3), the cardinal number of $\mathscr{P}(P)$ is bigger than $\aleph_0$. We shall now discover some more familiar sets which have cardinal number greater than $\aleph_0$.

Let R denote the set of real numbers.

Theorem D23 The half-open interval $[0, 1) = \{x : x \in R \wedge 0 \le x < 1\}$ is not denumerable.

Proof Assume that $[0, 1)$ is denumerable. Let $\Psi\colon P \xrightarrow[\text{onto}]{\text{1-1}} [0, 1)$. Denote $\Psi(n) = r_n$ for each n in P. Each r_n can be represented by an infinite decimal $.a_{n1}a_{n2}a_{n3}\cdots$. (Infinite decimals with an infinite tail of 9's, such as $.247999\cdots$, represent the same number as another decimal with an infinite tail of 0's; in the example, $.248000\cdots$. In our representation of r_n as an infinite decimal, we never choose a representation ending with an infinite tail of 9's.)

$$r_1 = .a_{11}a_{12}a_{13}\cdots$$
$$r_2 = .a_{21}a_{22}a_{23}\cdots$$
$$r_3 = .a_{31}a_{32}a_{33}\cdots$$
$$\vdots$$

Define a new decimal $.c_1c_2c_3\cdots$ as follows:

$$c_n = \begin{cases} 0 & \text{if } a_{n,n} \ne 0 \\ 1 & \text{if } a_{n,n} = 0. \end{cases}$$

Then the decimal $.c_1c_2c_3\cdots$ differs from each decimal r_n in the nth place. Hence, $.c_1c_2c_3\cdots$ does not appear in our list, contradicting the assumption that $\mathcal{R}(\Psi) = [0, 1)$. (Why can't $.c_1c_2c_3\cdots$ represent the same number as some r_n?) ∎

By a generalized interval, we shall mean a set of real numbers of any of the following forms.

$$[a, b] = \{x : x \in R \wedge a \le x \le b\} \quad \text{(closed interval)}$$
$$(a, b) = \{x : x \in R \wedge a < x < b\} \quad \text{(open interval)}$$
$$\left.\begin{array}{l}[a, b) = \{x : x \in R \wedge a \le x < b\} \\ (a, b] = \{x : x \in R \wedge a < x \le b\}\end{array}\right\} \text{(half-open intervals)}$$
$$[a, \infty) = \{x : x \in R \wedge a \le x\}$$
$$(a, \infty) = \{x : x \in R \wedge a < x\}$$
$$(-\infty, a] = \{x : x \in R \wedge x \le a\}$$
$$(-\infty, a) = \{x : x \in R \wedge x < a\}.$$

Theorem D24 All generalized intervals are equinumerous with each other and with the whole set R of real numbers.

Proof First, let us show that any two open intervals (a, b) and (c, d) are equinumerous. A simple geometric proof is pictured in Figure D.2.

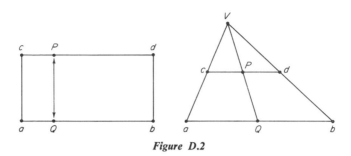

Figure D.2

If (a, b) and (c, d) have the same length, form a rectangle with (a, b) and (c, d) as opposite sides and associate each point in (a, b) with the corresponding point on the other side of the rectangle. If (c, d) is smaller than (a, b), place (c, d) over (a, b) and complete the triangle as in the diagram. For each point P in (c, d), draw the line L connecting P with the vertex V. The intersection Q of L with (a, b) is to be the point corresponding to P. This is clearly a one-one correspondence between (c, d) and (a, b). (A specific formula for a one-one correspondence between (a, b) and (c, d) is

$$\left(\frac{c - d}{a - b}\right) x + \frac{ad - bc}{a - b}.)$$

Thus, $(a, b) \approx (c, d)$.

Next, let us show that $(0, 1) \approx (0, \infty)$. This can be seen in the diagram in Figure D.3. We raise $(0, 1)$ to a 45° angle with the x axis. Let V be the point $(0, 1/\sqrt{2})$, which is above the origin and at the same height as the top of the interval $(0, 1)$. With each point P of the interval $(0, 1)$, we asso-

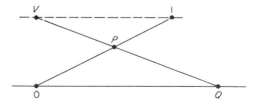

Figure D.3

ciate the point Q on $(0, \infty)$ obtained by extending the line connecting V and P until it intersects $(0, \infty)$. This clearly establishes a one-one correspondence between $(0, 1)$ and $(0, \infty)$. (A simple formula establishing a one-one correspondence between $(0, 1)$ and $(0, \infty)$ is $(1/(1 - x)) - 1$.)

We now can show that $(-1, 1) \approx R$. In fact, let Φ be a one-one correspondence between $(0, 1)$ and $(0, \infty)$. We need only "reflect" this in the origin, that is, for any u in $(-1, 1)$, let

$$\Psi(u) = \begin{cases} \Phi(u) & \text{if } u \in (0, 1) \\ 0 & \text{if } u = 0 \\ -\Phi(-u) & \text{if } u \in (-1, 0). \end{cases}$$

Then $\Psi \colon (-1, 1) \xrightarrow[\text{onto}]{1\text{--}1} R$. (A simple pictorial representation of a one-one correspondence between $(-1, 1)$ and R is presented in the diagram in Figure D.4.)

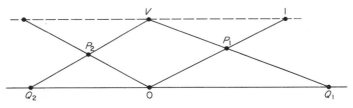

Figure D.4

Now we can show that, for any generalized interval I, $I \approx R$. There is some open interval (a, b) included in I. Thus, $(a, b) \subseteq I$. Since any two open intervals are equinumerous, $(a, b) \approx (-1, 1)$. Thus, $(-1, 1) \leq I$. Since $(-1, 1) \approx R$, $R \leq I$. Conversely, since $I \subseteq R$, $I \leq R$. Hence, by Theorem D4, $I \approx R$. ∎

EXERCISE

26. If $A \subseteq R$ and A includes some generalized interval, prove that $A \approx R$.

Notation   The symbol $\mathfrak{c}$ is used to denote $\mathscr{K}(R)$, the cardinal number of R. $\mathfrak{c}$ is called the **power of the continuum**.

From Theorem D24, we know that all generalized intervals have cardinal number $\mathfrak{c}$. From Theorem D23, we know that $\aleph_0 <_K \mathfrak{c}$. Whether there is a cardinal number between $\aleph_0$ and $\mathfrak{c}$ is a long-standing mathematical problem, called the **continuum problem**.

The fact proved in Theorem D23 that $P < R$ actually is not new. We already knew that $P < \mathscr{P}(P)$ and the following theorem tells us that that result is equivalent to $P < R$.

Theorem D25 $R \approx \mathscr{P}(P)$, that is, the set of real numbers is equinumerous with the set of all sets of natural numbers.

Proof By Theorem D16, $\mathscr{P}(P) \approx \{0, 1\}^P$. Hence, it suffices to prove that $R \approx \{0, 1\}^P$. Since $R \approx [0, 1)$, it suffices to prove that $[0, 1) \approx \{0, 1\}^P$. Any number in $[0, 1)$ has an infinite binary expansion (just as it has an infinite decimal representation); in the case of two binary representations for the same number, we discard the representation with an infinite tail of 1's. Every infinite binary representation can be considered to be an element of $\{0, 1\}^P$. For example, $.01101010\cdots$ is associated with the function f: $P \to \{0, 1\}$ such that $f(1) = 0, f(2) = 1, f(3) = 1, \ldots, f(n) =$ the nth place entry, etc. Thus, we have a one-one correspondence between $[0, 1)$ and a subset of $\{0, 1\}^P$. (Because those binary representations with infinite tails of 1's were not used, the range of this one-one correspondence is not all of $\{0, 1\}^P$.) Conversely, with each element g in $\{0, 1\}^P$, we associate the real number in $[0, 1)$ determined by the infinite ternary representation provided by g: $.g(0)g(1)g(2)\cdots$. In other words, g determines an infinite sequence of 0's and 1's, which can be interpreted as an infinite ternary representation of a real number in $[0, 1)$. This correspondence is one-one, since the infinite ternary representation obtained from g does not end in an infinite tail of 2's (in fact, it contains no 2's at all). Therefore, $\{0, 1\}^P \leq [0, 1)$. Since $[0, 1) \leq \{0, 1\}^P$ and $\{0, 1\}^P \leq [0, 1)$, it follows by Theorem D4 that $[0, 1) \approx \{0, 1\}^P$. ∎

It is possible to define on cardinal numbers "arithmetic" operations which are generalizations of the usual arithmetic operations on nonnegative integers.

Lemma D26 For any sets A and B, there exist sets A_1, B_1 such that $A_1 \approx A$, $B_1 \approx B$, and $A_1 \cap B_1 = \varnothing$.

Proof Let $A_1 = A \times \{0\}$ and $B_1 = B \times \{1\}$. ∎

Definition *Addition* Let $\mathfrak{m}$ and $\mathfrak{n}$ be cardinal numbers. Take any sets A, B such that $\mathscr{H}(A) = \mathfrak{m}$, $\mathscr{H}(B) = \mathfrak{n}$, and $A \cap B = \varnothing$. Then

$$\mathfrak{m} + \mathfrak{n} = \mathscr{H}(A \cup B).$$

By Lemma D14c, the result $\mathfrak{m} + \mathfrak{n}$ does not depend upon the choice of A and B.

Definitions   *Multiplication and Exponentiation*   Let $\mathfrak{m}$ and $\mathfrak{n}$ be cardinal numbers, and let A, B be any sets such that $\mathscr{H}(A) = \mathfrak{m}$, $\mathscr{H}(B) = \mathfrak{n}$. Then

$$\mathfrak{m} \times \mathfrak{n} = \mathscr{H}(A \times B),$$

and

$$\mathfrak{m}^{\mathfrak{n}} = \mathscr{H}(A^B).$$

By Lemma D14a,b, we know that $\mathfrak{m} \times \mathfrak{n}$ and $\mathfrak{m}^{\mathfrak{n}}$ do not depend upon the choice of A and B.

Theorem D27   Let $\mathfrak{m}$, $\mathfrak{n}$, $\mathfrak{p}$ be any cardinal numbers. Then

a. $\mathfrak{m} + \mathfrak{n} = \mathfrak{n} + \mathfrak{m}$.
b. $\mathfrak{m} + (\mathfrak{n} + \mathfrak{p}) = (\mathfrak{m} + \mathfrak{n}) + \mathfrak{p}$.
c. $\mathfrak{m} + 0 = \mathfrak{m}$.   (Remember that $0 = \mathscr{H}(\varnothing)$.)
d. $\mathfrak{m} \times \mathfrak{n} = \mathfrak{n} \times \mathfrak{m}$.
e. $\mathfrak{m} \times (\mathfrak{n} \times \mathfrak{p}) = (\mathfrak{m} \times \mathfrak{n}) \times \mathfrak{p}$.
f. $\mathfrak{m} \times 1 = \mathfrak{m}$.   (Remember that $1 = \mathscr{H}(P_1)$.)
g. $\mathfrak{m} \times (\mathfrak{n} + \mathfrak{p}) = (\mathfrak{m} \times \mathfrak{n}) + (\mathfrak{m} \times \mathfrak{p})$.
h. $\mathfrak{m}^0 = 1$.
i. $\mathfrak{n} \neq 0 \Rightarrow 0^{\mathfrak{n}} = 0$.
k. $\mathfrak{m}^{\mathfrak{n}+\mathfrak{p}} = \mathfrak{m}^{\mathfrak{n}} \times \mathfrak{m}^{\mathfrak{p}}$.
j. $(\mathfrak{m}^{\mathfrak{n}})^{\mathfrak{p}} = \mathfrak{m}^{\mathfrak{n} \times \mathfrak{p}}$.
l. $2 \leq_{\mathscr{H}} \mathfrak{m} \wedge 2 \leq_{\mathscr{H}} \mathfrak{n} \Rightarrow \mathfrak{m} + \mathfrak{n} \leq_{\mathscr{H}} \mathfrak{m} \times \mathfrak{n}$.

Proof   All of these results are direct consequences of the definitions and previous theorems. For example, (j) and (k) are reformulations of Lemma D15a,b, while (l) is essentially Lemma D21. The proofs are left as an exercise.   ■

EXERCISES

Assume $\mathfrak{m} \leq_K \mathfrak{n} \wedge \mathfrak{p} \leq_K \mathfrak{q}$. Prove:

27. $\mathfrak{m} + \mathfrak{p} \leq_K \mathfrak{n} + \mathfrak{q}$.
28. $\mathfrak{m} \times \mathfrak{p} \leq_K \mathfrak{n} \times \mathfrak{q}$.
29. $\mathfrak{m} + \mathfrak{p} + \mathfrak{n} >_K 0 \Rightarrow \mathfrak{m}^{\mathfrak{p}} \leq \mathfrak{n}^{\mathfrak{q}}$.

Theorem D28

a. $\aleph_0 \times \aleph_0 = \aleph_0$.

b. If n is a positive integer, $\aleph_0{}^n = \aleph_0$.

c. $0 <_K \mathfrak{m} \leq_K \aleph_0 \Rightarrow \mathfrak{m} \times \aleph_0 = \aleph_0$.

d. $\aleph_0 + \aleph_0 = \aleph_0$.

e. $\mathfrak{m} \leq_K \aleph_0 \Rightarrow \mathfrak{m} + \aleph_0 = \aleph_0$.

Proof   (a)   Theorem D19.

   (b)   Corollary D20b.

   (c)   Assume $0 <_K \mathfrak{m} \leq_K \aleph_0$. Then

$$\aleph_0 = 1 \times \aleph_0 \leq_K \mathfrak{m} \times \aleph_0 \leq_K \aleph_0 \times \aleph_0 = \aleph_0 \qquad \text{(by Part a).}$$

Hence, by Theorem D4, $\mathfrak{m} \times \aleph_0 = \aleph_0$.

   (d) $\aleph_0 \leq_K \aleph_0 + \aleph_0 \leq_K \aleph_0 \times \aleph_0 \qquad$ (by Theorem D27l)
$$= \aleph_0.$$

By Theorem D4, $\aleph_0 + \aleph_0 = \aleph_0$.

   (e)   Assume $\mathfrak{m} \leq_K \aleph_0$. Then,

$$\aleph_0 \leq_K \mathfrak{m} + \aleph_0 \leq_K \aleph_0 + \aleph_0 = \aleph_0.$$

By Theorem D4, $\mathfrak{m} + \aleph_0 = \aleph_0$.   ■

Theorem D29

a. $2^{\aleph_0} = \mathfrak{c}$.

b. $\mathfrak{c} \times \mathfrak{c} = \mathfrak{c}$.

c. $0 <_K \mathfrak{m} \leq_K \mathfrak{c} \Rightarrow \mathfrak{m} \times \mathfrak{c} = \mathfrak{c}$.

d. $\mathfrak{c} + \mathfrak{c} = \mathfrak{c}$.

e. $\mathfrak{m} \leq_K \mathfrak{c} \Rightarrow \mathfrak{m} + \mathfrak{c} = \mathfrak{c}$.

f. $0 <_K \mathfrak{m} \leq_K \aleph_0 \Rightarrow \mathfrak{c}^{\mathfrak{m}} = \mathfrak{c}$.

g. $\mathfrak{c} <_K 2^{\mathfrak{c}}$.

Proof   (a)   Theorem D25 and Theorem D16.

   (b) $\mathfrak{c} \times \mathfrak{c} = 2^{\aleph_0} \times 2^{\aleph_0} = 2^{\aleph_0 + \aleph_0} = 2^{\aleph_0}$.

   (c)   Assume $0 <_K \mathfrak{m} \leq_K \mathfrak{c}$. Then

$$\mathfrak{c} = 1 \times \mathfrak{c} \leq_K \mathfrak{m} \times \mathfrak{c} \leq_K \mathfrak{c} \times \mathfrak{c} = \mathfrak{c}.$$

By Theorem D4, $\mathfrak{m} \times \mathfrak{c} = \mathfrak{c}$.

(d) $c \leq_K c + c \leq_K c \times c$ (by Theorem D27l)
$$= c.$$

(e)   Assume $m \leq_K c$. Then

$$c \leq_K m + c \leq_K c + c = c.$$

(f)   Assume $0 <_K m \leq_K \aleph_0$. Then

$$c^m = (2^{\aleph_0})^m = 2^{\aleph_0 \times m} = 2^{\aleph_0} = c.$$

(g)   This is a special case of $m <_K 2^m$ (Theorems D3, D16).   ■

EXERCISES

30.   Show that the set of complex numbers has cardinal number c.

31.   Give a direct proof of Theorem D29d by using intervals of real numbers.

It turns out that, if either of m and n is an infinite cardinal number and neither is 0, then[†]

$$m + n = m \times n = \text{maximum}(m, n).$$

This result trivializes the arithmetic of addition and multiplication of cardinal numbers. However, exponentiation is a nontrivial operation and involves many unsolved problems. For further information about cardinal arithmetic, see Sierpinski [1958], Rubin [1967].

[†] The proof requires the Axiom of Choice.

AXIOMATIC SET THEORY AND THE EXISTENCE OF A PEANO SYSTEM

We shall carry through enough of the development of a system of axiomatic set theory to enable us to prove the existence of a Peano system. The axiomatic theory will be the classical Zermelo–Fraenkel–Skolem system, which we shall denote by ZFS.

In addition to the usual logical symbolism and axioms, ZFS contains a symbol $\in$ for a binary relation. (This relation is to be though of intuitively as the membership relation. The variables are to be thought of intuitively as ranging over sets. Thus, in axiomatic set theory, the only objects are sets.[†] Naturally, these intuitive ideas are to play absolutely no role in the derivation of theorems from the axioms.)

The equality relation is introduced by definition.

Definition $x = y \Leftrightarrow (\forall z)(z \in x \Leftrightarrow z \in y)$. Thus, two sets are equal if and only if they contain exactly the same members.

Axiom 1 *Extensionality*

$$(x = y \wedge y \in z) \Rightarrow x \in z.$$

From the definition of equality and Axiom 1, one can prove the basic properties of equality, for example, $x = x$, $x = y \rightarrow y = x$, $(x = y \wedge y = z) \Rightarrow x = z$, and the substitutivity property[‡]:

$$x = y \Rightarrow (\mathscr{A}(x, x) \Rightarrow \mathscr{A}(x, y))$$

[†] In pure mathematics, there is no need for non-sets.
[‡] See Appendix A or Mendelson [1964], pp. 75–79.

where $\mathscr{A}(x, x)$ is any formula and $\mathscr{A}(x, y)$ arises from $\mathscr{A}(x, x)$ by replacing some, but not necessarily all, occurrences of x by y.

Axiom 2 *Pairing*

$$(\forall x)(\forall y)(\exists z)(\forall u)(u \in z \Leftrightarrow (u = x \lor u = y)).$$

For any sets x, y, there is a set z whose only elements are x and y. The set z which is asserted to exist is unique; any other set containing just x and y must be equal to z by virtue of the definition of equality. We denote the unique set z containing x and y by $\{x, y\}$. Thus,

$$(\forall u)(u \in \{x, y\} \Leftrightarrow (u = x \lor u = y)).$$

Notation $\{x\} = \{x, x\}$. Then $\{x\}$ is a set whose only element is x.

Definition   *Ordered Pair*

$$(x, y) = \{\{x\}, \{x, y\}\}.$$

Theorem ZFS 1

$$(x, y) = (u, v) \Rightarrow (x = u \land y = v).$$

Proof Verify that the proof of this result in Chapter 1, p. 26, uses only the basic definitions of equality and ordered pairs.   ∎

Axiom 3   *Union*

$$(\forall x)(\exists y)(\forall u)(u \in y \Leftrightarrow (\exists v)(u \in v \land v \in x)).$$

For any set x, there exists a set y which contains precisely those sets which belong to some set in x. By the definition of equality, there is only one such set y. It is called the **sum set** of x and is denoted $\bigcup x$. Traditionally, $\bigcup x$ is denoted $\bigcup_{v \in x} v$ and is called the **union** of the elements of x.

Definition $a \cup b = \bigcup \{a, b\}$.

Theorem ZFS 2

$$(\forall u)(u \in a \cup b \Leftrightarrow (u \in a \lor u \in b)).$$

Proof $u \in a \cup b \Leftrightarrow (\exists v)(u \in v \wedge v \in \{a, b\})$
$\Leftrightarrow (\exists v)(u \in v \wedge (v = a \vee v = b))$
$\Leftrightarrow u \in a \vee u \in b.$ ∎

EXERCISES

Prove:

1. $a \cup b = b \cup a.$

2. $(a \cup b) \cup c = a \cup (b \cup c).$

3. $a \cup a = a.$

4. $\bigcup \{a\} = a.$

Definitions $a \subseteq b \Leftrightarrow (\forall u)(u \in a \Rightarrow u \in b).$
$a \subsetneqq b \Leftrightarrow a \subseteq b \wedge a \neq b.$

$a \subseteq b$ is read: **a is included in b** or **a is a subset of b**.
$a \subsetneqq b$ is read: **a is a proper subset of b**.

EXERCISES

Prove:

5. $a = b \Leftrightarrow (a \subseteq b \wedge b \subseteq a).$

6. $a \subseteq a.$

7. $(a \subseteq b \wedge b \subseteq c) \Rightarrow a \subseteq c.$

8. $a \subseteq a \cup b \wedge b \subseteq a \cup b.$

9. $a \subsetneqq b \Rightarrow b \nsubseteq a.$[†]

10. $(a \subsetneqq b \wedge b \subseteq c) \Rightarrow a \subsetneqq c.$

11. $(a \subseteq c \wedge b \subseteq c) \Rightarrow a \cup b \subseteq c.$

12. $v \in x \Rightarrow v \subseteq \bigcup x.$

13. $(\forall v)(v \in x \Rightarrow v \subseteq b) \Rightarrow \bigcup x \subseteq b.$

14. $a \subseteq b \Leftrightarrow a \cup b = b.$

Axiom 4 *Separation* Let $\mathscr{A}(x)$ be any formula of ZFS.

$$(\forall x)(\exists y)(\forall u)(u \in y \Leftrightarrow (u \in x \wedge \mathscr{A}(u))).$$

For any set x, there is a set y containing precisely those members of x which

[†] $b \nsubseteq a$ stands for: not($b \subseteq a$).

satisfy $\mathscr{A}$. By the definition of equality, there is exactly one such set y. As notation for the set y, we shall use

$$\{u : u \in x \wedge \mathscr{A}(u)\}.$$

Theorem ZFS 3 For any sets a, b, there is a unique set y such that

$$(\forall u)(u \in y \Leftrightarrow (u \in a \wedge u \in b)).$$

Proof Let $\mathscr{A}(u)$ be $u \in b$, and let $y = \{u : u \in a \wedge \mathscr{A}(u)\}$. ∎

Notation The unique set y of Theorem ZFS 3 will be denoted $a \cap b$. Thus,

$$(\forall u)(u \in a \cap b \Leftrightarrow (u \in a \wedge u \in b)).$$

$a \cap b$ is called the **intersection** of a and b.

EXERCISES

Prove:

15. $a \cap b = b \cap a$.

16. $a \cap (b \cap c) = (a \cap b) \cap c$.

17. $a \cap a = a$.

18. $a \cap b \subseteq a \wedge a \cap b \subseteq b$.

19. $(c \subseteq a \wedge c \subseteq b) \Rightarrow c \subseteq a \cap b$.

20. $a \subseteq b \Leftrightarrow a \cap b = a$.

Theorem ZFS 4 There is a unique set y such that $(\forall u)(u \notin y)$.

Proof Take any set x.[†] Let $y = \{u : u \in x \wedge u \neq u\}$. Clearly, $u \notin y$ for all u. The uniqueness of such a set y follows from the definition of equality. ∎

Notation The unique set y of Theorem ZFS 4 is denoted $\varnothing$ and is called the **empty set**. Thus, $(\forall u)(u \notin \varnothing)$.

[†] It follows from the axioms of logic that there must be at least one set.

EXERCISES

Prove:

21. For any set x, $\varnothing \subseteq x$.

22. $a \cup \varnothing = a$.

23. $a \cap \varnothing = \varnothing$.

24. (Russell's Paradox) $\neg (\exists y)(\forall x)(x \in y \Leftrightarrow x \notin x))$. (*Hint*: Assume there is such a set y and then consider $y \in y$. Notice that no set-theoretic axioms are necessary for the proof of this exercise.)

25. There cannot be a universal set, that is, $\neg (\exists y)(\forall x)(x \in y)$. (*Hint*: Use Exercise 4 and the Separation Axiom.)

Theorem ZFS 5

$$(\forall x)(x \neq \varnothing \Rightarrow (\exists ! y)(\forall u)(u \in y \Leftrightarrow (\forall v)(v \in x \Rightarrow u \in v))).$$

For any nonempty set x, there is a unique set y consisting of those sets which belong to every member of x.

Proof Let $z \in x$. Let $y = \{u : u \in z \wedge (\forall v)(v \in x \Rightarrow u \in v)\}$.   ■

Notation   The unique set y of Theorem ZFS 5 is denoted $\bigcap x$, or, sometimes in more traditional form, $\bigcap_{v \in x} v$. It is called the **intersection** of all the sets in x. Notice that the existence of such a set depends upon x being nonempty.

EXERCISES

Prove:

26. $v \in x \rightarrow \bigcap x \subseteq v$.

27. $\bigcap \{a, b\} = a \cap b$.

Axiom 5 *Power Set*

$$(\forall x)(\exists y)(\forall u)(u \in y \Leftrightarrow u \subseteq x).$$

For any set x, there is a set y whose elements are the subsets of x.

Notation   By the definition of equality, the set y of Axiom 5 is unique. It is denoted $\mathscr{P}(x)$ and is called the **power set** of x.

EXERCISES

Prove:

28. $\mathscr{P}(\varnothing) = \{\varnothing\}$.

29. $\mathscr{P}(\{x\}) = \{\varnothing, \{x\}\}$.

Theorem ZFS 6

$$(\forall x)(\forall y)(\exists z)(\forall u)(u \in z \Leftrightarrow (\exists a)(\exists b)(u = (a, b) \wedge a \in x \wedge b \in y)).$$

For any sets x and y, there is a set z whose elements are precisely those ordered pairs (a, b) with first component a in x and second component b in y.

Proof If $a \in x$ and $b \in y$, then $(a, b) = \{\{a\}, \{a, b\}\} \in \mathscr{P}(\mathscr{P}(x \cup y)))$. Let

$$z = \{u : u \in \mathscr{P}(\mathscr{P}(x \cup y)) \wedge (\exists a)(\exists b)(u = (a, b) \wedge a \in x \wedge b \in y)\}. \qquad \blacksquare$$

Notation The set z of Theorem ZFS 6 is unique, by the definition of equality. It is denoted $x \times y$ and is called the **Cartesian product** of x and y.

EXERCISE

30. Prove: $x \times y = y \times x \Leftrightarrow (x = y \vee x = \varnothing \vee y = \varnothing)$.

Definition By a **relation** we mean any set of ordered pairs. Thus, x is a relation $\Leftrightarrow (\forall u)(u \in x \Rightarrow (\exists a)(\exists b)(u = (a, b)))$.

We say that a set f is a **function** if and only if f is a relation and

$$(\forall u)(\forall v)(\forall w)(((u, v) \in f \wedge (u, w) \in f) \Rightarrow v = w).^{\dagger}$$

Notice that, if $(u, v) \in f$, then $\{\{u\}, \{u, v\}\} \in f$. Hence, $\{u\} \in \bigcup f$ and $\{u, v\} \in \bigcup f$. Therefore, $u \in \bigcup (\bigcup f)$ and $v \in \bigcup (\bigcup f)$. This clarifies the following definitions.

Definitions

$$\mathscr{D}(f) = \{u : u \in \bigcup (\bigcup f) \wedge (\exists v)((u, v) \in f)\}.$$
$$\mathscr{R}(f) = \{v : v \in \bigcup (\bigcup f) \wedge (\exists u)((u, v) \in f)\}.$$

---

† An intuitive discussion of relations and functions, with numerous examples, may be found in Chapter 1, Sections 14–21.

$\mathscr{D}(f)$ is called the **domain** of f and consists of all first components of ordered pairs in f. $\mathscr{R}(f)$ is called the **range** of f and consists of all second components of ordered pairs in f.

If f is a function and $u \in \mathscr{D}(f)$, then there is a unique v such that $(u, v) \in f$. This unique v will be denoted in the usual way by $f(u)$.

We say that f is a **function from x into y**, and we shall write $f \colon x \to y$, if and only if $\mathscr{D}(f) = x$ and $\mathscr{R}(f) \subseteq y$.

Notation   Let $\mathscr{B}(u, v)$ be a formula of ZFS. Then

$\mathrm{Un}(\mathscr{B})$ will stand for: $(\forall u)(\forall v)(\forall w)(\mathscr{B}(u, v) \wedge \mathscr{B}(u, w) \Rightarrow v = w)$.

Axiom 6   *Replacement*   Let $\mathscr{B}(u, v)$ be a formula of ZFS

$$(\forall x)(\mathrm{Un}(\mathscr{B}) \Rightarrow (\exists y)(\forall v)(v \in y \Leftrightarrow (\exists u)(u \in x \wedge \mathscr{B}(u, v)))).$$

Theorem ZFS 7   The Separation Axiom (Axiom 4) is provable from the Replacement Axiom (Axiom 6).

Proof   Let $\mathscr{A}(u)$ be a formula of ZFS. We must prove that

$$(\exists y)(\forall v)(v \in y \Leftrightarrow (v \in x \wedge \mathscr{A}(v))).$$

Let $\mathscr{B}(u, v)$ stand for $u = v \wedge \mathscr{A}(u)$. It is easy to check that $\mathrm{Un}(\mathscr{B})$. Hence, by Axiom 6, there is a set y such that

$$(\forall v)(v \in y \Leftrightarrow (\exists u)(u \in x \wedge u = v \wedge \mathscr{A}(u))).$$

Hence,

$$(\forall v)(v \in y \Leftrightarrow (v \in x \wedge \mathscr{A}(v))).   \blacksquare$$

Axiom 7   *Infinity*

$$(\exists x)(\varnothing \in x \wedge (\forall u)(u \in x \Rightarrow u \cup \{u\} \in x)).$$

Definition   Let us call a set x **Z-inductive**[†] if and only if

$$\varnothing \in x \wedge (\forall u)(u \in x \Rightarrow u \cup \{u\} \in x).$$

Axiom 7 asserts that there exists a Z-inductive set. Notice that the intersection of two Z-inductive sets is again a Z-inductive set.

[†] We use *Z-inductive* rather than the more customary *inductive* because the latter has been used already with a different (although analogous) meaning.

We want to define the nonnegative integers to be the intersection of all Z-inductive sets. This cannot be done directly on the basis of our axioms because the collection of Z-inductive sets is not a set. However, a more circuitous approach is possible.

Theorem ZFS 8   There is a unique set Nn such that

$$(\forall u)(u \in Nn \Leftrightarrow (\forall v)(v \text{ is } Z\text{-inductive} \Rightarrow u \in v)).$$

Thus, the elements of Nn are those sets which belong to every Z-inductive set.

Proof   By Axiom 7, there exists a Z-inductive set x_0. Let $y = \{z : z \in \mathscr{P}(x_0)$ $\wedge\, z$ is Z-inductive$\}$. y is the set of all Z-inductive sets which are included in x_0. Clearly, $x_0 \in y$, and, therefore, $y \neq \varnothing$. Define $Nn = \bigcap_{z \in y} z$. Nn consists of all sets which belong to every set in y. Consider any set u. Assume $(\forall v)(v$ is Z-inductive $\Rightarrow u \in v)$. Therefore, $(\forall z)$ $(z$ is Z-inductive $\wedge\, z \subseteq x_0$ $\Rightarrow u \in z)$. Thus, $u \in \bigcap_{z \in y} z = Nn$. Conversely, assume $u \in Nn$, and let v be any Z-inductive set. Then $v \cap x_0$ is Z-inductive. Hence, $v \cap x_0 \in y$. Thus, since $u \in Nn = \bigcap_{z \in y} z$, it follows that $u \in v \cap x_0$. Hence, $u \in v$. Therefore, $(\forall v)(v$ is Z-inductive $\Rightarrow u \in v)$. We have shown that $u \in Nn \Leftrightarrow (\forall v)(v$ is Z-inductive $\Rightarrow u \in v)$. That Nn is the only such set follows from the definition of equality.   ∎

Theorem ZFS 9

a. Nn is Z-inductive.
b. Nn is a subset of every Z-inductive set.

Proof   (a) $\varnothing$ is a member of every Z-inductive set. Hence, $\varnothing \in Nn$. Now, assume $u \in Nn$. Then u is a member of every Z-inductive set. Consider any Z-inductive set v. Then $u \in v$. Since v is Z-inductive, $u \cup \{u\} \in v$. Hence, $u \cup \{u\}$ is a member of every Z-inductive set, and, therefore, $u \cup \{u\} \in Nn$. We have shown: $u \in Nn \Rightarrow u \cup \{u\} \in Nn$.

(b)   Let v be any Z-inductive set. Assume $u \in Nn$. Then, by Theorem ZFS 8, $u \in v$. Thus, $Nn \subseteq v$.   ∎

Theorem ZFS 10   *Mathematical Induction*

$$[A \subseteq Nn \wedge \varnothing \in A \wedge (\forall u)(u \in A \Rightarrow u \cup \{u\} \in A)] \Rightarrow A = Nn.$$

Proof Assume $A \subseteq Nn \wedge \varnothing \in A \wedge (\forall u)(u \in A \Rightarrow u \cup \{u\} \in A)$. Then, A is Z-inductive. By Theorem ZFS 9b, $Nn \subseteq A$. But $A \subseteq Nn$. Hence, $A = Nn$. ∎

Definition A set z is said to be **transitive** if and only if

$$(\forall u)(\forall v)((u \in v \wedge v \in z) \Rightarrow u \in z),$$

that is, every member of a member of z is a member of z.

EXERCISES

Prove:

31. z is transitive $\Leftrightarrow \bigcup z \subseteq z$.

32. $\varnothing$ is transitive.

Lemma ZFS 11 *Every member of Nn is transitive.*

Proof Let $A = \{u : u \in Nn \wedge u$ is transitive$\}$. Clearly, $\varnothing \in A$, since $\varnothing$ is transitive. Assume now that $u \in A$. Hence, $u \in Nn$ and u is transitive. Since $u \in Nn$ and Nn is Z-inductive, $u \cup \{u\} \in Nn$. In addition, $u \cup \{u\}$ is transitive. (For, if we assume that $w \in v \wedge v \in u \cup \{u\}$, then either $v \in u$ or $v = u$. If $v \in u$, then $w \in u$, since u is transitive. If $v = u$, then, since $w \in v$, $w \in u$. Thus, $(w \in v \wedge v \in u \cup \{u\}) \Rightarrow w \in u \cup \{u\}$.) We have shown: $u \in A \Rightarrow u \cup \{u\} \in A$. By Theorem ZFS 10, $A = Nn$, and, therefore, every member of Nn is transitive. ∎

Lemma ZFS 12 *Assume $u \in Nn \wedge v \in Nn$. Then*

$$u \cup \{u\} = v \cup \{v\} \Rightarrow u = v.$$

Proof Assume $u \in Nn \wedge v \in Nn \wedge u \cup \{u\} = v \cup \{v\}$. Assume for the sake of contradiction that $u \neq v$. Now, $u \in u \cup \{u\}$. Hence, $u \in v \cup \{v\}$. Since $u \neq v$, $u \notin \{v\}$. Hence, $u \in v$. Therefore, since v is transitive (by Lemma ZFS 11), every member of u is also a member of v, that is, $u \subseteq v$. Similarly, $v \subseteq u$. Hence, $u = v$. ∎

We define a function $S: Nn \to Nn$ such that $S(u) = u \cup \{u\}$ for every u in Nn. More precisely,

$$S = \{z : z \in Nn \times Nn \wedge (\exists u)(z = (u, u \cup \{u\}))\}.$$

Theorem ZFS 13 $(Nn, S, \varnothing)$ is a Peano system.

Proof   We already know that $\varnothing \in Nn$ and that $S: Nn \to Nn$. We must prove Axioms (P1)–(P3) for Peano systems.

(P1) $(\forall u)(u \in Nn \Rightarrow S(u) \neq \varnothing)$.
This is obvious, since $u \in u \cup \{u\} = S(u)$.

(P2) $(\forall u)(\forall v)((u \in Nn \wedge v \in Nn \wedge S(u) = S(v)) \Rightarrow u = v)$.
This is Lemma ZFS 12.

(P3) $(\forall A)([A \subseteq Nn \wedge \varnothing \in A \wedge (\forall u)(u \in A \Rightarrow S(u) \in A)] \Rightarrow A = Nn)$.
This is Theorem ZFS 10.   ∎

Theorem ZFS 13 is a proof of the Basic Existence Assumption for Peano systems. It may be reasonably objected that the Axiom of Infinity (Axiom 7) amounts to the assumption of the Basic Existence Axiom. Axiom 7 is equivalent to the assumption of the existence of an infinite set. Such an axiom cannot be proved from the other axioms in any axiomatic set theory similar to the one described above.[†] What we have shown is that the assumption of the existence of an infinite set implies the existence of a special kind of infinite set, a Peano system.

For further information on various systems of axiomatic set theory, the reader is referred to Fraenkel and Bar-Hillel [1958] and Hatcher [1968].

[†] There are entirely different kinds of set theories in which the existence of an infinite set is provable from certain basic set-theoretic principles which seem to have nothing to do with the notion of infinity. An example is Quine's system NF (see Rosser [1953], Specker [1953], Hatcher [1968]). However, NF has a drawback (namely, the falsity of the Axiom of Choice) which *seems* to make it unsuitable as a foundation for mathematics.

CONSTRUCTION OF THE REAL NUMBERS VIA DEDEKIND CUTS

Our purpose is to give an alternative construction of a complete ordered field. The method was invented by Richard Dedekind.

Let us start with the ordered field of rational numbers

$$\mathcal{Q} = (Q, +, \times, <).^{\dagger}$$

The idea behind Dedekind's construction is that the gaps in Q can be thought of as representing the "missing" irrational numbers.[‡] For example, $\sqrt{2}$ will be represented by the cut (A, B) that it determines in Q:

$$A = \{x : x \in Q \wedge (x < 0 \vee x^2 < 2)\},$$
$$B = \{x : x \in Q \wedge 0 < x \wedge x^2 > 2\}.$$

To make for a uniform treatment, every rational number also will be replaced by a corresponding cut; instead of the rational number r, one can think of the cut (A_r, B_r), where

$$A_r = \{x : x \in Q \wedge x < r\} \quad \text{and} \quad B_r = \{x : x \in Q \wedge x > r\}.$$

Of course, we could have used the cut (A_r^*, B_r^*), where

$$A_r^* = \{x : x \in Q \wedge x \leq r\} \quad \text{and} \quad B_r^* = \{x : x \in Q \wedge x > r\},$$

but we wish to associate a unique cut with each rational number. Finally, for further simplification in technical details, instead of cuts one uses the lower half of each cut. Now we shall present the precise definition.

[†] For brevity, we write simply $+$, $\times$, $<$ instead of $+_Q$, $\times_Q$, $<_Q$.
[‡] For definitions of "gap" and "cut", see p. 188.

Definition By a **lower cut** we shall mean any nonempty subset A of Q such that

i.   For any x, y in Q, $(x \in A \wedge y < x) \Rightarrow y \in A$;

ii. A has no maximum element, that is, $x \in A \Rightarrow (\exists y)(y \in A \wedge x < y)$;

iii. $A \neq Q$.

Let LC denote the set of lower cuts.

Remark If A is a lower cut, $a \in A$, and $b \notin A$, it follows from Part i of the definition that $a < b$. (For, if $b \leq a$, then $b \in A$.) Thus, every element not in A is an upper bound of A.

Definitions For any rational number u, let

$$u_{LC} = \{x : x \in Q \wedge x < u\}.$$

In particular,

$$0_{LC} = \{x : x \in Q \wedge x < 0\} = \text{the set of negative rational numbers,}$$

and

$$1_{LC} = \{x : x \in Q \wedge < 1\}.$$

Lemma F1 For any u in Q, $u_{LC} \in LC$.

Proof It is clear that u_{LC} is a nonempty subset of Q.

(i)   Assume $x \in u_{LC} \wedge y < x$. Then $x < u$. Hence, $y < u$, that is, $y \in u_{LC}$.

(ii) Assume $x \in u_{LC}$. Then $x < u$. Hence, $x < (x + u)/2$ and $(x + u)/2 \in u_{LC}$. Thus, u_{LC} has no maximum.

(iii) Take any rational number z greater than u. Then $z \notin u_{LC}$. Thus, $u_{LC} \neq Q$.   ∎

Definitions Consider any A and B in LC.

$$A < B \text{ means } A \subsetneq B;$$
$$A \leq B \text{ means } A < B \text{ or } A = B, \text{ that is, } A \subseteq B.$$

Lemma F2 For any A, B, C in LC:

a. $A \not< A$;

b. $(A < B \wedge B < C) \Rightarrow A < C$;

c. Either $A < B$ or $A = B$ or $B < A$.

Proof (a) and (b) are obvious from the definitions.

(c) Assume $A \nless B$ and $A \neq B$. Hence, there is some rational number z in A but not in B. We wish to prove that $B \subsetneqq A$. Since $A \neq B$, it suffices to show that $B \subseteq A$. Take any b in B. We must prove that $b \in A$. Now, either $b < z$ or $z = b$ or $z < b$. Clearly, $z \neq b$, since $z \notin B$. In addition, $z \nless b$. (For, if $z < b$, then, since B is a lower cut, property i of the definition of lower cuts would imply that $z \in B$, which is impossible.) Hence, the only possibility left is that $b < z$. But $z \in A$, and, therefore, by property i, $b \in A$. ∎

EXERCISE

Assume $A \in LC$. Prove: $0_{LC} < A$ if and only if A contains at least one positive rational number.

Definition *Addition* Let A and B be lower cuts. Then

$$A \boxplus B = \{a + b : a \in A \wedge b \in B\}.$$

Lemma F3 For any A, B, C in LC:

a. $A \boxplus B \in LC$;

b. $A \boxplus B = B \boxplus A$;

c. $(A \boxplus B) \boxplus C = A \boxplus (B \boxplus C)$;

d. $A \boxplus 0_{LC} = A$.

Proof (a) Clearly, $A \boxplus B$ is a nonempty set of rational numbers. In addition:

i. Assume $a + b \in A \boxplus B$, where $a \in A$ and $b \in B$. In addition, assume $z < a + b$. Let $w = (a + b) - z$. Thus, $w > 0$. Therefore, $a - w < a$, and so, $a - w \in A$. Hence, $z = (a - w) + b \in A \boxplus B$.

ii. Assume $a + b \in A \boxplus B$, where $a \in A$ and $b \in B$. Then, by property ii for A, there is a rational number a' such that $a < a'$ and $a' \in A$. Hence, $a + b < a' + b \in A \boxplus B$. Thus, $A \boxplus B$ has no maximum.

iii. Take any u in $Q - A$, and any v in $Q - B$. Then, for any $a \in A$, $b \in B$, we have $a < u$ and $b < v$. Therefore, $a + b < u + v$. Thus, $u + v \notin A \boxplus B$.

Thus, $A \boxplus B$ is a lower cut.

(b) and (c) are obvious from the definitions.

(d)   Assume $a + b \in A \boxplus 0_{LC}$, where $a \in A$ and $b \in 0_{LC}$. Hence, $b < 0$. Therefore, $a + b < a$, and so, $a + b \in A$. Thus, $A \boxplus 0_{LC} \subseteq A$. Conversely, assume $a \in A$. By property (ii), there is some z in A such that $a < z$. Then, $a - z < 0$, and so, $a - z \in 0_{LC}$. But, $a = z + (a - z) \in A \boxplus 0_{LC}$. Thus, $A \subseteq A \boxplus 0_{LC}$.   ∎

Lemma F4   Assume $A \in LC$ and $c > 0$. Then there exists a in A and there exists an upper bound b of A which is not a least upper bound of A such that $b - a = c$.

Proof   Choose any element a_1 in A and any upper bound b_1 of A which is not a lub of A. Moreover, if there exists a lub x of A, choose $a_1 > x - c$. By the Archimedean property of Q, there exists a natural number n such that $nc > b_1 - a_1$.[†] Therefore, $a_1 + nc > b_1$, and so, $a_1 + nc$ is an upper bound of A. Let k be the least natural number such that $a_1 + kc$ is an upper bound of A. Then, $a_1 + (k - 1)c \in A$. Now, $a_1 + kc$ is not a lub of A. (For, if $a_1 + kc$ is a lub of A, then, by our choice of a_1, $a_1 > (a_1 + kc)$ $- c = a_1 + (k - 1)c \geq a$, which is impossible.) Let $a = a_1 + (k - 1)c$ and let $b = a_1 + kc$. Then $b - a = c$.   ∎

Definition   *Subtraction*   If $A, B \in LC$, let

$$-A = \{-z : z \text{ is an upper bound of } A \text{ but } z \text{ is not a lub of } A\},$$

and $A \boxminus B = A \boxplus (-B)$ (see Figure F.1).

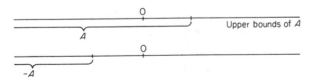

Figure F.1

† Observe that we have used the Archimedean property of Q. The method of Dedekind cuts does not seem to be applicable to a non-Archimedean ordered field K. However, if we restricted ourselves to cuts satisfying Lemma F4, we would obtain an ordered field $\hat{K}$ which is an extension of the original field K. Although $\hat{K}$ is not complete, it can be considered to be a *generalized completion* of K (see Baer [1970], Scott [1967]).

Lemma F5 Assume $A \in LC$. Then

a. $-A \in LC$.
b. $A \boxplus (-A) = 0_{LC}$.
c. $--A = A$.
d. $-(A \boxplus B) = (-A) \boxplus (-B)$.

Proof (a) Clearly, $-A$ is a nonempty subset of Q.

(i) Assume $-z \in -A$, where z is an upper bound of A but not a lub of A. Assume $w < -z$. Then $z < -w$. Hence $-w$ is an upper bound of A and $-w$ is not a lub of A. Therefore, $w = -(-w) \in -A$.

(ii) Assume $-z \in -A$, where z is an upper bound of A but not a lub of A. Then there is an upper bound v of A which is less than z. Since $v < (v + z)/2 < z$, $(v + z)/2$ is an upper bound of A, which is not a lub of A. Therefore, $-((v + z)/2) \in -A$ and $-((v + z)/2) > -z$.

(iii) Let $w \in A$. Since A has no maximum, w is not an upper bound of A. Hence, $-w \notin -A$.

(b) Assume $u + v \in A \boxplus (-A)$, where $u \in A$ and $v \in (-A)$. Then $-v$ is an upper bound of A, but not a lub of A. Therefore, $u < -v$. Hence, $u + v < 0$. Thus, $A \boxplus (-A) \subseteq 0_{LC}$. Conversely, assume $z \in 0_{LC}$. Then $z < 0$. Let $w = -z > 0$. By Lemma F4, there exists some $a \in A$ and some upper bound b of A which is not a lub of A such that $w = b - a$. Then $z = a + (-b) \in A \boxplus (-A)$. Hence, $0_{LC} \subseteq A \boxplus (-A)$, and, therefore, $A \boxplus (-A) = 0_{LC}$.

(c) By Part b, $(-A) \boxplus (-(-A)) = 0_{LC}$. Adding A to both sides, we obtain $-(-A)) = A$.

(d) $(A \boxplus B) \boxplus ((-A) \boxplus (-B)) = (A \boxplus (-A)) \boxplus (B \boxplus (-B))$
$$= 0_{LC} \boxplus 0_{LC} = 0_{LC}.$$

Now, add $-(A \boxplus B)$ to both sides. ∎

Lemma F6 Assume $A, B, C \in LC$. Then

a. $A < B \Leftrightarrow A \boxplus C < B \boxplus C$.
b. $A < B \Leftrightarrow 0_{LC} < B \boxminus A$.
c. $A < 0_{LC} \Leftrightarrow 0_{LC} < -A$.

Proof (a) Assume $A < B$. Then $A \subsetneqq B$. Clearly, $A \boxplus C \subseteq B \boxplus C$, since every sum $a + c$, where $a \in A \wedge c \in C$, is also in $B \boxplus C$. Now take some b'

in B such that $b' \notin A$. Choose some b^* in B such that $b' < b^*$. By Lemma F4, there exists c^* in C and there exists an upper bound d of C which is not a lub of C such that

$$d - c^* = \frac{b^* - b'}{2} < b^* - b'.$$

Now, for any a in A and any c in C, $c - c^* < d - c^* < b^* - b' < b^* - a$. Hence, $a + c < b^* + c^*$. Therefore, $b^* + c^* \in B \boxplus C$, but $b^* + c^* \notin A \boxplus C$ Hence, $A \boxplus C \subsetneq B \boxplus C$, that is, $A \boxplus C < B \boxplus C$.

Conversely, assume $A \boxplus C < B \boxplus C$. By what we have just shown, $A = A \boxplus C \boxplus (-C) < B \boxplus C \boxplus (-C) = B$.

(b) $A < B \Leftrightarrow A \boxplus (-A) < B \boxplus (-A)$

$\qquad\quad \Leftrightarrow 0_{LC} < B \boxminus A$.

(c) Put $0_{LC} = B$ in Part b. ∎

Now we must tackle the problem of defining multiplication in LC.

Definition Let $A, B \in LC$.

Case 1 $0_{LC} < A \wedge 0_{LC} < B$.

Let $A \boxtimes B = \{x : x \leq 0\} \cup \{ab : 0 < a \wedge 0 < b \wedge a \in A \wedge b \in B\}$.

Case 2 $A = 0_{LC}$ or $B = 0_{LC}$.

Let $A \boxtimes B = 0_{LC}$.

Case 3 $A < 0_{LC} \wedge B < 0_{LC}$.

Let $A \boxtimes B = (-A) \boxtimes (-B)$.

Case 4 $A < 0_{LC} \wedge 0_{LC} < B$.

Let $A \boxtimes B = -((-A) \boxtimes B)$.

Case 5 $B < 0_{LC} \wedge 0_{LC} < A$.

Let $A \boxtimes B = -(A \boxtimes (-B))$.

Lemma F7 Let $A, B \in LC$. Then $A \boxtimes B \in LC$.

Proof It clearly suffices to prove this result when $0_{LC} < A$ and $0_{LC} < B$, by virtue of Lemma F5a. Thus, we assume that $0_{LC} < A$ and $0_{LC} < B$. Then,

$$A \boxtimes B = \{x : x \leq 0\} \cup \{ab : 0 < a \wedge 0 < b \wedge a \in A \wedge b \in B\}.$$

Obviously, $A \boxtimes B$ is a nonempty set of rational numbers.

(i)   Assume $u \in A \boxtimes B$ and $v < u$. We must prove that $v \in A \boxtimes B$. If $v \leq 0$, then, by definition, $v \in A \boxtimes B$. Hence, we may assume $0 < v$. Then $0 < u$ and $u = ab$, where $a \in A$, $b \in B$, and a and b are positive. Now, $v/u < 1$. Hence $(v/u)a < a$. Therefore,

$$v = \frac{v}{u}\, u = \frac{v}{u}\,(ab) = \left(\frac{v}{u}\, a\right) b \in A \boxtimes B.$$

(ii)   Assume $u \in A \boxtimes B$. We must find an element of $A \boxtimes B$ greater than u. This is obvious if $u \leq 0$. Hence, we may assume that $0 < u$. Then $u = ab$, where a and b are positive elements of A and B, respectively. Since A has no maximum, there is an element a^* of A greater than a. Hence, $a^*b \in A \boxtimes B$ and $u = ab < a^*b$.

(iii)   Let $c \notin A$ and $d \notin B$. Then c and d are positive and greater than any elements of A and B, respectively. Therefore, if a and b are positive elements of A and B, respectively, $ab < cd$. Therefore, $cd \notin A \boxtimes B$.   ∎

Observe that it follows from the definition of multiplication that the product of positive lower cuts is a positive lower cut, and, then, that the product of two negative lower cuts is positive, and the product of a positive and a negative lower cut is negative. (Use Lemma F6c.)

Lemma F8   Assume $A, B, C \in LC$.

a. $A \boxtimes B = B \boxtimes A$.
b. $(A \boxtimes B) \boxtimes C = A \boxtimes (B \boxtimes C)$.
c. $A \boxtimes 1_{LC} = A$.

Proof   (a)   *Case 1* $0_{LC} < A \wedge 0_{LC} < B$.
Then $A \boxtimes B = B \boxtimes A$ follows easily from the definition.

Case 2 $A = 0_{LC}$ or $B = 0_{LC}$.
Then $A \boxtimes B = 0_{LC} = B \boxtimes A$.

Case 3 $A < 0_{LC} \wedge B < 0_{LC}$.
Then $A \boxtimes B = (-A) \boxtimes (-B) = (-B) \boxtimes (-A) = B \boxtimes A$.

Case 4 $A < 0_{LC} \wedge 0_{LC} < B$.
Then $A \boxtimes B = -((-A) \boxtimes B) = -(B \boxtimes (-A)) = B \boxtimes A$.

Case 5 $B < 0_{LC} \wedge 0_{LC} < A$.
Then $A \boxtimes B = -(A \boxtimes (-B)) = -((-B) \boxtimes A) = B \boxtimes A$.

(b)   This is proved by another tedious case-by-case analysis.

Case 1 A, B, and C are positive. Then the result is clear from the definition and the obvious fact that the product of positive lower cuts is a positive lower cut.

Case 2   At least one of A, B, C is 0_{LC}. Then both sides are 0_{LC}.

Case 3 A, B, C are all negative. Then

$$(A \boxtimes B) \boxtimes C = ((-A) \boxtimes (-B)) \boxtimes C = -(((-A) \boxtimes (-B)) \boxtimes (-C)).$$
$$A \boxtimes (B \boxtimes C) = A \boxtimes ((-B) \boxtimes (-C)) = -((-A) \boxtimes ((-B) \boxtimes (-C))).$$

The right-hand sides are equal by virtue of Case 1.

Case 4 A and B positive; C negative.

$$(A \boxtimes B) \boxtimes C = -((A \boxtimes B) \boxtimes (-C)).$$
$$A \boxtimes (B \boxtimes C) = A \boxtimes (-(B \boxtimes (-C))) = -(A \boxtimes (--(B \boxtimes (-C))))$$
$$= -(A \boxtimes (B \boxtimes (-C))).$$

But, $(A \boxtimes B) \boxtimes (-C) = A \boxtimes (B \boxtimes (-C))$ by Case 1.

Case 5 A and C positive; B negative.

$$(A \boxtimes B) \boxtimes C = C \boxtimes (A \boxtimes B) = (C \boxtimes A) \boxtimes B \quad \text{by Case 4.}$$
$$A \boxtimes (B \boxtimes C) = A \boxtimes (C \boxtimes B) = (A \boxtimes C) \boxtimes B \quad \text{by Case 4}$$
$$= (C \boxtimes A) \boxtimes B.$$

Case 6 B and C positive; A negative.

$$(A \boxtimes B) \boxtimes C = (B \boxtimes A) \boxtimes C = B \boxtimes (A \boxtimes C) \quad \text{by Case 5.}$$
$$A \boxtimes (B \boxtimes C) = (B \boxtimes C) \boxtimes A = B \boxtimes (C \boxtimes A) \quad \text{by Case 4}$$
$$= B \boxtimes (A \boxtimes C).$$

Case 7 A and B are negative; C positive.

$$A \boxtimes (B \boxtimes C) = A \boxtimes (-((-B) \boxtimes C)) = (-A) \boxtimes ((-B) \boxtimes C)$$
$$= ((-A) \boxtimes (-B)) \boxtimes C \quad \text{by Case 1}$$
$$= (A \boxtimes B) \boxtimes C.$$

Case 8 A and C negative; B positive.

$$A \boxtimes (B \boxtimes C) = A \boxtimes (C \boxtimes B) = (A \boxtimes C) \boxtimes B \quad \text{by Case 7.}$$
$$(A \boxtimes B) \boxtimes C = C \boxtimes (A \boxtimes B) = (C \boxtimes A) \boxtimes B \quad \text{by Case 7}$$
$$= (A \boxtimes C) \boxtimes B.$$

Case 9 B and C negative; A positive.

$A \boxtimes (B \boxtimes C) = (B \boxtimes C) \boxtimes A = B \boxtimes (C \boxtimes A)$ by Case 7.

$(A \boxtimes B) \boxtimes C = (B \boxtimes A) \boxtimes C = B \boxtimes (A \boxtimes C)$ by Case 8

$$= B \boxtimes (C \boxtimes A).$$

(c) *Case 1* A is positive.

$A \boxtimes 1_{LC} = \{x : x \leq 0\} \cup \{ab : 0 < a \wedge 0 < b \wedge a \in A \wedge b < 1\}.$

Since A is positive, $\{x : x \leq 0\} \subseteq A$. If $a \in A \wedge 0 < b < 1 \wedge 0 < a$, then $ab < a$, and, therefore, $ab \in A$. Thus, $A \boxtimes 1_{LC} \subseteq A$. Conversely, assume $a \in A$. If $a \leq 0$, then $a \in A \boxtimes 1_{LC}$. If $a > 0$, then there must be a larger element a^* in A. Hence, $a = a^*(a/a^*) \in A \boxtimes 1_{LC}$. Hence, $A \boxtimes 1_{LC} = A$.

Case 2 $A = 0_{LC}$.

Then $A \boxtimes 1_{LC} = 0_{LC} \boxtimes 1_{LC} = 0_{LC} = A$.

Case 3 A is negative.

$A \boxtimes 1_{LC} = -((-A) \boxtimes 1_{LC}) = -(-A)$ by Case 1

$$= A \quad \text{by Lemma F5c.} \quad \blacksquare$$

Lemma F9 Let $A, B, C \in LC$. Then

$$A \boxtimes (B \boxplus C) = (A \boxtimes B) \boxplus (A \boxtimes C).$$

Proof This is another proof by a tedious case-by-case analysis.

Case 1 A, B, C are all positive. Then $A \boxtimes B, A \boxtimes C, B \boxplus C,$ $A \boxtimes (B \boxplus C)$, and $(A \boxtimes B) \boxplus (A \boxtimes C)$ are all positive. Anything in $A \boxtimes (B \boxplus C)$ is either ≤ 0 or of the form $a(b + c)$, where $a > 0, b + c > 0,$ $a \in A, b \in B, c \in C$. But, $a(b + c) = ab + ac$. Furthermore, $ab \in A \boxtimes B$ and $ac \in A \boxtimes C$. (For, if $b \leq 0$, then $ab \leq 0$, and so, $ab \in A \boxtimes B$. If $b > 0$, then by definition, $ab \in A \boxtimes B$. Similarly, $ac \in A \boxtimes C$.) Therefore,

$$A \boxtimes (B \boxplus C) \subseteq (A \boxtimes B) \boxplus (A \boxtimes C).$$

Conversely, an object in $(A \boxtimes B) \boxplus (A \boxtimes C)$ is of one of the following forms:

I. $ab + a_1 c$, where $a, b, c, a_1 > 0$, $a, a_1 \in A$, $b \in B$, $c \in C$.

II. $ab + u$, where $a, b > 0$, $a \in A$, $b \in B$, $u \leq 0$.

III. $u + ac$, where $a, c > 0$, $a \in A$, $c \in C$, $u \leq 0$.

IV. $u + u'$, where $u \leq 0$ and $u' \leq 0$.

In Case I, let $a_2 = \max(a, a_1)$. Then

$$0 < ab + a_1c \le a_2b + a_2c = a_2(b + c) \in A \boxtimes (B \boxplus C).$$

Hence, $ab + a_1c \in A \boxtimes (B \boxplus C)$.

In Case II, $ab + u \le ab$. Therefore, $ab + u \in A \boxtimes B$. If $ab + u \le 0$, then $ab + u \in A \boxtimes (B \boxplus C)$. If $ab + u > 0$, then, since $ab + u \in A \boxtimes B$, $ab + u = a'b'$, where $a', b' > 0$, $a' \in A$, $b' \in B$. But $b' \in B \boxplus C$, since $b' = b' + 0$. Hence, $a'b' \in A \boxtimes (B \boxplus C)$.

Case III is treated similarly to Case (II).

In Case IV, $u + u' \le 0$. Hence, $u + u' \in A \boxtimes (B \boxplus C)$. (For, A and $B \boxplus C$ are positive, and, therefore, $A \boxtimes (B \boxplus C)$ contains all nonpositive elements.)

In all cases, we have shown that $(A \boxtimes B) \boxplus (A \boxtimes C) \subseteq A \boxtimes (B \boxplus C)$. Hence, $(A \boxtimes B) \boxplus (A \boxtimes C) = A \boxtimes (B \boxplus C)$.

Case 2 $A = 0_{LC}$, or $B = 0_{LC}$, or $C = 0_{LC}$. Obvious.

Case 3 A negative, B and C positive.

$$
\begin{aligned}
A \boxtimes (B \boxplus C) &= -((-A) \boxtimes (B \boxplus C)) \\
&= -(((-A) \boxtimes B) \boxplus ((-A) \boxtimes C)) \quad \text{by Case 1} \\
&= -((-(A \boxtimes B)) \boxplus (-(A \boxtimes C))) \\
&= (--(A \boxtimes B)) \boxplus (--(A \boxtimes C)) \quad \text{by Lemma F5d.} \\
&= (A \boxtimes B) \boxplus (A \boxtimes C) \quad \text{by Lemma F5c.}
\end{aligned}
$$

Case 4 A positive, and one of B and C negative (say B).

Case 4a $C = -B$. Then $A \boxtimes (B \boxplus C) = A \boxtimes 0_{LC} = 0_{LC}$ and

$$
\begin{aligned}
(A \boxtimes B) \boxplus (A \boxtimes C) &= (A \boxtimes B) \boxplus (A \boxtimes (-B)) \\
&= (-(A \boxtimes (-B))) \boxplus (A \boxtimes (-B)) = 0_{LC}.
\end{aligned}
$$

Case 4b $C > -B$. Then $B \boxplus C > 0_{LC}$ and $C = (-B) \boxplus (B \boxplus C)$.

$$
\begin{aligned}
(A \boxtimes B) \boxplus (A \boxtimes C) &= (A \boxtimes B) \boxplus (A \boxtimes ((-B) \boxplus (B \boxplus C))) \\
&= -(A \boxtimes (-B)) \boxplus ((A \boxtimes (-B)) \boxplus (A \boxtimes (B \boxplus C))) \\
&= A \boxtimes (B \boxplus C). \quad \text{by Case 1}
\end{aligned}
$$

Case 4c $C < -B$. Then $B \boxplus C < 0_{LC}$, $-B \boxplus (-C) > 0_{LC}$, and $-B = C \boxplus ((-B) \boxplus (-C))$.

$$
\begin{aligned}
(A \boxtimes B) \boxplus (A \boxtimes C) &= -(A \boxtimes (-B)) \boxplus (A \boxtimes C) \\
&= -(A \boxtimes (C \boxplus ((-B) \boxplus (-C)))) \boxplus (A \boxtimes C) \\
&= -((A \boxtimes C) \boxplus (A \boxtimes ((-B) \oplus (-C)))) \boxplus (A \boxtimes C) \\
&= (-(A \boxtimes C)) \boxplus (-(A \boxtimes ((-B) \boxplus (-C)))) \boxplus (A \boxtimes C) \\
&= -(A \boxtimes ((-B) \boxplus (-C))) \\
&= -(A \boxtimes (-(B \boxplus C))) = A \boxtimes (B \boxplus C).
\end{aligned}
$$

Case 5 A positive, B and C negative. Then $B \boxplus C$ is negative, by Lemma F6a.

$$A \boxtimes (B \boxplus C) = -(A \boxtimes (-(B \boxplus C)))$$
$$= -(A \boxtimes ((-B) \boxplus (-C))) \quad \text{by Lemma F5d}$$
$$= -((A \boxtimes (-B)) \boxplus (A \boxtimes (-C))) \quad \text{by Case 1}$$
$$= (-(A \boxtimes (-B))) \boxplus (-(A \boxtimes (-C))) \quad \text{by Lemma F5d}$$
$$= (A \boxtimes B) \boxplus (A \boxtimes C).$$

Case 6 A negative, one of B and C negative (say B).

Case 6a $B = -C$.

$$A \boxtimes (B \boxplus C) = A \boxtimes 0_{LC} = 0_{LC}.$$
$$(A \boxtimes B) \boxplus (A \boxtimes C) = (A \boxtimes (-C)) \boxplus (A \boxtimes C)$$
$$= ((-A) \boxtimes C) \boxplus (-((-A) \boxtimes C)) = 0_{CL}.$$

Case 6b $B > -C$. Then $B \oplus C > 0_{LC}$.

$$A \boxtimes (B \boxplus C) = -((-A) \boxtimes (B \boxplus C))$$
$$= -(((-A) \boxtimes B) \boxplus ((-A) \otimes C)) \quad \text{by Case 4}$$
$$= (-((-A) \boxtimes B)) \boxplus (-((-A) \boxtimes C)) \quad \text{by Lemma F5d}$$
$$= (-(-((-A) \boxtimes (-B)))) \boxplus (A \boxtimes C)$$
$$= ((-A) \boxtimes (-B)) \boxplus (A \boxtimes C) = (A \boxtimes B) \boxplus (A \boxtimes C).$$

Case 6c $B < -C$. Then $B \boxplus C < 0_{LC}$.

$$A \boxtimes (B \boxplus C) = (-A) \boxtimes (-(B \boxplus C))$$
$$= (-A) \boxtimes ((-B) \boxplus (-C)) \quad \text{by Lemma F5d}$$
$$= ((-A) \boxtimes (-B)) \boxplus ((-A) \boxtimes (-C)) \quad \text{by Case 4}$$
$$= (A \boxtimes B) \boxplus (-((-A) \boxtimes (--C))) = (A \boxtimes B) \boxplus (A \otimes C).$$

Case 7 A, B, and C all negative.

$$A \boxtimes (B \boxplus C) = (-A) \boxtimes (-(B \boxplus C))$$
$$= (-A) \boxtimes ((-B) \boxplus (-C)) \quad \text{by Lemma F5d}$$
$$= ((-A) \boxtimes (-B)) \boxplus ((-A) \boxtimes (-C)) \quad \text{by Case 1}$$
$$= (A \boxtimes B) \boxplus (A \boxtimes C). \quad \blacksquare$$

We have shown so far that $(LC, \boxplus, \boxtimes, <)$ is a commutative ring with unit element. In particular, all theorems proved about commutative rings with unit element in Chapter 3 are available for use.

Lemma F10   Let $A \in LC$ and $A \neq 0_{LC}$. Then there exists B in LC such that $A \boxtimes B = 1_{LC}$.

Proof   *Case 1* $A > 0_{LC}$. Let

$$B = \{x : x \leq 0\} \cup \{1/u : u \text{ is an upper bound of } A \text{ but not a lub of } A\}.$$

First, let us show that $B \in LC$. Clearly, B is a nonempty subset of Q.

(i)   Assume $z \in B \wedge w < z$. We must prove that $w \in B$. If $w \leq 0$, then $w \in B$. So, we may assume that $w > 0$. Hence, $0 < z$. Then $1/z$ is an upper bound of A but not a lub of A. But $1/w > 1/z$. Hence, $1/w$ is an upper bound of A which is not a lub of A. Therefore, $w \in B$.

(ii)   Assume $z \in B$. If $z \leq 0$, then there is an element of B greater than z. So, we may assume that $z > 0$. Hence, $1/z$ is an upper bound of A which is not a lub of A. Therefore, there is an upper bound w of A which is less than $1/z$. Therefore, $v = (w + (1/z))/2$ is an upper bound of A which is not a lub of A, and $v < 1/z$. Hence, $1/v > z$ and $1/v \in B$. Therefore, B has no maximum.

(iii)   Let a be a positive element of A. Then $1/a \notin B$, since $a = 1/(1/a)$ is not an upper bound of A.

Thus, $B \in LC$. It is clear that $0_{LC} < B$. Now we must prove that $A \boxtimes B = 1_{LC}$. Assume $z \in A \boxtimes B$. If $z \leq 0$, then $z \in 1_{LC}$. So, we assume $z > 0$. Then $z = ab$, where $a, b > 0$, $a \in A$, $b \in B$. Then $1/b$ is an upper bound of A. Hence, $a < 1/b$, and, therefore, $ab < 1$, that is, $z = ab \in 1_{LC}$. Thus, $A \boxtimes B \subseteq 1_{LC}$. Conversely, assume $z \in 1_{LC}$. If $z \leq 0$, then $z \in A \boxtimes B$. Hence, we may assume $0 < z < 1$. Let a_1 be any positive element of A. Then $(1 - z)a_1 > 0$. Hence, by Lemma F4, there is an element a in A and an upper bound w of A which is not a lub of A such that $w - a = (1 - z)a_1$. Then $w - a < (1 - z)w$. Hence, $zw < a$. Then $w = (1/z)(zw) < a/z$. Thus, a/z is an upper bound of A and is not a lub of A. Hence, $z = a(1/(a/z)) \in A \boxtimes B$. Thus, $1_{LC} \subseteq A \boxtimes B$. We have shown that $A \boxtimes B = 1_{LC}$.

Case 2 $A < 0_{LC}$.

Then $-A > 0_{LC}$. By Case 1, there exists B in LC such that $(-A) \boxtimes B = 1_{LC}$. Hence,

$$A \boxtimes (-B) = -(A \boxtimes B) = (-A) \boxtimes B = 1_{LC}. \qquad \blacksquare$$

Lemma F11   Let A, B, $C \in LC$. Then

$$(0_{LC} < C \wedge A < B) \Rightarrow A \boxtimes C < B \boxtimes C.$$

Proof Since $A < B$, we have $B = A \boxplus (B \boxminus A)$, where $B \boxminus A > 0_{LC}$ (by Lemma F6b). Then $(B \boxminus A) \boxtimes C > 0_{LC}$ and

$$B \boxtimes C = (A \boxplus (B \boxminus A)) \boxtimes C = (A \boxtimes C) \boxplus (B \boxminus A) \boxtimes C) > A \boxtimes C. \quad \blacksquare$$

Corollary F12 $(LC, \boxplus, \boxtimes, <)$ is an ordered field.

Proof We have, in the course of Lemmas F1–F11, verified all the axioms for an ordered field.   ∎

Theorem F13 $(LC, \boxplus, \boxtimes, <)$ is a complete ordered field.

Proof To prove completeness, we shall prove the lub property and use Theorem 5.3.3. Assume that $\mathscr{B}$ is a nonempty subset of LC which is bounded above. Let D be a lower cut which is an upper bound of $\mathscr{B}$. We must show that $\mathscr{B}$ has a lub in LC. Let $A = \bigcup_{B \in \mathscr{B}} B =$ the union of all the lower cuts in $\mathscr{B}$. First, let us prove that $A \in LC$. Since $\mathscr{B}$ is nonempty, A is a nonempty subset of Q.

(i)   Assume $z \in A \wedge w < z$. We must prove that $w \in A$. Now, $z \in B$ for some B in $\mathscr{B}$. Since B is a lower cut, $w \in B$. Hence, $w \in A$.

(ii)   Assume $z \in A$. Then $z \in B$ for some B in $\mathscr{B}$. Since B is a lower cut, there is some w in B such that $z < w$. Hence, $w \in A$, and $z < w$.

(iii)   Let d be an upper bound of the lower cut D. Let $a \in A$. Then $a \in B$ for some B in $\mathscr{B}$. Since D is an upper bound of $\mathscr{B}$, $B \leq D$. Hence, $B \subseteq D$. Therefore, $a \in D$. Then $a < d$. Thus, every element of A is less than d. Therefore, $d \notin A$.

We have shown that $A \in LC$. From the definition of A, $B \subseteq A$ for all B in $\mathscr{B}$. Hence, $B \leq A$ for all B in $\mathscr{B}$. Thus, A is an upper bound of $\mathscr{B}$. We must show that A is less than every other upper bound of $\mathscr{B}$. Assume C is a lower cut which is an upper bound of $\mathscr{B}$. Then $B \leq C$ for all B in $\mathscr{B}$, and, therefore, $B \subseteq C$ for all B in $\mathscr{B}$. Hence, $A = \bigcup_{B \in \mathscr{B}} B \subseteq C$. So, $A \leq C$. Thus, A is a lub of $\mathscr{B}$.   ∎

The tedious case-by-case proofs of some of the theorems above (for example, Lemma F9) can be avoided to some extent if we start with the positive rational numbers instead of all rational numbers. Lower cuts would then contain only positive rational numbers. The system of lower cuts would correspond to the system of positive real numbers. The set of all real numbers can then be generated in the same way that we generated the integers from the natural numbers. For details of this approach, see Landau [1951] or Hafstrom [1967].

COMPLEX NUMBERS

Let $\mathscr{R} = (R, +, \times, <)$ be the system of real numbers. The system $\mathscr{R}$ is not an **algebraically closed field,** that is, there are nonconstant polynomials with real coefficients which have no real roots. Examples of such polynomials include

$$x^2 + 1, \qquad x^2 + 2, \qquad x^2 - 2x + 2, \qquad x^4 - 6x^3 - x^2 - 18x + 10.$$

Historically, a "number" i was assumed to exist with the property that $i^2 = -1$, that is, $i = \sqrt{-1}$. Then the "complex numbers" $a + bi$, where a and b are real, could be shown to form an algebraically closed field which is an extension of the real number field. We shall show how this construction of the complex number field can be accomplished in a logically unobjectionable way. The approach we shall use was discovered by the Irish mathematician W. R. Hamilton [1837].

Let C be the set $R \times R$ of all ordered pairs of real numbers. Intuitively, think of (a, b) as representing $a + bi$.

Definition Assume $(a, b) \in C$ and $(u, v) \in C$. Then

$$(a, b) \oplus (u, v) = (a + u, b + v),$$

and

$$(a, b) \otimes (u, v) = (au - bv, av + bu).^\dagger$$

These definitions are our versions of the traditional operations:

$$(a + bi) + (u + vi) = (a + u) + (b + v)i,$$

---

† As usual, if a and u are real numbers, we sometimes write au instead of $a \times u$.

and

$$(a + bi) \times (u + vi) = (au - bv) + (av + bu)i.$$

These were "derived" in the traditional treatment on the basis of the assumption that the "complex numbers" $a + bi$ obeyed the usual arithmetic laws (such as commutativity and associativity of $+$ and $\times$, and the distributivity of $\times$ with respect to $+$) and the assumption that $i^2 = -1$.

Definitions

$$0_C = (0, 0).$$
$$1_C = (1, 0).$$
$$-(a, b) = (-a, -b).$$

Theorem G1 $(C, \oplus, \otimes)$ is a field.

Proof   One must verify the axioms for a field. Assume that (a, b), (u, v), (x, y) are any elements of C.

1. $((a, b) \oplus (u, v)) \oplus (x, y) = (a + u, b + v) \oplus (x, y)$
$$= (a + u + x, b + v + y).$$

   $(a, b) \oplus ((u, v) \oplus (x, y)) = (a, b) \oplus (u + x, v + y)$
$$= (a + u + x, b + v + y).$$

2. $(a, b) \oplus (u, v) = (a + u, b + v) = (u + a, v + b) = (u, v) \oplus (a, b).$

3a. $(a, b) \oplus 0_C = (a, b) \oplus (0, 0) = (a + 0, b + 0) = (a, b).$

3b. $(a, b) \oplus (-(a, b)) = (a, b) \oplus (-a, -b) = (a + (-a), b + (-b))$
$$= (0, 0) = 0_C.$$

4. $(a, b) \otimes ((u, v) \otimes (x, y))$
$$= (a, b) \otimes (ux - vy, uy + vx)$$
$$= (a(ux - vy) - b(uy + vx), a(uy + vx) + b(ux - vy))$$
$$- (aux - avy - buy - bvx, auy + avx + bux - bvy).$$

   $((a, b) \otimes (u, v)) \otimes (x, y)$
$$= (au - bv, av + bu) \otimes (x, y)$$
$$= ((au - bv)x - (av + bu)y, (au - bv)y + (av + bu)x)$$
$$= (aux - bvx - avy - buy, auy - bvy + avx + bux).$$

5. $(a, b) \otimes ((u, v) \oplus (x, y))$
$$= (a, b) \otimes (u + x, v + y)$$
$$= (a(u + x) - b(v + y), a(v + y) + b(u + x))$$
$$= (au + ax - bv - by, av + ay + bu + bx).$$

$((a, b) \otimes (u, v)) \oplus ((a, b) \otimes (x, y))$
$$= (au - bv, av + bu) \oplus (ax - by, ay + bx)$$
$$= (au - bv + ax - by, av + bu + ay + bx).$$

6. $(a, b) \otimes 1_C = (a, b) \otimes (1, 0) = (a \cdot 1 - b \cdot 0, a \cdot 0 + b \cdot 1) = (a, b).$

7. $(a, b) \otimes (u, v) = (u, v) \otimes (a, b)$. This is left as an exercise for the reader.

9. (Remember that (8) is superfluous.) Assume $(a, b) \neq 0_C$. Thus $a \neq 0$ or $b \neq 0$. We must show that there exists (w, z) in C such that $(a, b) \otimes (w, z) = 1_C$. Going back to the traditional notation, we have

$$\frac{1}{a + bi} = \frac{1}{a + bi} \cdot \frac{a - bi}{a - bi} = \frac{a - bi}{a^2 + b^2} = \frac{a}{a^2 + b^2} - \frac{b}{a^2 + b^2} i.$$

Hence, we consider the pair $(a/(a^2 + b^2), -b/(a^2 + b^2))$. Notice that $a^2 + b^2 > 0$ since $a \neq 0$ or $b \neq 0$. Then

$$(a, b) \otimes \left(\frac{a}{a^2 + b^2}, \frac{-b}{a^2 + b^2} \right)$$
$$= \left(a \cdot \frac{a}{a^2 + b^2} - b \cdot \frac{-b}{a^2 + b^2}, \; a \cdot \frac{-b}{a^2 + b^2} + b \cdot \frac{a}{a^2 + b^2} \right)$$
$$= \left(\frac{a^2}{a^2 + b^2} + \frac{b^2}{a^2 + b^2}, \; \frac{-ab}{a^2 + b^2} + \frac{ab}{a^2 + b^2} \right)$$
$$= \left(\frac{a^2 + b^2}{a^2 + b^2}, \; 0 \right) = (1, 0) = 1_C.$$

10. $0_C = (0, 0) \neq (1, 0) = 1_C.$ ∎

Let $i = (0, 1)$. Then:
$$i^2 = i \otimes i = (0, 1) \otimes (0, 1) = (0 \cdot 0 - 1 \cdot 1, 0 \cdot 1 + 1 \cdot 0)$$
$$= (-1, 0) = -(1, 0) = -1_C.$$

Since $i^2 = -1_C$, it follows that $(C, \oplus, \otimes)$ cannot be made into an ordered field. For, in an ordered field, a square is always nonnegative, while -1 is negative.

Theorem G2 For any u in R, let $\Phi(u) = (u, 0)$. Then Φ is an isomorphism from $(R, +, \times)$ into $(C, \oplus, \otimes)$.

Proof Clearly, $\Phi: R \xrightarrow{\text{1-1}} C$. Now, take any u and v in R. Then

$$\Phi(u) \oplus \Phi(v) = (u, 0) \oplus (v, 0) = (u + v, 0) = \Phi(u + v).$$
$$\Phi(u) \otimes \Phi(v) = (u, 0) \otimes (v, 0) = (uv - 0 \cdot 0, u \cdot 0 + 0 \cdot v)$$
$$= (uv, 0) = \Phi(uv). \quad \blacksquare$$

Let $R^{\#} = \{(u, 0) : u \in R\}$. Then $R^{\#}$ is the range of the isomorphism Φ of Theorem G2. Hence, $R^{\#}$ determines a subfield of C isomorphic to the real number field. Let $(a, b) \in C$. Then

$$(a, 0) \oplus ((b, 0) \otimes i) = (a, 0) \oplus ((b, 0) \otimes (0, 1))$$
$$= (a, 0) \oplus (b \cdot 0 - 0 \cdot 1, b \cdot 1 + 0 \cdot 0)$$
$$= (a, 0) \oplus (0, b)$$
$$= (a, b).$$

If, as is customary, we write $(b, 0)i$ instead of $(b, 0) \otimes i$, then

$$(*) \quad (a, b) = (a, 0) \oplus (b, 0)i.$$

Now, let us agree to write u instead of $(u, 0)$. In a sense, this amounts to substituting the real number field for its isomorphic copy $R^{\#}$ in C. Then, (*) becomes

$$(a, b) = a \oplus bi,$$

which brings us back to the traditional notation for complex numbers. Notice that

$$(a \oplus bi) \oplus (c \oplus di) = (a + c) \oplus (b + d)i,$$

and

$$(a \oplus bi) \otimes (c \oplus di) = (ac - bd) \oplus (ad + bc)i.$$

EXERCISE

Show that the field $(C, \oplus, \otimes)$ has characteristic zero.

In the complex number field, there is a special operation called **conjugation,** which turns out to be extremely useful.

Definition   Let $(a, b) \in C$. Then $(a, -b)$ is called the **conjugate** of (a, b) and is denoted $\overline{(a, b)}$. In traditional notation,

$$\overline{a \oplus bi} = a - bi.$$

For,

$$\begin{aligned}
a - bi &= a \oplus (-(bi)) = a \oplus (-((b, 0) \otimes (0, 1))) \\
&= a \oplus (-(0, b)) = a \oplus (0, -b) = (a, 0) \oplus (0, -b) \\
&= (a, -b) = \overline{(a, b)}.
\end{aligned}$$

Theorem G3   Let α and β be any elements of C. Then

a. $\overline{\alpha \oplus \beta} = \bar{\alpha} \oplus \bar{\beta}.$

b. $\overline{-\alpha} = -\bar{\alpha}.$

c. $\overline{\alpha - \beta} = \bar{\alpha} - \bar{\beta}.$

d. $\overline{\alpha\beta} = \bar{\alpha}\bar{\beta}.$

e. $\overline{\left(\dfrac{\alpha}{\beta}\right)} = \dfrac{\bar{\alpha}}{\bar{\beta}}$ if $\beta \neq 0_C.$

f. $\bar{\bar{\alpha}} = \alpha.$

Proof   Let $\alpha = a \oplus bi$ and $\beta = c \oplus di$.

(a) $\overline{\alpha \oplus \beta} = \overline{(a \oplus bi) \oplus (c \oplus di)} = \overline{(a + c) \oplus (b + d)i}$
    $= (a + c) - (b + d)i = (a + c) - (bi \oplus di)$
    $= (a - bi) \oplus (c - di) = \overline{a + bi} \oplus \overline{c + di} = \bar{\alpha} \oplus \bar{\beta}.$

(b) $\overline{-\alpha} = \overline{(-a) \oplus (-b)i} = (-a) - (-b)i = (-a) \oplus bi$
    $= -(a - bi) = -\overline{(a + bi)}.$

(c) This follows from (a) and (b).

(d)–(f)   are exercises for the reader.   ∎

Although $(C, \oplus, \otimes)$ cannot be made into an ordered field, there is an operation in C analogous to the absolute value operation in ordered fields.

Definition $|a, b)| = \sqrt{a^2 + b^2}$. Thus, $|a \oplus bi| = \sqrt{a^2 + b^2}$.

Notice that $|0_C| = 0$ and, if $(a, b) \neq 0_C$, then $|(a, b)|$ is a positive real number. If u is a real number, then $|u|$ is the ordinary absolute value in the real number system. (Remember that u is identified with the complex number $(u, 0)$.)

Theorem G4 For any α, β in C:

a. $\alpha\bar{\alpha} = |\alpha|^2$.

b. $|\bar{\alpha}| = |\alpha|$.

c. $|\alpha\beta| = |\alpha||\beta|$.

d. If $\beta \neq 0_C$, $\left|\dfrac{\alpha}{\beta}\right| = \dfrac{|\alpha|}{|\beta|}$.

e. $|-\alpha| = |\alpha|$.

Proof Let $\alpha = a \oplus bi$, $\beta = c \oplus di$.

(a) $\alpha\bar{\alpha} = (a \oplus bi)(a - bi) = a^2 + b^2 = |\alpha|^2$.

(b) $|\bar{\alpha}| = |\overline{a \oplus bi}| = |a - bi| = \sqrt{a^2 + (-b)^2} = \sqrt{a^2 + b^2} = |\alpha|$.

(c) $|\alpha\beta|^2 = (\alpha\beta)\overline{(\alpha\beta)}$ by Part a
$$= \alpha\beta\,\bar{\alpha}\bar{\beta}   \text{by Theorem G3d.}$$
$$= (\alpha\bar{\alpha})(\beta\bar{\beta}) = |\alpha|^2|\beta|^2.$$

Therefore, $|\alpha\beta|^2 - |\alpha|^2|\beta|^2 = 0$. Hence,

$$(|\alpha\beta| - |\alpha||\beta|)(|\alpha\beta| + |\alpha||\beta|) = 0.$$

If $\alpha \neq 0_C$ and $\beta \neq 0_C$, $|\alpha\beta| + |\alpha||\beta| > 0$, and therefore,

$$|\alpha\beta| - |\alpha||\beta| = 0,   \text{that is,} |\alpha\beta| = |\alpha||\beta|.$$

If $\alpha = 0_C$ or $\beta = 0_C$, it is clear that $|\alpha\beta| = |\alpha||\beta|$.

(d) $\left|\dfrac{\alpha}{\beta}\right||\beta| = \left|\dfrac{\alpha}{\beta}\beta\right|$ by Part c
$$= |\alpha|.$$

Therefore, $|\alpha/\beta| = |\alpha|/|\beta|$.

(e) Use Part c, with $\beta = -1$.   ∎

Theorem G5

a. *Triangle Inequality* For any α, β in C:

$$|\alpha \oplus \beta| \leq |\alpha| + |\beta|.$$

b. $|\alpha - \beta| \geq ||\alpha| - |\beta||$.

Proof   (a)   Let $\alpha = a \oplus bi$ and $\beta = c \oplus di$.

$0 \le (ad - bc)^2 \Rightarrow 0 \le a^2d^2 - 2abcd + b^2c^2$

$\Rightarrow 2abcd \le a^2d^2 + b^2c^2$

$\Rightarrow a^2c^2 + 2abcd + b^2d^2 \le a^2c^2 + a^2d^2 + b^2c^2 + b^2d^2$

$\Rightarrow (ac + bd)^2 \le (a^2 + b^2)(c^2 + d^2)$

$\Rightarrow ac + bd \le \sqrt{a^2 + b^2} \ \sqrt{c^2 + d^2}$

$\Rightarrow 2ac + 2bd \le 2\sqrt{a^2 + b^2} \ \sqrt{c^2 + d^2}$

$\Rightarrow a^2 + 2ac + c^2 + b^2 + 2bd + d^2 \le$
$$(a^2 + b^2) + 2\sqrt{a^2 + b^2} \ \sqrt{c^2 + d^2} + (c^2 + d^2)$$

$\Rightarrow (a + c)^2 + (b + d)^2 \le$
$$(a^2 + b^2) + 2\sqrt{a^2 + b^2} \ \sqrt{c^2 + d^2} + (c^2 + d^2)$$

$\Rightarrow \sqrt{(a + c)^2 + (b + d)^2} \le \sqrt{a^2 + b^2} + \sqrt{c^2 + d^2}$

$\Rightarrow |\alpha \oplus \beta| \le |\alpha| + |\beta|$.

(The logical sequence of steps in this proof is, of course, the reverse of the way in which they were discovered.)

(b) $|\alpha| = |(\alpha - \beta) \oplus \beta| \le |\alpha - \beta| + |\beta|$ by Part a.

Hence, $|\alpha| - |\beta| \le |\alpha - \beta|$. Exchanging α and β, we also obtain $|\beta| - |\alpha| \le |\beta - \alpha| = |\alpha - \beta|$. Hence, $|\alpha - \beta| \ge |\alpha| - |\beta|$ and $|\alpha - \beta| \ge |\beta| - |\alpha|$.   Therefore, $|\alpha - \beta| \ge ||\alpha| - |\beta||$,   since $||\alpha| - |\beta||$ is either $|\alpha| - |\beta|$ or $|\beta| - |\alpha|$.   ■

For further development of the theory of complex numbers, the reader is referred to any standard textbook in the theory of functions of a complex variable (for example, Ahlfors [1966] or Heins [1968]). Simple proofs of the Fundamental Theorem of Algebra (that is, every nonconstant polynomial with complex coefficients has a complex root) may be found in van der Waerden [1949], Fine [1901], or Feferman [1964].

EXERCISES

We say that a complex number z is a **limit** of a sequence $\langle \alpha_n \rangle$ of complex numbers if and only if, for any positive real number ε, there is a positive integer n_0 such that $n \ge n_0$ $\Rightarrow |\alpha_n - z| < \varepsilon$.

1. Prove that a sequence $\langle \alpha_n \rangle$ of complex numbers has at most one limit. If such a limit exists, the sequence is said to be **convergent** and the limit is denoted Lim α_n.

    By a **Cauchy sequence in** C we mean a sequence $\langle \alpha_n \rangle$ of complex numbers such that, for any positive real ε, there is a positive integer n_0 such that $k, n \geq n_0 \Rightarrow |\alpha_n - \alpha_k| < \varepsilon$.

2. Let $\langle \alpha_n \rangle$ be a sequence of complex numbers, where $\alpha_n = a_n \oplus b_n i$.

    a. Prove that $\langle \alpha_n \rangle$ is a Cauchy sequence in C if and only if $\langle a_n \rangle$ and $\langle b_n \rangle$ are Cauchy sequences of real numbers.

    b. Prove that every Cauchy sequence of complex numbers is convergent.

BIBLIOGRAPHY

This list contains not only books and papers referred to in the text but also other references which the author recommends to the reader for further study.

Adamson, I. (1965) "Introduction to Field Theory," Wiley (Interscience), New York.
Ahlfors, L. (1966) "Complex Analysis," 2nd ed., McGraw-Hill, New York.
Anderson, K. and D. Hall (1963) "Sets, Sequences, and Mappings," Wiley, New York.
Apostol, T. (1957), "Advanced Calculus," Addison-Wesley, Reading, Massachusetts.
Apostol, T. (1961–1962) "Calculus," 2 Vols., Ginn (Blaisdell), Boston, Massachusetts.
Artin, E. (1959) "Theory of Algebraic Numbers," Göttingen.
Artin, E. and O. Schreier (1926) Algebraische Konstruktion reeler Körper, *Abh. Math. Sem. Univ. Hamburg* **5**, 85–99.
Auslander, L. (1969) "What Are Numbers?" Scott-Foresman,
Bachmann, G. (1964) "Introduction to p-adic Numbers and Valuations," Academic Press, New York.
Baer, R. (1927) Über nicht-archimedisch geordnete Körper, *S.-B. Heidelberger, Akad. Wiss. Math. - Natur. Kl.* **8**, 3–13.
Baer, R. (1970) Dichte, Archimedizität und Starrheit geordneter Körper, *Math. Ann.* **188**, 165–205.
Bar Hillel, J. See Fraenkel, A. A., and J. Bar Hillel.
Barker, S. (1964) "Philosophy of Mathematics," Prentice-Hall, Englewood Cliffs, New Jersey.
Beck, A., M. Bleicher, and D. Crowe (1969) "Excursions into Mathematics," Chapters 2 and 6, Worth,
Bell, E. T. (1937) "Men of Mathematics," Simon & Schuster, New York.
Bell, E. T. (1945) "The Development of Mathematics," McGraw-Hill, New York.
Benecerraf, P. (1965) What numbers could not be, *Philos. Rev.* **74**, 47–74.
Beth, E. (1959) "The Foundations of Mathematics," North-Holland Publ., Amsterdam.
Birkhoff, G. See Maclane, S., and G. Birkhoff.
Bishop, E. (1967) "Foundations of Constructive Analysis," McGraw-Hill, New York.
Bleicher, M. See Beck, A., M. Bleicher, and D. Crowe.
Boas, R., Jr. (1960) "A Primer of Real Functions," Math. Ass. Amer.,
Borofsky, S. (1950) "Elementary Theory of Equations," Macmillan, New York.
Borovskii, Yu. E. (1956) The independence of the Archimedean axiom, *Uspehi Mat. Nauk* **11**, 161–167 (in Russian).
Bosch, W., and P. Krajkiewicz (1970) A categorical system of axioms for the complex numbers, *Math. Mag.* **43**, 67–70.

Bourbaki, N. (1950–1952) "Algèbre," Vol. II, Chapters IV-VI, Hermann, Paris.

Bourbaki, N. (1960) "Éléments d'Histoire des Mathématiques," Hermann, Paris.

Boyer, C. (1959) "The History of the Calculus and its Conceptual Development," Dover, New York.

Boyer, C. (1968) "A History of Mathematics," Wiley, New York.

Breusch, R. (1954) A proof of the irrationality of π, *Amer. Math. Monthly* **61**, 631–632.

Cajori, F. (1919) "A History of Mathematics," Macmillan, New York.

Cajori, F. (1929) "A History of Mathematical Notations," 2 Vols, Open Court, La Salle, Illinois.

Cantor, G. (1872) Über die Ausdehnung eines Satzes aus der Theorie der trigonometrischen Reihen, *Math. Ann.* **5**, 123–132.

Chinn, W., and N. Steenrod (1969) "First Concepts of Topology," Random House, New York.

Cohen, L. and G. Ehrlich (1963) "The Structure of the Real Number System," Van Nostrand—Reinhold, Princeton, New Jersey.

Cohen, L. and C. Goffman (1949) A theory of transfinite convergence, *Trans. Amer. Math. Soc.* **66**, 65–74.

Courant, R. and F. John (1965) "Introduction to Calculus and Analysis," Vol. 1, Wiley (Interscience), New York.

Courant, R. and H. Robbins (1941) "What is Mathematics?," Oxford Univ. Press, London and New York.

Crowe, D. See Beck, A., M. Bleicher, and D. Crowe.

Dantzig, T. (1954) "Number, The Language of Science," Macmillan, New York.

Dedekind, R. (1963) "Essays on the Theory of Numbers," Dover, New York. (Reprint of 1901 Open Court ed.; Translations of "Stetigkeit und Irrationalzahlen," 1872, and "Was Sind und Was Sollen Die Zahlen," 1888.)

Devinatz, A. (1968) "Advanced Calculus," Holt, New York.

Dieudonne, J. (1960) "Foundations of Modern Analysis," Academic Press, New York.

Efimov, N. V. (1953) "Higher Geometry," 3rd ed., Moscow, Leningrad (in Russian).

Ehrlich, G. See Cohen, L., and G. Ehrlich.

Ehrlich, G. See Goldhaber, J., and G. Ehrlich.

Eves, H. (1969) "An Introduction to the History of Mathematics," 3rd ed. Holt, New York.

Feferman, S. (1964) "The Number Systems," Addison-Wesley, Reading, Massachusetts.

Fine, H. B. (1901) "College Algebra," Ginn, Boston, Massachusetts.

Fraenkel, A. A. (1961) "Abstract Set Theory," North-Holland Publ., Amsterdam.

Fraenkel, A. A., and J. Bar Hillel (1958) "Foundations of Set Theory," North-Holland Publ., Amsterdam.

Frege, G. (1884) "Die Grundlagen der Arithmetik," Breslau.

Frege, G. (1893) "Grundgesetze der Arithmetik," Vol. I, Jena.

Frege, G. (1903) "Grundgesetze der Arithmetik," Vol. II, Jena.

Fuchs, L. (1963) "Partially Ordered Algebraic Systems," Pergamon, Oxford.

Gaal, L. (1971) "Classical Galois Theory with Examples," Markham, Chicago, Illinois.

Gaal, S. (1964) "Point Set Topology," Academic Press, New York.

Gelbaum, B. and J. Olmsted (1964) "Counterexamples in Analysis," Holden-Day, San Francisco, California.

Gelfand, S., et al. (1969) "Sequences, Combinations, Limits," MIT Press, Cambridge, Massachusetts.

Gericke, H. (1970) "Geschichte des Zahlbegriffs," Bibliographisches Inst., Mannheim.

Gillman, L. and M. Jerison (1960) "Rings of Continuous Functions," Van Nostrand—Reinhold, Princeton, New Jersey.

Gleason, A. (1966) "Fundamentals of Abstract Analysis," Addison-Wesley, Reading, Massachusetts.

Gleyzal, A., (1937) Transfinite numbers, *Proc. Nat. Acad. Sci. U.S.A.* **23**, 281–287.

Gödel, K. (1947) What is Cantor's continuum problem, *Amer. Math. Monthly* **54**, 515–525.

Goffman, C. See Cohen, L. and C. Goffman.

Goldhaber, J. and G. Ehrlich (1970) "Algebra," Macmillan, New York.

Goodstein, R. L. (1948) "A Textbook of Mathematical Analysis," Oxford Univ. Press, London and New York.

Hafstrom, J. (1967) "Introduction to Analysis and Abstract Algebra," Saunders, Philadelphia, Pennsylvania.

Hahn, H. (1907) Über die nichtarchimedischen Grössensysteme, *S.-B. Kaiserlichen Akad. Wiss. Vienna, Sect. IIa* **116**, 601–653.

Hall, D. See Anderson, K. and D. Hall.

Halmos, P. (1960) "Naive Set Theory," Van Nostrand—Reinhold, Princeton, New Jersey.

Hamilton, N., and J. Landin (1961) "Set Theory: The Structure of Arithmetic," Allyn & Bacon, Rockleigh, New Jersey.

Hamilton, W. R. (1837) Theory of conjugate functions, or algebraic couples; with a preliminary and elementary essay on algebra as the science of pure time, *Trans. Royal Irish Acad.* **17**, 283–422. (Also in "The Mathematical Papers of Sir W. R. Hamilton," Vol. III, Algebra, pp. 3-100. Cambridge Univ. Press, London and New York.)

Hardy, G. H. (1938) "A Course of Pure Mathematics," Cambridge Univ. Press, London and New York.

Hardy, G. H., and E. Wright (1954) "An Introduction to the Theory of Numbers," Oxford Univ. Press, London and New York.

Hatcher, W. (1968) "The Foundations of Mathematics," Saunders, Philadelphia, Pennsylvania.

Hauschild, K. (1966) Über die Konstruktion von Erweiterungskörpern zu nichtarchimedisch angeordneten Körpern mit Hilfe Hölderschen Schnitte, *Wiss. Z. Humboldt-Univ. Berlin Math.-Natur. Reihe* **5**.

Hauschild, K. (1967) Cauchyfolgen höheren Typus in angeordneten Körpern, *Z. Math. Logik Grundlagen Math.* **13**, 55–66.

Hayden, S. and J. Kennison (1968) "Zermelo-Fraenkel Set Theory," Merrill, Columbus, Ohio.

Heine, E. (1872) Die Elemente der Functionenlehre, *J. Reine Angew. Math.* **74**, 172–188.

Heins, M. (1968) "Complex Function Theory," Academic Press, New York.

Henkin, L., W. Smith, V. Varineau, and M. Walsh (1962) "Retracing Elementary Mathematics," Macmillan, New York.

Herstein, N. (1964) "Topics in Algebra," Ginn (Blaisdell), Boston, Massachusetts.

Hilbert, D. (1902) "Foundations of Geometry," Open Court, La Salle, Illinois.

Hobson, E. W. (1926) "Functions of a Real Variable," 2 Vols. Cambridge Univ. Press, London and New York.

Huntington, E. (1955) "The Continuum and Other Types of Serial Order," Dover, New York. (Reprint of 1917 Ed.)

Jacobson, N. (1964) "Lectures in Abstract Algebra," Vol. III, Van Nostrand—Reinhold, Princeton, New Jersey.

Jerison, M. See Gillman, L., and M. Jerison.

John, F. See Courant, R., and F. John.

Kaczmarz, S. M. (1932) Axioms for arithmetic, *J. London Math. Soc.* **7,** 179–182.

Kelley, J. (1955) "General Topology," Van Nostrand—Reinhold, Princeton, New Jersey.

Kennison, J. See Hayden, S., and J. Kennison.

Kershner, R. B., and L. R. Wilcox (1950), "The Anatomy of Mathematics," Ronald Press, New York.

Knopp, K. (1951) "Theory and Application of Infinite Series," Hafner, New York.

Krajkiewicz, P. See Bosch, W., and P. Krajkiewicz.

Landau, E. (1951) "Foundations of Analysis," Chelsea, Bronx, New York.

Landin, J. See Hamilton, N., and J. Landin.

LeVeque, W. (1956) "Topics in Number Theory," Vol. I, Addison-Wesley, Reading, Massachusetts.

Lightstone, A. H. (1965) "Symbolic Logic and the Real Number System," Harper, New York.

Lorenzen, P. (1971) "Differential and Integral, A Constructive Introduction to Classical Analysis," Univ. of Texas Press, Austin.

Luxemburg, W. A. J. (1964) "Non-Standard Analysis," California Inst. of Technol., Pasadena, California.

McCarthy, P. (1966) "Algebraic Extensions of Fields," Ginn (Blaisdell), Boston, Massachusetts.

Maclane, S. (1939) The universality of formal power series, *Bull. Amer. Math. Soc.* **45,** 888–890.

Maclane, S., and G. Birkhoff (1967) "Algebra," Macmillan, New York.

McShane, E. J. (1962) "A Theory of Limits, Studies in Modern Analysis," pp. 7–29, Math. Ass. Amer.

Machover, M. and J. Hirschfeld (1969) "Lectures on Non-Standard Analysis," Springer, Berlin.

Manheim, J. (1964) "The Genesis of Point Set Topology," Macmillan, New York.

Mendelson, E. (1964) "Introduction to Mathematical Logic," Van Nostrand—Reinhold, Princeton, New Jersey.

Mendelson, E. (1970) "Introduction to Boolean Algebra and Switching Circuits," Schaum-McGraw-Hill, New York.

Meyer, J.-P. See Mostow, G., J. Sampson, and J.-P. Meyer.

Monk, J. D. (1969) "Introduction to Set Theory," McGraw-Hill, New York.

Moss, R., and G. Roberts (1968) A creeping lemma, *Amer. Math. Monthly* **75,** 649–652.

Mostow, G., J. Sampson, and J.-P. Meyer (1963) "Fundamental Structures of Algebra," McGraw-Hill, New York.

Neumann, B. H. (1949) On ordered division rings, *Trans. Amer. Math. Soc.* **66,** 202–252.

Niven, I. (1956) "Irrational Numbers," Math. Ass. Amer.

Niven, I. (1961) "Numbers: Rational and Irrational," Random House, New York.

Oberschelp, A. (1968) "Aufbau des Zahlensystems." Göttingen,

Olmsted, J. (1962) "The Real Number System," Appleton, New York.

Olmsted, J. See Gelbaum, B., and J. Olmsted.

Parsons, C. (1965) Frege's theory of numbers, *in* "Philosophy in America" (M. Black ed.), pp. 180–203, Cornell Univ. Press, Ithaca, New York.

Peano, G. (1891) Sul concetto di numero, *Riv. mat. Turin*, **1**, 87–102, 256–267.

Peano, G. (1895–1908) "Formulaire des Mathématiques," Vols. I–V, Turin,

Pollard, H. (1950) "The Theory of Algebraic Numbers," Math. Ass. Amer.

Robbins, H. See Courant, R., and H. Robbins.

Roberts, G. See Moss, R., and G. Roberts.

Robertson, J. (1970) A comparison of the Archimedean and completeness properties, *Math. Mag.* **43**, 92–93.

Robinson, A. (1963) "Introduction to Model Theory and the Metamathematics of Algebra," North-Holland. Publ., Amsterdam.

Robinson, A. (1966) "Non-Standard Analysis," North-Holland Publ., Amsterdam.

Rohrbach, H. (1950) Das Axiomensystem von Erhard Schmidt für die Menge der natürlichen Zahlen, *Math. Nachrichten* **4**, 315–321.

Rosser, J. B. (1953) "Logic for the Mathematician," McGraw-Hill, New York.

Rubin, J. (1967) "Set Theory for the Mathematician," Holden-Day, San Francisco, California.

Rudin, W. (1953) "Principles of Mathematical Analysis," McGraw-Hill, New York.

Russell, B. (1903) "The Principles of Mathematics," Cambridge Univ. Press, London.

Russell, B. (1919) "Introduction to Mathematical Philosophy," Macmillan, New York.

Sampson, J. See Mostow, G., and J.-P. Meyer.

Šanin, N. (1968) Constructive real numbers and function spaces, *Amer. Math. Soc. Transl.* Vol. 21, Providence, Rhode Island.

Scott, D. (1969) On completing ordered fields, Applications of Model Theory, *Proc. Int. Symp. California Inst. Technol.* (1967), pp. 274–278, Holt, New York.

Seidenberg, A. (1954) A new decision method for elementary algebra, *Ann. of Math.* **60**, 365–374.

Sierpinski, W. (1958) Cardinal and Ordinal Numbers, "Monografie Matematyczne," Vol. 34, Warsaw.

Sierpinski, W. (1964) Elementary Theory of Numbers, "Monografie Matematyczne," Vol. 42, Warsaw.

Sikorski, R. (1948) On an ordered algebraic field, *C. R. Soc. Sci. Lett. Varsovie Cl. III*, **41**, 69–96.

Smith, W. See Henkin, L., W. Smith, V. Varineau, and M. Walsh.

Specker, E. (1953) The axiom of choice in Quine's "New Foundations for Mathematical Logic ", *Proc. Nat. Acad. Sci. U.S.A.* **39**, 972–975.

Spivak, M. (1967) "Calculus," Benjamin, New York.

Steenrod, N. See Chinn, W. and N. Steenrod.

Stein, S. (1963) "Mathematics: The Man-Made Universe," Freeman, San Francisco, California.

Stolyar, A. (1970) "Introduction to Elementary Mathematical Logic," MIT Press, Cambridge, Massachusetts.

Struik, D. (1969) "A Source Book in Mathematics, 1200-1800," Harvard Univ. Press, Cambridge, Massachusetts.

Suppes, P. (1961) "Axiomatic Set Theory," Van Nostrand—Reinhold, Princeton, New Jersey.

Tarski, A. (1948) "A Decision Method for Elementary Algebra and Geometry," Rand, Santa Monica, California.

Tarski, A. (1965) "An Introduction to Logic and the Methodology of Deductive Sciences," 3 ed. Oxford Univ. Press, London and New York.

van der Waerden, B. L. (1954) "Science Awakening," Noordhoff, Groningen.

van der Waerden, B. L. (1949) "Modern Algebra, I," Ungar, New York.

Varineau, V. See Henkin, L., W. Smith, V. Varineau, and M. Walsh.

Waismann, F. (1951) "Introduction to Mathematical Thinking," Harper, New York.

Walsh, M. See Henkin, L., W. Smith, V. Varineau, and M. Walsh.

Wang, H. (1957) The axiomatization of arithmetic, *J. Symbolic Logic*, **22**, 145–158.

Weiss, E., (1963) "Algebraic Number Theory," McGraw-Hill, New York.

Weyl, H. (1934) "Das Kontinuum," Chelsea, Bronx, New York. (Reprint of 1918 Ed.)

Whyburn, G. (1964) "Topological Analysis," Princeton Univ. Press, Princeton, New Jersey.

Wilcox, L. R. See Kershner, R. B. and L. R. Wilcox.

Wilder, R. L. (1952) "The Foundations of Mathematics," Wiley, New York.

Wilder, R. L. (1968) "The Evolution of Mathematical Concepts," Wiley, New York.

Wright, E. See Hardy, G. H., and E. Wright.

Zippin, L. (1962) "Uses of Infinity," Random House, New York.

INDEX OF SPECIAL SYMBOLS

INDEX